Advanced Statistics: Statistical Distributions

Advanced Statistics: Statistical Distributions

Editor: Alice Morgan

MURPHY & MOORE
www.murphy-moorepublishing.com

Published by Murphy & Moore Publishing
1 Rockefeller Plaza,
New York City, NY 10020, USA
www.murphy-moorepublishing.com

Advanced Statistics: Statistical Distributions
Edited by Alice Morgan

International Standard Book Number: 978-1-63987-013-4 (Hardback)

Cataloging-in-Publication Data

Advanced statistics : statistical distributions / edited by Alice Morgan.
p. cm.
Includes bibliographical references and index.
ISBN 978-1-63987-013-4
1. Distribution (Probability theory). 2. Statistics. 3. Mathematical statistics. I. Morgan, Alice.
QA273.6 .A38 2022
519.24--dc23

Contents

Preface

The discipline that is concerned with the collection, organization, analysis, interpretation, and presentation of data is referred to as statistics. The two main statistical methods that are used in data analysis are descriptive statistics and inferential statistics. Descriptive statistics summarizes data from a sample using indexes such as the mean or standard deviation, whereas inferential statistics draws conclusions from data that are subject to random variation. Statistics is applied in machine learning and data mining, academia, computing, business, mathematics, and arts. Statistical distribution refers to the parameterized mathematical function which provides the probabilities of different outcomes for a random variable. This book is compiled in such a manner, that it will provide in-depth knowledge about the theory and practice of statistical distributions. Also included herein is a detailed explanation of the various concepts and applications of this field. This book aims to equip students and experts with advanced topics and upcoming concepts in this area.

The researches compiled throughout the book are authentic and of high quality, combining several disciplines and from very diverse regions from around the world. Drawing on the contributions of many researchers from diverse countries, the book's objective is to provide the readers with the latest achievements in the area of research. This book will surely be a source of knowledge to all interested and researching the field.

In the end, I would like to express my deep sense of gratitude to all the authors for meeting the set deadlines in completing and submitting their research chapters. I would also like to thank the publisher for the support offered to us throughout the course of the book. Finally, I extend my sincere thanks to my family for being a constant source of inspiration and encouragement.

Editor

1

Testing the Rasch model with the conditional likelihood ratio test: sample size requirements and bootstrap algorithms

Rainer W. Alexandrowicz[1*] and Clemens Draxler[2]

*Correspondence: rainer.alexandrowicz@aau.at
[1] Department of Psychology, Applied Psychology and Methods Research Unit, Universitaetsstr. 65, 9020 Klagenfurt, Austria
Full list of author information is available at the end of the article

Abstract

Background: The Rasch model allows for a conditional likelihood ratio goodness of fit test. The speed of approximation of the test statistic to the limiting distribution as a function of sample size and test length has not been analyzed so far. Three bootstrap simulation methods are analyzed with respect to their performance in providing a proper distribution of the test statistic under the null- and the alternative hypothesis.

Results: We found a stable approximation to the limiting χ^2-distribution for sample sizes of at least 500 and 10 items. The three bootstrap algorithms rendered consistent results for the H_0-cases but not for the H_1-cases.

Conclusion: A sequential probability sampling scheme proves sufficiently apt for generating samples under the alternative hypothesis. This superiority can be justified from a theoretical point of view.

Keywords: Rasch model, Conditional likelihood ratio test, Bootstrap analysis, Sequential importance sampling, Number of bootstrap samples

AMS Subject Classification: Primary 62F40; 62G09; 62G10; 62H10; 62E17; secondary 91B70

1 Introduction

The dichotomous logistic model according to Rasch (1960, 1966; henceforth denoted as Rasch Model, RM) allows for assessing its adequacy for describing a given data set by means of a conditional Likelihood Ratio Test (LRT; Andersen 1973). The test statistic is approximately χ^2-distributed if the sample size $n \to \infty$. Hence, small samples will deteriorate inference, i.e. the limiting distribution will not provide sufficiently precise quantiles for a reasoned decision and we have to switch to the bootstrap (cf. Efron and Tibshirani 1998), which is computationally demanding.

However, no systematic investigation has been undertaken so far to analyze the rate of approximation of the test statistic to the limiting distribution. It is therefore difficult to decide when it is safe to use the χ^2-distribution or when a bootstrap is required. This question shall be tackled in a simulation study. Moreover, if we switch to the bootstrap, precision depends on the number of bootstrap samples. A concrete guideline will be given, how many bootstrap replications are required to fulfill a desired precision criterion.

The following outline shall guide the reader through the details of this study:

Theoretical Background We start with explaining the fundamentals of the Rasch Model (Section 2.1) and the essential basics of model parameter estimation (Section 2.2) to an extent required to understand the simulation procedures applied in the study. Section 2.3 shows the basics of the LRT, the test statistic of which the study focusses. The task of determining the speed of approximation of the test statistic to its limiting χ^2-distribution breaks down into three separate questions, which are formulated in Section 2.4.

Methods In order to perform the simulation study, bootstrap samples in line with the RM have to be generated. For that purpose, several algorithms are at our disposal, which are introduced in Section 3.2. The study considers the distribution of the test statistic under both the null and the alternative hypothesis. These two scenarios require different simulation strategies, which are explained in Section 3.1. The simulation study covers numerous different scenarios, which may arise in practical application. Section 3.3 lists the simulation parameters considered for that purpose.

Results The complex details of the study are split into results concerning the H_0-case (Section 4.1) and the H_1-case (Section 4.2). Finally, Section 4.3 introduces a flexible formula to compute an adequate number of bootstrap samples, if this procedure is required.

2 Theoretical background

2.1 The Rasch model (RM)

The RM is a discrete probability model of a Bernoulli variable, $X_{vi} \in \{0, 1\}$, assuming two real valued parameters θ_v ($v = 1 \ldots n$) and β_i ($i = 1 \ldots k$),

$$P(X_{vi} = x_{vi}) = \frac{e^{x_{vi}(\theta_v - \beta_i)}}{1 + e^{(\theta_v - \beta_i)}}. \tag{1}$$

A typical application of model (1) is psychometrics, with θ_v describing respondent's v ability to solve a task (or item) and β_i describing the difficulty of task (or item) i. Both parameters are unbounded in value, i.e. $\theta_v, \beta_i \in \mathbb{R}$. By means of the substitutions $\xi_v = \exp(\theta_v)$ and $\epsilon_i = \exp(-\beta_i)$ we yield the so-called multiplicative notation of the model equation,

$$P(X_{vi} = x_{vi}) = \frac{(\xi_v \epsilon_i)^{x_{vi}}}{1 + \xi_v \epsilon_i}. \tag{2}$$

Due to the exponentiation, ξ_v and ϵ_i take positive values only, and ϵ_i is interpreted as an item easiness parameter. Conditional on both parameter vectors $\boldsymbol{\theta} = (\theta_1, \theta_2, \ldots, \theta_v, \ldots, \theta_n)^T$ and $\boldsymbol{\beta} = (\beta_1, \beta_2, \ldots, \beta_i, \ldots, \beta_k)^T$ the binary responses are assumed to be independent so that the joint distribution of all n responses to all k items ist given by the product of (1) (or (2), respectively) over v and i. This assumption is usually termed conditional or local independence.

The RM is a member of the exponential family (cf. Molenaar 1995, p. 41) with the sums $R_v = \sum_i X_{vi}$ and $S_i = \sum_v X_{vi}$ being the sufficient statistics for the parameters θ_v and β_i, respectively. The separability theorem (Fisher 1922) applies (Rasch 1966, p. 95; Rost 2001, p. 28), hence items can be compared independently of the ability parameters occurring in the sample and abilities can be estimated independently of the items used (given the items are in line with the model and the model holds for all respondents).

2.2 Parameter estimation

Several parameter estimation methods have been developed. Most straightforward from maximum likelihood theory is the Unconditional Maximum Likelihood approach (UML; or Joint Maximum Likelihood, JML; cf. Baker and Kim 2004, ch. 5.6). Here we determine estimates for both parameter vectors simultaneously by finding the maximum of the unconditional likelihood function as a function of $\boldsymbol{\theta}$ and $\boldsymbol{\beta}$. This is achieved by setting the partial derivatives equal to zero and applying the pertinent numeric methods to solve a system of nonlinear equations (cf. ibid., p. 136). However, this approach suffers from the so-called incidental parameter problem as expressed in Neyman and Scott (1948). While the item parameters appear as structural (or fixed) parameters, the person parameters constitute a random draw from the population and are therefore incidental (or nuisance) parameters. The simultaneous appearance of both kinds of parameters may cause inconsistent item parameter estimates. Corrective procedures have been proposed (cf. Molenaar 1995, p. 43), but there was dispute concerning their effect (cf. Baker and Kim 2004, ch. 5.6.2).

Generally, this incidental parameter problem may be overcome by marginalization or conditional inference (cf. Pawitan 2001, p. 274). In the first case (Marginal Maximum Likelihood estimation, MML, cf. Baker and Kim 2004, ch. 6), the incidental parameters θ_v are replaced by assuming a proper distribution $G(\theta)$ in the population (e.g. the normal), requiring only the hyperparameters $\boldsymbol{\tau}$ of $G(\theta)$ to be determined (i.e. the mean and the variance of $G(\theta)$ in our example). Although this solves the incidental parameter problem, the correct choice of $G(\cdot)$ is decisive for obtaining correct estimates (cf. Molenaar 1995, p. 47).

The second approach is the Conditional Maximum Likelihood estimation method (CML; Andersen 1970), directly involving the parameter separability feature. At its heart, this method overcomes the incidental parameter problem by conditioning on each respondent's observed value of the nuisance parameter's sufficient statistic, i.e. the score r_v, when estimating the item parameters. The conditional likelihood function L_C of the item parameter vector $\boldsymbol{\epsilon} = (\epsilon_1, \epsilon_2, \ldots, \epsilon_i, \ldots, \epsilon_k)^T$ given the vector of scores $\boldsymbol{r} = (r_1, r_2, \ldots, r_v, \ldots, r_n)^T$ can be written as

$$L_C(\boldsymbol{\epsilon}|\boldsymbol{r}) = \prod_v \left(\prod_i \frac{\epsilon_i^{x_{vi}}}{\gamma_{r_v}} \right). \tag{3}$$

The term γ_r denotes the elementary symmetric function of order r, covering a complex combinatorical task (cf. Andersen 1972; Formann 1986; Gustafsson 1980). Expression (3) plays a crucial role in the conditional Likelihood Ratio Test (LRT; see next Section).

In the CML context, the person parameters are determined in a separate step, where the $\hat{\beta}_i$ are assumed to be the true item parameters and the $\hat{\theta}_v$ are obtained using maximum likelihood estimation (cf. Hoijtink and Boomsma 1995). Because the score r_v is a sufficient statistic for the person parameter, all respondents yielding the same score will obtain the same ability estimate, which will be termed $\hat{\theta}_r$ or $\hat{\xi}_r$, respectively.

In terms of the Rasch Model, items being never or always solved (i.e. $s_i = 0$ and $s_i = n$) are infinitely difficult or easy, respectively. The same applies to respondents solving either no item or all items (i.e. $r_v = 0$ and $r_v = k$). While this is seldom a problem for the items (it is unlikely that in a sample of reasonable size an item is never or always solved), it may constitute a problem for person parameter estimation, especially, when the instrument is short (i.e. k is small). However, practicioners demand estimates for all respondents, so we have to make further assumptions in order to obtain parameter estimates for such cases as well. These may be obtained with the Weighted Maximum Likelihood Estimation Method (WLE; Warm 1989). Based on a Bayesian argument, person parameter estimates are decreased in their absolute value, thus attenuating their unbounded growth.

2.3 Assessing model fit

Numerous methods for assessing the adequacy of the RM have been proposed, an overview of which can be found in Glas and Verhelst (1995). The present study focusses on the conditional Likelihood Ratio Test (LRT, Andersen 1973; Kreiner and Christensen 2013), which relies on the CML estimation method.

If the model holds, item parameter estimates do not differ across subsamples but for random variation (invariability assumption, cf. Engelhard Jr. 2013). The LRT allows for an assessment of this assumption by comparing the conditional likelihood of the entire dataset according to Eq. (3), henceforth denoted L_0, with the product of the conditional likelihoods obtained from subsets $j = 1 \ldots g$ of the data set,

$$L_1 = \prod_{j=1}^{g} L_C^{(j)} \left(\boldsymbol{\epsilon}^{(j)} | \boldsymbol{r}^{(j)} \right). \tag{4}$$

Andersen (1973) has shown that the quantity

$$Z = -2 \log \frac{L_0}{L_1} \tag{5}$$

follows asymptotically a central χ^2-distribution with

$$df = (k-1)(g-1), \tag{6}$$

given that the Rasch Model is the true model and the subsample sizes $n^{(j)} \longrightarrow \infty$ (ibid., p. 128). Andersen referred to a split according to the score r_v, however, one may apply a criterion of substantive interest, like sex, treatment group, or a random split. In many applications, two groups are formed at the median of the score distribution. Without loss of generality, we will consider this median split in the present study.

2.4 Study purpose

The present study targets the following three questions:

Q1 How close fits the sampling distribution of (5) the central χ^2-distribution for small values of n and k, and in which cases a bootstrap simulation might be preferable due to lack of approximation? This question is analyzed for both the H_0-case of model fit (Results, Section 4.1) and for model violations under a given H_1 (Results, Section 4.2).

Q2 Second, do three pertinent bootstrap methods differ with respect to their preciseness in providing appropriate approximations of the type-I-error probability? These results are part of the tables of Sections 4.1 and 4.2.

Q3 And third, if a bootstrap is applied, which number of bootstrap replicates is required to obtain a sufficiently stable estimate of the desired quantile for the H_0 case (Results, Section 4.3)?

3 Methods

The three questions shall be tackled by means of a simulation study, determining the sampling distribution of the test statistic (5) for various combinations of n and k. Usually, a simulation study starts with fixing the population parameters of interest and drawing samples from this population. In our case, this would comprise fixing a set of k item parameters and n person parameters (or $k-1$ person parameters associated with each score r, respectively). However, the CML approach relies on the sufficient statistics of the person parameters. We would, therefore, have to find those r_v, which are associated with a given set of person parameters and item parameters. This task is difficult to achieve, hence we developed the following procedure:

- First, a set of k item parameters $\boldsymbol{\beta}^*$ and n person parameters $\boldsymbol{\theta}^*$ is fixed, representing the population of interest. The item parameters $\boldsymbol{\beta}^*$ were chosen equidistantly from the interval $[-1, 1]$ and person parameters $\boldsymbol{\theta}^*$ were randomly sampled from the $N(0, 1)$.
- Then, an initial sample $\boldsymbol{X}_0$ of size $n \times k$ in line with the assumptions of the Rasch Model is drawn from this population, yielding the realized values of the initially chosen parameters *and* the according sufficient statistics. The parameter estimates $\hat{\boldsymbol{\beta}}^0$ and $\hat{\boldsymbol{\theta}}^0$ of this initial sample $\boldsymbol{X}_0$ supersede the initially chosen $\boldsymbol{\beta}^*$ and $\boldsymbol{\theta}^*$. We now dispose of both the parameter values and the accompanying sufficient statistics, which are required for the bootstrap algorithm introduced in Section 3.2.3.

The sample $\boldsymbol{X}_0$ serves as the basis for the generation of bootstrap samples providing the distribution of the test statistic (5).

3.1 Sampling under the null and the alternative hypothesis

In order to obtain an inital data set $\boldsymbol{X}_0$ providing for the distribution of the test statistic under the null hypothesis of model fit, we take the overall parameter vector $\boldsymbol{\beta}^0$. This choice assumes no subgroup characteristics to be present.

In contrast, a data set $\boldsymbol{X}_0$ providing for the distribution of the test statistic under the alternative hypothesis is attained by separately bootstrapping $j = 1 \ldots g$ subsamples of size $n^{(j)}$ using the original subsamples' item parameter estimates $\hat{\beta}_i^{(j)}$. These subsample parameter vectors will necessarily differ, at least by chance, i.e. $\hat{\boldsymbol{\beta}}^{(1)} \neq \hat{\boldsymbol{\beta}}^{(2)} \neq \ldots \neq \hat{\boldsymbol{\beta}}^{(j)} \neq \ldots \neq \hat{\boldsymbol{\beta}}^{(g)}$. Merging these subsamples to one bootstrap sample of size $n = \sum_j n^{(j)}$ will therefore result in a sample violating the item invariance assumption. Hence, such a dataset constitutes a random draw from a population realizing the alternative hypothesis fixed at a model deviation, which is constituted by the item parameter differences of the $\hat{\boldsymbol{\beta}}^{(j)}$. If we now apply the LRT in the usual manner, the bootstrap distribution of the test statistic represents its distribution under the alternative.

3.2 Generating bootstrap samples

Several methods for generating bootstrap samples in the context of the RM have been proposed, two of which have gained some popularity. A third method, which to the authors' knowledge has not been described in the context of the Rasch Model before, is introduced in Section 3.2.3. These methods may cause differing distributions of the LR test statistic for reasons outlined below, what might affect the conclusions in an unpredictable manner (cf. Q2 in Section 2.4). Therefore, all three methods will be applied parallel in order to evaluate their impact upon the resulting distribution of the LR test statistic.

Note that the Nonparametric (or "Naïve") Bootstrap (cf. Davison and Hinkley 1997) is not suited for generating bootstrap samples in the CML context. This has theoretical reasons, which are elucidated in the light of the present findings in Section 5.3.

3.2.1 A "Normal" Approach

In the conditional estimation approach, unbiased item parameter estimates will be attained irrespective of the actual ability distribution. Therefore, the first simulation method uses only the item parameter estimates $\hat{\boldsymbol{\beta}}^{0}$, while the person parameters are sampled from a freely chosen distribution. In our case, this was the $N(0, \sigma^2)$, with randomly chosen but not too extreme values of σ^2. For notational ease, the hat will be omitted in the following.

The normal distribution has been chosen for it is arguably a proper candidate for numerous characteristics frequently assessed in those areas of social science, where the Rasch-Model is typically applied. This approach will be termed *normal marginals*, although, of course, the row marginal sums $r_1 \ldots r_n$ are discrete by nature; it is the underlying parameters that are sampled from the normal. The method has been described in van den Wollenberg (1982) and has gained some popularity for generating data compliant with the RM, wherefore it is considered in the present analysis.

3.2.2 Remaining with the observed

Here we use both the person parameter estimates and the item parameter estimates obtained from $\boldsymbol{X}_0$. The probability of a positive response is determined by using Eq. (1) and the according parameter estimates $\hat{\theta}_v$ and $\hat{\beta}_i$. However, this method raises two issues:

First, the CML method only allows for obtaining the item parameter estimates $\hat{\beta}_i$. For the estimation of the person parameters θ_v, the item parameter estimates $\hat{\beta}_i$ are taken as if they were the true parameters β_i. Hence, the random error associated with the item parameter estimates remains unconsidered, possibly rendering the person parameter estimates deficient. This could deteriorate the bootstrap procedure to an unknown extent.

Second, the ML estimate for respondents solving no item or all items would tend towards plus or minus infinity. Three ways of handling this situation could be thought of:

(i) The modified estimates (WLE) according to Warm (1989) could be applied instead. But, as has been elucidated in the last paragraph of Section 2.2, this method

systematically attenuates the person parameter estimates, making the implications for our bootstrap procedure imponderable.

(ii) The WLE could be inserted only for respondents with $r = 0$ and $r = k$, and the ML estimates otherwise. This would probably reduce the problem largely, as in most cases only few respondents (compared to the total sample size) will realize such scores. This method is implemented in the `WinMIRA` software of von Davier (2001), for example.

(iii) One can deliberately use arbitrary values for respondents with $r = 0$ and $r = k$, for example ± 15, so that the resulting score will almost surely be equal to 0 or k. Such a method is applied in the software package `M-Plus` (Muthén 1998–2004; p. 35).

However, any of these three approaches is heuristical and thus has to be considered as unsatisfactory from a statistical point of view.

Because the intention is to maintain the original ability distribution as far as possible, method *(iii)* was applied in the present study. Nevertheless, this approach will not preserve the individual scores r_v. Therefore, this approach will be termed *free marginals*, because the marginal scores are likely to differ from the original ones.

3.2.3 The Rasch point of view

In contrast, a sequential importance sampling procedure following a truly conditional approach will be taken into consideration. It merely regards the conditional pattern probabilities in the way they are used in the CML estimation method. Here, the sufficient statistics r_v for the person parameter estimates are conditioned upon, making any distributional assumptions superfluous. The probability of a response vector $\boldsymbol{x}_v = (x_{v1}, \ldots, x_{vk})$ conditional on the score equals

$$P(\boldsymbol{x}_v | r_v) = \gamma_{r_v}^{-1} \prod_i \epsilon_i^{x_{vi}}. \tag{7}$$

The algorithm starts with a respondent's observed score and computes his or her probability of solving the first item

$$P(x_{v1} = 1 | r_v) = \gamma_{r_v}^{-1} \epsilon_1 \gamma_{r_v - 1}. \tag{8}$$

Then, we transform this response probability to a manifest response by comparing it to a random number u drawn from the standard uniform distribution, i.e. $u \in U(0, 1)$. If $P(x_{v1} = 1 | r_v)$ exceeds u, the bootstrap respondent's v first manifest response x_{v1} is set to one and otherwise to zero (cf. van den Wollenberg 1982, p. 88). In case the response evaluates to one, this person's score r_v is reduced by one, otherwise not, yielding the modified score after step one, $r_v^{(1)}$. The procedure continues with the second item in the same manner and proceeds until all k items have been processed. As soon as the modified score after i steps, $r_v^{(i)}$, equals zero, the probability of solving one of the remaining items has zero probability and the corresponding responses are set to zero. If $r_v^{(i)}$ at any step i equals the number of the remaining items, all remaining responses are set to one.

That way, each individuals original score r_v is maintained, which is equivalent to fixing the row marginals of the observed data set $\boldsymbol{X}_0$. Therefore, this procedure will be termed *fixed marginals* in the following. Nevertheless, the items' sufficient statistics still vary according to the probability distribution described by the RM, which is the information the LRT relies upon.

3.3 Simulation parameters

The simulation comprised $k = 5$, 10, and 15 items and $n = 100$, 250, 500, 750, 1000, 2500, and 5000 observations. To each of the 21 possible combinations arising, which will be denoted designs, the three bootstrap algorithms (normal, free, and fixed) for both the H_0 and the H_1 case were applied. According to Eq. (6), the degrees of freedom were 4, 9, and 14, respectively.

One crucial aspect of the present study is to differentiate between inaccuracies due to a lack of approximation of the actual distribution of the test statistic to the limiting distribution (i.e. a truly statistical problem) on the one hand and an inaccurate bootstrap distribution due to an insufficient number of bootstrap samples (i.e. a merely technical problem) on the other hand. Preliminary trial runs suggested that $m = 200{,}000$ bootstrap replications seem to suffice for the required distinction. Assuming that this number of bootstrap replicates makes the bootstrap caused (i.e. technical) error negligible, any remaining deviation from the limiting density will be attributable to a true lack of approximation.

In order to determine the minimum number of bootstrap replicates required for a sufficiently good approximation of the bootstrap densities to the true ones under the null hypothesis (i.e. Q3 in Section 2.4), random samples of decreasing size $m^* < m$ have repeatedly been drawn with replacement from the original 200,000 samples of each design. The following values were chosen for m^*: 500, 1000, 1500, 2000, 2500, 5000, 7500, 10,000, 15,000, 20,000, 25,000, 50,000, 100,000, and 150,000. Each draw of size m^* was repeated 1000 times (m^{**}) in order to obtain a distribution of the LR test statistic for each m^*.

The densities obtained by means of the bootstrap will be depicted using kernel density estimators with bandwidth parameters of 0.2 to 0.5. The simulation itself was performed with the program `Ganz Rasch` (Alexandrowicz 2012), which supports all three simulation techniques introduced in Section 3.2. The simulation results were analyzed with R (R Core Team 2015).

4 Results

The results of the simulation study regarding questions Q1 and Q2 are presented separately for the H_0- and the H_1-case (Sections 4.1 and 4.2). Section 4.3 covers the results regarding question Q3, the required number of bootstrap samples.

4.1 Approximation under the null hypothesis

In order to describe the approximation of the bootstrap distributions to the limiting ones, we opposed the first four moments and selected quantiles of the estimated and the theoretical distribution of the test statistic; This step is accompanied by a density plot of the respective distributions. Second, the p-values of the LRT evaluated using both the bootstrap and the limiting distributions were opposed to each other.

4.1.1 Descriptive approach

Tables 4, 5 and 6 in the Appendix show the sample statistics of the 21 different designs for the fixed, the free, and the normal marginals case. The simulated values are opposed to the respective values of the χ^2-distribution with the according degrees of freedom (first row in each block). Each pair of columns denotes the estimated value of the statistic along with the relative deviation (in %) compared to the exact value of the limiting distribution. Two tendencies are discernible across all designs: The largest deviations can be found for the smallest sample size ($n = 100$) combined with the fewest items ($k = 5$, i.e. $df = 4$). For example, in the fixed marginals case (Table 4), the mean deviates by 10.4 % for 5 items and 100 observations, by 6.0 % for 10/100, and by 4.3 % for 15/100. In comparison, with 5000 observations, the respective deviations were -0.1 % ($k = 5$), 0.1 % ($k = 10$), and < 0.1 % ($k = 15$).

Comparing the three simulation algorithms reveals the smallest relative deviation to appear for the bootstrap technique using normal marginals. For example, the relative error regarding the 95 %-quantile (being most important for hypothesis testing) is 2.2 % ($df = 4$), 1.2 % ($df = 9$), and 0.9 % ($df = 14$). The respective figures for the fixed marginals case are 2.6 %, 1.2 %, and 1.0 %, while the normal marginals produces deviations of 0.8 %, 0.5 %, and 0.6 %.

The density plots (Fig. 1) allow for a rough assessment of the overall fit of the bootstrap distributions. The plot is threefold (normal, free, and fixed marginals) with three clusters of densities according to the *df*. Each line represents a certain sample size (i.e. seven per cluster). As can be seen, the seven lines per method and *df* cannot be kept apart in any of the plots, therefore no attempt was made to label the lines. Also, three lines indicating the limiting densities with the respective degrees of freedom have been superimposed (red lines). However, they mostly disappear in the three clusters, indicating overall agreement of the bootstrap generated distributions with the limiting distributions.

In the normal marginals case (Fig. 1, left hand plot), we would virtually identify no deviation of the bootstrap densities from the limiting ones. This observation was independent of the degrees of freedom and the sample size. In the other two cases (free and fixed marginals), one line per cluster seems somewhat dislocated, indicating a reduced probability of smaller χ^2-values and slightly heavier tails (the latter is hardly

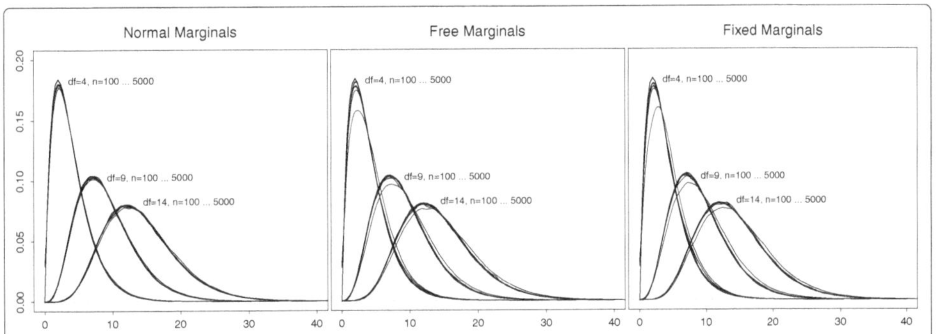

Fig. 1 Bootstrap densities of the test statistic using normal, free, and fixed marginals. Each density cluster represents same degrees of freedom; lines cover the sample sizes involved. The red lines indicate the limiting distribution with the according degrees of freedom

discernible). These lines represent the $n = 100$ cases, in which the approximation seems deficient.

For both the free and the fixed marginals method (Fig. 1, middle and right hand plot), the bootstap densities for $df = 4$ are located beyond the limiting distributions. The densities covering $df = 9$ show the same tendency but to a much slighter extent, and when $df = 14$, this effect vanishes entirely. The free and the fixed marginals method seem not to differ with respect to this issue. Therefore, we can conclude that the approximation improves with the degrees of freedom, which is in line with theory. Roughly ten degrees of freedom seem sufficient for the approximation to be considered satisfactory, given a sample size of at least 250.

4.1.2 Inferential Approach

In order to compare the bootstrap distributions with the according limiting ones, we used the Kolmogorov-Smirnov (K-S) test (cf. Thode 2002, ch. 5.1.1) to test the null hypothesis that the bootstrap distributions follow a χ^2-distribution with the respective degrees of freedom (i.e. 4 in the 5-items designs, 9 in the 10-items designs, and 14 in the 15-items designs). The results are given in Table 7 in the Appendix.

Many of the tests yield a significant result using a type-I error risk of 5 %. However, we recognize some non-significant results for *(a)* larger samples and *(b)* larger instruments: Non-significant results were obtained for the combinations (k/n) 5/5000 (fix), 10/1000 (free + nv), 10/5000 (nv), 15/750 (nv), 15/2500 (fix + free), and all three combinations 15/5000. The normal marginals appeared slightly better, as 4 out of 21 tests were not significant, opposed to 3 out of 21 for both the fixed and the free marginals cases. This is in line with the findings based on the descriptive statistics above.

Note that the K-S-tests rely on 200,000 bootstrap samples each, hence they are by far overpowered. Assuming that the probability of an error of the second kind almost vanishes with such large samples, a non-significant result corroborates our supposition that the bootstrap distribution in fact realizes the limiting distribution. Therefore, the mere fact that at least some of the tests were not significant is in fact a remarkable result. If we further look at the values of the K-S test statistics D: None of them exceeds 0.07 (which appeared with 5/100, fixed marginals algorithm). The K-S test evaluates the maximum difference of the CDFs of the bootstrap generated distributions and the limiting ones, which diverge at most by 7 % in the cases considered here.

4.1.3 Comparison of the p-values

Two comparisons shall enhance conclusions from a practical point of view: First, we emulate the action a person ignorant of approximation problems might take. This means, we simply use the seeked quantile of the limiting χ^2-distribution (e.g. 9.49 for $\alpha = 0.05$ and $df = 4$) and decide whether or not to retain the H_0. For that purpose, we calculated the p-value the χ^2-quantile of the limiting distribution would yield when applied to the according bootstrap distribution (reflecting the proper probability measure). The (relative) difference of the p-value to the nominal α quantifies how misleading such a procedure would be. Table 8 in the Appendix presents this comparison for the three simulation algorithms, considering critical values of $\alpha = 0.10$, 0.05, and 0.01.

The p-values were considerably increased for small samples, yielding differences of up to 52 % ($\alpha = 0.01$, $n = 100$, $k = 10$, nv). Generally, sample sizes 100 and 250 and (in some instances) 500 would lead to substantially more significant results if we decided to use the quantiles of the limiting distribution rather than the bootstrap-based ones. Comparing the three simulation algorithms revealed that the normal marginals procedure performed somewhat better than the free and the fixed marginals algorithm. However, the discrepancies of the three methods are mostly moderate. Further, errors are more pronounced for low values of the type-I-error risk. As soon as samples exceed 500 observations, the deviations become increasingly smaller.

Remember that each bootstrap analysis is based on an actual realization of the test statistic (5), allowing for a second check: Table 1 compares the observed values of these test statistics applied to both the limiting χ^2-distribution and to the respective bootstrap generated distribution. As can be seen, most differences seem negligible. This is not surprising, because we now consider values far away from the regions relevant to inferential decisions (i.e. the distributions' tails). Hence too heavy tails of the bootstrap distributions

Table 1 Comparison of p-values of the observed test statistics evaluated at both the limiting and the bootstrap distribution

	Normal marginals			Free marginals			Fixed marginals		
k = 5/df = 4	χ_l^2	χ_b^2	Diff.	χ_l^2	χ_b^2	Diff.	χ_l^2	χ_b^2	Diff.
100	0.01	0.01	0.000	0.01	0.01	0.001	0.01	0.01	0.000
250	0.97	0.97	0.001	0.97	0.96	0.008	0.97	0.96	0.008
500	0.03	0.03	0.000	0.03	0.03	0.001	0.03	0.03	0.000
750	0.39	0.39	0.000	0.39	0.39	0.004	0.39	0.38	0.007
1000	0.69	0.69	0.001	0.69	0.68	0.004	0.69	0.68	0.004
2500	0.44	0.44	−0.001	0.44	0.44	0.000	0.44	0.44	0.001
5000	0.96	0.96	0.001	0.96	0.96	0.000	0.96	0.96	−0.001
k = 10/df = 9									
100	0.48	0.46	0.017	0.48	0.43	0.046	0.48	0.43	0.049
250	0.32	0.32	0.006	0.32	0.31	0.010	0.32	0.31	0.010
500	0.96	0.95	0.002	0.96	0.96	0.000	0.96	0.95	0.002
750	0.63	0.63	0.002	0.63	0.63	0.004	0.63	0.63	0.000
1000	0.79	0.79	−0.002	0.79	0.79	0.000	0.79	0.79	0.001
2500	0.26	0.26	0.000	0.26	0.25	0.000	0.26	0.25	0.001
5000	0.31	0.31	0.000	0.31	0.31	0.002	0.31	0.31	0.002
k = 15/df = 14									
100	0.08	0.07	0.006	0.08	0.07	0.011	0.08	0.07	0.010
250	0.60	0.59	0.007	0.60	0.58	0.012	0.60	0.59	0.009
500	0.07	0.07	0.001	0.07	0.07	0.002	0.07	0.07	0.003
750	0.58	0.58	0.001	0.58	0.57	0.002	0.58	0.58	0.000
1000	0.96	0.96	0.002	0.96	0.96	−0.001	0.96	0.96	0.001
2500	0.34	0.34	0.003	0.34	0.34	0.000	0.34	0.34	0.001
5000	0.63	0.63	0.001	0.63	0.63	0.000	0.63	0.63	0.000

Notes: χ_l^2 : limiting distribution of the test statistic; χ_b^2: bootstrap distribution of the test statistic; Diff. = Difference of p-values

are compensated by regions of decreased probabilities for lower values of the test statistic.

4.2 Approximation under the alternative hypothesis

The analysis of the three bootstrap methods under the alternative refers to comparing the observed differences between the three methods with the given sample sizes. At this point, we have to keep in mind that each original sample $\boldsymbol{X}_0$ constitutes a random realization of a population distribution of its own, hence the bootstrap generated distributions of the test statistic differ across the various designs (i.e. each combination of number of items k and sample size n) and are therefore incomparable. Any attempt to achieve the same subgroup parameter would go beyond the objectives of the present study.

Table 2 shows the descriptive statistics for the three bootstrap methods. We see considerable differences in some cases: For example, in the $k = 5/n = 250$ design, the mean of the bootstrap distribution generated with the normal marginals method is more than three times larger than in the free or the fixed marginals case (the latter two being fairly similar). A similar tendency occurs in the $k = 5/n = 750$ design, also for the normal marginals method, yet to a weaker extent. In contrast, the fixed and the free marginals method yielded fairly similar distributions for all designs.

In order to rule out technical reasons for the unexpected distributions, simulations of the 5-items designs have been repeated twice. However, the results were virtually identical, highly deviating distributions appeared repeatedly, with no apparent pattern regarding sample size (in the first repetition, the phenomenon occurred with $n = 500$, $n = 750$, and $n = 2500$, and in the second with $n = 250$, $n = 500$, and, to a lesser extent, $n = 500$, $n = 2500$, and $n = 5000$). In no case, such pecularities were to observe with any of the other two algorithms, i.e. fixed or free marginals.

4.3 Number of bootstrap samples

When applying the bootstrap, we have to decide on the number m^* of bootstrap samples, required to obtain sufficiently precise results in justifiable time, as bootstrapping may consume a considerable amount of time for large data sets and/or many items. First of all, a means is required to summarize the (loss of) precision when m^* decreases. Because it is a common choice to evaluate the test statistic with respect to the 95 %-quantile of the appropriate limiting distribution, we will concentrate on this measure.

Rather than starting a new simulation, we resorted on the vast amount of data already at hand: The original simulation covered 7 sample sizes times 3 item counts times 3 algorithms, which totals in $7 \times 3 \times 3 = 63$ vectors, each containing $m = 200,000$ realizations of the test statistic (5). In order to evaluate the effect of less than 200,000 bootstrap samples ($m^* < m$), we drew random subsamples of 14 different sizes m^* from each of the 63 vectors, each repeated 1000 times. For each of these 882×1000 samples, we determined the empirical 95 %-quantile, yielding 882 vectors containing 1000 estimates of the quantile under consideration, $\hat{q}_{.95}$. (For notational ease, the index will be omitted.)

The minimum and maximum value per vector would express a worst case appraisal of error to be found in the simulated data sets. However, in order to avoid singular outliers to detract from a more general perspective, the five most extreme values in each direction were averaged, a procedure which can be considered a stabilized minimum and maximum. The difference of these two figures is divided by the corresponding quantile of the according limiting distribution, which makes the measure comparable across all designs. It will be termed *relative range*, *rr*:

$$rr = \frac{\frac{1}{5}\sum_{i=n-4}^{n} \hat{q}_{(i)} - \frac{1}{5}\sum_{i=1}^{5} \hat{q}_{(i)}}{\chi^2_{[.95;df]}}, \tag{9}$$

with $\hat{q}_{(i)}$ denoting the sorted values of the quantile estimates per combination of n, k, algorithm, and m^*.

Figure 2 shows the relative range by number of bootstrap replications m^*. Two clear structures are discernible: First, all lines exhibit a (negative) logarithmic shape without exemption, with deviations rapidly decreasing with increasing number of bootstrap samples. Second, the larger the number of items k, the faster the deviations decrease together with increasing m^*. However, the latter phenomenon is considerably smaller than the first one.

Due to the clear shape of the curves, we tried to formulate a general model predicting the required number of bootstrap replicates given a desired precision criterion in terms of rr. In order to apply a linear model, the logarithm of the relative range rr and the negative logarithm of the bootstrap replication number m^* were taken. The algorithm was dummy coded (fr serving as reference category) and the number of variables k and the sample size n were directly entered into the model equation

$$y = \beta_0 + \beta_1 \log(m^*) + \beta_2 k + \beta_3 n + \beta_{4[1]} fx + \beta_{4[2]} nv, \tag{10}$$

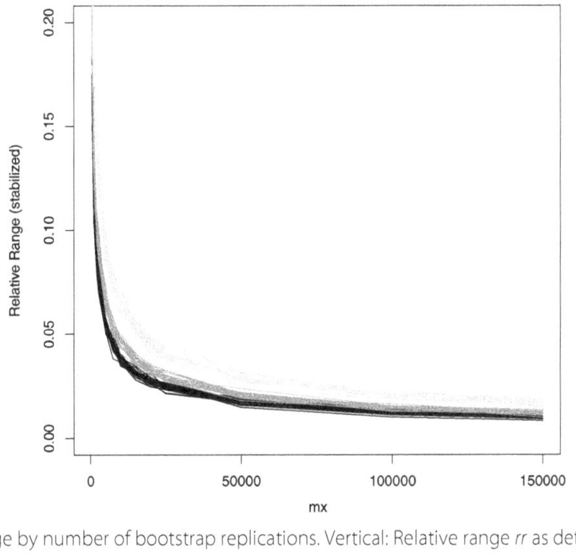

Fig. 2 Relative range by number of bootstrap replications. Vertical: Relative range *rr* as defined in Eq. (9); Horizontal: number of bootstrap replications, m^*. The 63 lines indicate the 3 algorithms × 3*k* × 7*n* different simulation designs. Grey shadings indicate *k*: light = 5 items, medium = 10 items, and dark = 15 items

with $y = -\log(rr)$ and $\beta_{(\cdot)}$ denoting the regression coefficients. Their estimates are given in Table 3 along with the respective significance measures.

The model R^2 equalled .994, which indicates a good fit of model (10). Aside of the intercept β_0, the coefficients regarding the bootstrap replication number, β_1, and the number of items, β_2, were significantly different from zero, but not those for the sample size, β_3, or the simulation algorithm, $\beta_{4[\cdot]}$.

This affirms the impression already derived from Fig. 2 that m^* and (to a much lesser extent) k suffice for the determination of the precision of the bootstrap analysis. From the coefficients indicated in Table 3, a rule of thumb has been developed to obtain a rough estimation of the required number of bootstrap samples (the coefficients were rounded):

$$\widehat{m}^* = \exp\big(4 - 0.1k - 2\log(rr)\big). \tag{11}$$

If, for example, one wants to test 8 items with the LRT using two split groups, then the critical value $\chi^2_{[.95;7]} = 14.067$. The 95 % quantile of the bootstrap distribution shall not exceed the interval $[13, 15]$ (which complies with the probabilities .964 and .927, respectively), the range is two and the relative range is $rr = 2/14.1 = 0.142$ (note that the deviations do not behave symmetrically, but this seems negligible in order to obtain a rough estimation of m^*). Then, the optimal number of bootstrap replicates according to (11) amounts to 1219, hence 1200 bootstrap samples will be a good choice.

5 Discussion

The present study focuses on practical issues when applying the Likelihood Ratio statistic according to Andersen (1973) for testing the binary Rasch Model. If the model holds, the Likelihood Ratio test statistic approaches the limiting distribution to a sufficient extent even in cases where samples were small or items were few. The most problematic combination of 5 items and 100 observations revealed moderate deviations from the limiting distribution. But even in the most problematic cases the CDFs of the bootstrap and the according limiting distribution differed by no more than 7 %, which seems justifiable to us.

Generally, the approximation of the test statistic under the null-hypothesis shows sufficient approximation to the theoretical distribution if samples comprise at least 500 respondents and an instrument with more than ten items is considered. For studies considering smaller samples or fewer items we recommend the more expensive bootstrap method. However, this is little a drawback as bootstrapping small samples takes only a reasonable amount of time. In order to further control the required time, Eq. (11) provides an easily applicable rule of thumb allowing to limit the number of bootstrap samples warranting a precision criterion of interest.

5.1 Size Matters

But so far, only the type-I error probability of falsely rejecting the null hypothesis has been taken into consideration. It will be overpowered if samples are large, hence irrelevant model deviations will become significant, although they might be acceptable from a substantial point of view, which, in turn, might give rise to generally scorn the LRT as such.

However, a significance test has its merits as well, as it allows to rely on the decision criterion of statistical significance, which is fundamental in scientific reasoning. In order to avoid the propagation of unsubstantiated rules of thumb (cf. Maxwell 2000), a prospective power analysis (in the sense of Cohen 1988) is required. For that purpose, the simulation technique presented here provides a reliable means to obtain the required non-central distributions of the test statistic.

5.2 Simulation technique

In the present study, three pertinent bootstrap algorithms, which we termed normal marginals, free marginals, and fixed marginals method, have been compared. While there was hardly any difference in the null hypothesis cases, some striking differences were encountered for the non-central distributions, deserving further inspection: The LRT assumes the rowsums r_v to be fixed at their observed values, therefore the fixed marginals bootstrap adopts this assumption. Any deviation from the observed scores inevitably yields a different likelihood and the sampling distribution of the test statistic will change.

Interestingly, the present study revealed that such a change primarily occurs in the non-central case, which can be explained: If we let the rowsums vary freely (as has been done in the normal marginals and the free marginals case), score frequencies change as well. Now, in the central case, the same item parameter estimates $\hat{\boldsymbol{\beta}}^{(0)}$ are used for all subsamples, which supports the assumptions made in the null-hypothesis. But in the non-central case, possibly differing subgroup estimates are used for generating the bootstrap samples. If, say, score group two yields highly deviating estimates $\hat{\boldsymbol{\beta}}^{(2)}$, but the score two has (by chance) only sometimes occurred, the deviation will not be much reflected in the test statistic. But if the same score group would have appeared with a high frequency, the deviating estimates will considerably change the product of the subgroup likelihoods in Eq. (4) and the the test statistic will reflect the model violation. Hence, the test statistic (and, in turn, its bootstrap distribution) depends on the relative frequencies of the scores, which explains the observed differences of the three bootstrap methods considered.

Hence, the fixed marginals method has to be considered superior, not only for the theoretical reasons outlined above, but also for the present study revealed in certain cases the differences to be striking. The normal marginals method yielded problematic distributions of the test statistic (5) when simulating the H_1-distribution, which may also be explained: If we split along the score r_v (using the median, for example), score distributions in the subsamples will inevitably differ, causing the observed differences. Therefore, the normal marginals method is not eligible for that purpose.

5.3 Don't be naïve!

As has been mentioned above, the naïve bootstrap (i.e. drawing response vectors with replacement from the original sample, cf. Davison and Hinkley 1997) cannot be applied to the present problem: In the specific case of the LRT, this would cause the split group membership to be drawn at random as well, thus changing the subgroup frequencies n_j of each bootstrap sample in an entirely unpredictable manner (the likewise argument exists for the case of regression analysis, cf. Enders 2010, p. 150).

One might therefore consider to draw the observations separately from the original subsamples. However, this would not yield the desired results: Only the response patterns of the possibly differing subgroup members form each group then, hence we end up with the distribution of the test statistic under the alternative hypothesis.

5.4 Subgroup frequencies

The similarities of the three algorithms in the H_0-case could be traced back to the fact that the original samples $\boldsymbol{X}_0$ have been generated with $\boldsymbol{\theta} \sim N(0, 1)$: All bootstrap procedures reproduced the marginals' distributions exceptionally well, what will not be necessarily the case in a practical application. Hence, we should not use the free or the normal marginals method but the fixed marginals method to perform a power analysis in the sense of Cohen (1988).

However, this algorithm has far-reaching consequences for further research: If one wanted to perform a power analysis of the LRT by means of a simulation study using the fixed marginals method, he or she would have to consider both the item parameters and the score group frequencies. But unfortunately, we face a technical complication here: In order to vary the sample size seamlessly (which is necessary to obtain the optimal sample size), the relative subgroup frequencies have to be maintained. For twice the original sample (or any other integer multiple), each observation can be drawn twice (or three times, and so on). But for any other sample size, the relative frequencies n_j/n would have to be carefully approximated, allowing for a sufficient reproduction of the marginals with increasing n.

5.5 How many bootstrap samples?

One question, which always has to be considered when applying of the bootstrap, is the number of bootstrap replicates that have to be generated. For this purpose, a very general solution has been found in the present study. Within the parameters considered, the expected maximum deviation of the 95 % quantile of the true distribution can be determined using the number of samples and the number of items. Of course, Eq. (11) could be extended to any other measure of interest, like another quantile, for example. The present approach demonstrated the feasibility of a means to generally determine the required number of bootstrap samples.

5.6 Limitations and outlook

One limitation of the present study can be seen in the fact that only the two groups sample split has been considered. However, the procedure presented here allows for a straightforward extension to any number of split groups. Further, this seems to be only a minor obstactle for the practical application of the present results, as available sample sizes seldom allow for splitting into more than two groups.

The LRT can also be applied to polytomous extensions of the RM, like, for example, the Rating Scale Model (Andrich 1978), or the Partial Credit Model (Masters 1982). Power considerations for these models have to be tackled separately, as the number of parameters to be estimated changes with the number of response categories involved. Further, the LRT plays a crucial role in testing the linear logistic extension of the RM (Linear Logistic Test Model, LLTM, Fischer 1973), where, the likelihood of the (empirically more restrictive) LLTM is opposed to that of the RM (cf. Alexandrowicz 2011). Again, the methods described here can be adapted accordingly.

The results obtained in the present study have two important implications: First, we are able to obtain the distribution of the test statistic under a fixed alternative by means of the recommended bootstrap method. This allows for determining the required sample size to detect a model violation which is considered relevant from a substantive point of view with fixed risks for the type-I and the type-II errors. And second, the necessary number of bootstrap replications for warranting a desired precision can be obtained. One might object that Eq. (11) only covers up to 15 items and may therefore not be used for larger instruments. But as we have seen, the larger the number of items, the fewer bootstrap samples are necessary given everything else is held constant. Therefore, one is on the safe side using a minimum of 500 bootstrap samples for data sets comprising more items. The same applies to sample sizes beyond those considered here.

One reason impairing the applicability of the LRT is that the power of the test for a given model deviation could not be determined. As a consequence, no sample planning was possible, leaving the researcher in the dark whether a significant result indicates a model deviation of substantial interest or was merely the consequence of too large a sample. However, this fundamental problem has been overcome by Draxler (2010) for the Wald-test and generalized to the LRT and the Rao-Score-test by Draxler and Alexandrowicz.

These present results allow to determine the optimal sample size required to detect a model deviation considered relevant from a substantial point of view with given risks of an error of the first and the second kind. We believe that the LRT is a valuable tool for testing whether an instrument allows for establishing a measurement and the present findings will facilitate its liable utilization.

6 Conclusion

The test statistic of the conditional Likelihood Ratio Test approximates its limiting distribution very fast. Only the combination of 5 items and 100 respondents revealed slight deviations, however, increasing either the number of items or the sample size will allow for employing the quantiles of the respective χ^2-distribution in the usual manner. Hence, the cLRT may be applied with confidence in many situations. All three bootstrap algorithms perform well under the null hypothesis and provide reliable estimates of the quantiles required for testing the null hypothesis of model fit taking the error of the first kind into consideration. In order to take the error of the second kind into account as well, we have to find the according non-central distribution of the test statistic given a specified model deviation. Here, the three simulation methods differed considerably. We recommend the newly introduced technique warranting the row marginals to remain at their observed values for this technique has to be considered superior on theoretical reasons. Finally, we provide an easy to apply formula for identifying the necessary number of bootstrap samples allowing to limit the bootstrap-related error to a freely definable degree. Especially in studies involving a large set of items or sample, this formula will prove useful to perform the bootstrap in a reasonable amount of time.

Appendix

Table 2 Moments and quantiles of the non-central bootstrap distributions

k/df	*n*	meth	mean	var	cv	skew	kurt	q50	q90	q95	q99
5/4	100	nv	4.46	9.80	0.70	1.39	2.90	3.75	8.65	10.54	14.70
		free	4.65	10.07	0.68	1.31	2.64	4.01	8.89	10.73	14.82
		fix	4.72	9.78	0.66	1.24	2.29	4.10	8.92	10.75	14.70
	250	nv	50.60	123.32	0.22	0.25	0.02	50.14	65.15	69.58	78.32
		free	14.89	27.97	0.36	0.75	0.86	14.19	21.97	24.66	30.01
		fix	13.43	23.23	0.36	0.78	0.82	12.66	20.03	22.35	27.03
	500	nv	5.07	12.34	0.69	1.34	2.63	4.30	9.81	11.83	16.42
		free	4.65	10.64	0.70	1.38	2.87	3.91	9.01	10.95	15.29
		fix	4.66	10.75	0.70	1.39	2.87	3.92	9.06	11.02	15.41
	750	nv	9.36	29.00	0.58	0.98	1.32	8.45	16.61	19.47	25.63
		free	6.77	18.59	0.64	1.16	1.89	5.93	12.61	15.03	20.20
		fix	6.77	18.47	0.63	1.15	1.90	5.94	12.58	14.99	19.98
	1000	nv	11.61	38.43	0.53	0.89	1.08	10.65	19.97	23.14	29.83
		free	8.64	26.15	0.59	1.05	1.57	7.75	15.51	18.34	24.17
		fix	8.80	26.28	0.58	1.03	1.50	7.93	15.66	18.47	24.53
	2500	nv	9.82	31.20	0.57	1.00	1.46	8.91	17.33	20.24	26.67
		free	6.95	19.47	0.63	1.14	1.83	6.10	12.91	15.39	20.71
		fix	7.01	19.70	0.63	1.14	1.79	6.15	13.03	15.51	20.78
	5000	nv	20.44	73.41	0.42	0.67	0.63	19.49	31.91	35.99	44.45
		free	13.48	44.64	0.50	0.82	0.93	12.56	22.48	25.80	32.77
		fix	13.95	45.78	0.48	0.79	0.88	13.06	23.05	26.33	33.54
10/9	100	nv	16.63	48.66	0.42	0.76	0.84	15.73	25.98	29.40	36.58
		free	16.93	49.66	0.42	0.71	0.67	16.07	26.36	29.88	36.84
		fix	17.76	53.02	0.41	0.70	0.66	16.90	27.50	31.02	38.17
	250	nv	15.23	43.00	0.43	0.80	0.98	14.37	24.01	27.26	34.21
		free	15.23	42.59	0.43	0.78	0.85	14.36	23.97	27.25	34.03
		fix	15.83	45.04	0.42	0.77	0.83	14.97	24.82	28.11	35.05
	500	nv	24.60	80.08	0.36	0.60	0.47	23.66	36.55	40.73	49.11
		free	25.09	82.19	0.36	0.60	0.52	24.17	37.13	41.32	50.00
		fix	26.38	88.22	0.36	0.61	0.52	25.42	38.92	43.31	52.47
	750	nv	17.83	53.23	0.41	0.73	0.77	16.95	27.59	31.15	38.46
		free	18.13	54.32	0.41	0.73	0.76	17.21	28.00	31.63	39.10
		fix	18.83	57.38	0.40	0.72	0.77	17.90	28.96	32.67	40.32
	1000	nv	19.13	58.74	0.40	0.71	0.71	18.22	29.33	33.11	40.76
		free	20.03	62.19	0.39	0.69	0.67	19.13	30.53	34.31	42.26
		fix	21.12	66.66	0.39	0.65	0.57	20.24	32.04	35.97	43.88
	2500	nv	14.21	38.70	0.44	0.82	0.91	13.34	22.57	25.73	32.28
		free	14.47	39.67	0.44	0.80	0.89	13.63	22.97	26.10	32.56
		fix	14.96	42.22	0.43	0.80	0.93	14.11	23.64	26.98	33.72
	5000	nv	15.19	42.61	0.43	0.79	0.90	14.31	23.92	27.23	33.92
		free	15.14	42.54	0.43	0.80	0.92	14.27	23.86	27.15	33.93
		fix	15.49	43.84	0.43	0.77	0.76	14.62	24.38	27.65	34.50
15/14	100	nv	21.08	56.32	0.36	0.68	0.66	20.23	31.10	34.73	42.13
		free	21.59	58.86	0.36	0.66	0.61	20.74	31.86	35.45	42.99
		fix	21.98	59.95	0.35	0.62	0.50	21.15	32.32	35.97	43.36
	250	nv	27.02	80.26	0.33	0.60	0.52	26.13	38.94	43.13	51.73
		free	28.16	84.17	0.33	0.57	0.44	27.30	40.38	44.68	53.11
		fix	28.84	87.07	0.32	0.56	0.38	27.92	41.25	45.48	54.39

Table 2 Moments and quantiles of the non-central bootstrap distributions (*Continued*)

k/df	*n*	meth	mean	var	cv	skew	kurt	q50	q90	q95	q99
	500	nv	20.85	55.05	0.36	0.66	0.62	20.05	30.72	34.36	41.78
		free	20.85	55.55	0.36	0.67	0.64	20.01	30.76	34.34	41.62
		fix	21.13	56.11	0.35	0.69	0.71	20.25	31.10	34.78	42.27
	750	nv	27.01	80.95	0.33	0.60	0.51	26.08	39.02	43.18	51.58
		free	27.86	82.61	0.33	0.59	0.60	27.01	39.87	44.06	52.72
		fix	28.37	85.27	0.33	0.58	0.49	27.50	40.65	45.00	53.45
	1000	nv	36.14	115.14	0.30	0.49	0.32	35.28	50.30	55.18	65.02
		free	37.22	121.78	0.30	0.51	0.34	36.31	51.83	56.89	66.78
		fix	38.39	124.94	0.29	0.49	0.35	37.49	53.20	58.22	68.15
	2500	nv	24.07	68.46	0.34	0.62	0.54	23.20	35.08	38.98	47.12
		free	24.78	71.02	0.34	0.62	0.55	23.90	36.02	39.95	48.09
		fix	25.39	73.60	0.34	0.61	0.54	24.51	36.81	40.85	48.82
	5000	nv	27.56	81.96	0.33	0.58	0.45	26.69	39.65	43.83	52.29
		free	28.52	85.93	0.33	0.56	0.42	27.65	40.82	45.09	53.90
		fix	29.17	89.11	0.32	0.58	0.47	28.26	41.71	46.07	55.00

Table 3 Coefficients of the linear model predicting the negative log of the relative error from the log of the number of bootstrap samples, the number of items, the sample size and the bootstrap algorithm

Par.	Pred.	Est.	S.E.	*t*-value	*Pr* (> |*t*|)
β_0	Intcpt	−2.079	0.014	−144.488	< .001
β_1	m^*	0.498	0.001	361.291	< .001
β_2	k	0.052	0.001	91.534	< .001
β_3	n	0.000	0.000	1.009	0.313
$\beta_{4[1]}$	$fr : fx$	−0.004	0.006	− 0.719	0.472
$\beta_{4[2]}$	$fr : nv$	−0.002	0.006	− 0.291	0.771

Table 4 Descriptive statistics for the fixed marginals case under the null hypothesis

	mean	%	var	%	cv	%	skew	%	kurt	%	q50	%	q90	%	q95	%	q99	%
limiting: df = 4	4.00		8.00		0.71		1.41		3.00		3.36		7.78		9.49		13.28	
k = 5/n = 100	4.42	10.4	8.86	10.7	0.67	−4.7	1.23	−12.8	2.17	−27.7	3.81	13.5	8.45	8.6	10.17	7.2	13.82	4.1
k = 5/n = 250	4.19	4.7	8.88	10.9	0.71	0.6	1.39	−1.9	2.74	−8.6	3.47	3.4	8.22	5.7	9.99	5.3	13.91	4.8
k = 5/n = 500	4.10	2.4	8.53	6.7	0.71	0.9	1.44	1.8	3.10	3.2	3.42	1.9	7.98	2.6	9.81	3.4	13.70	3.2
k = 5/n = 750	4.06	1.5	8.26	3.3	0.71	0.2	1.43	1.2	3.12	3.9	3.40	1.3	7.90	1.5	9.60	1.2	13.46	1.4
k = 5/n = 1000	4.02	0.6	8.05	0.7	0.71	−0.3	1.40	−0.7	2.96	−1.3	3.38	0.7	7.81	0.4	9.54	0.6	13.29	0.1
k = 5/n = 2500	4.02	0.5	8.19	2.3	0.71	0.7	1.43	1.2	3.03	1.2	3.37	0.4	7.83	0.6	9.57	0.9	13.51	1.8
k = 5/n = 5000	4.00	−0.1	7.99	−0.2	0.71	−0.0	1.43	0.9	3.19	6.4	3.36	0.1	7.76	−0.2	9.47	−0.2	13.14	−1.0
avg. (df = 4)	4.12	2.9	8.39	4.9	0.70	−0.4	1.39	−1.5	2.90	−3.3	3.46	3.0	7.99	2.7	9.74	2.6	13.55	2.1
limiting: df = 9	9.00		18.00		0.47		0.94		1.33		8.34		14.68		16.92		21.67	
k = 10/n = 100	9.54	6.0	20.05	11.4	0.47	−0.4	0.89	−5.9	1.10	−17.7	8.88	6.4	15.55	5.9	17.87	5.6	22.77	5.1
k = 10/n = 250	9.14	1.6	18.67	3.7	0.47	0.3	0.95	1.2	1.36	2.3	8.47	1.5	14.93	1.7	17.23	1.8	22.07	1.9
k = 10/n = 500	9.08	0.9	18.32	1.8	0.47	−0.0	0.95	0.6	1.42	6.4	8.43	1.0	14.80	0.8	17.07	0.9	21.83	0.8
k = 10/n = 750	9.02	0.2	18.22	1.2	0.47	0.4	0.95	0.3	1.30	−2.6	8.33	−0.2	14.75	0.5	16.97	0.3	21.81	0.7
k = 10/n = 1000	9.01	0.1	17.93	−0.4	0.47	−0.3	0.94	0.1	1.37	2.8	8.34	−0.0	14.69	0.0	16.92	0.0	21.64	−0.1
k = 10/n = 2500	9.02	0.2	17.97	−0.2	0.47	−0.3	0.93	−1.1	1.26	−5.5	8.36	0.2	14.73	0.3	16.95	0.2	21.48	−0.9
k = 10/n = 5000	9.01	0.1	17.87	−0.7	0.47	−0.5	0.93	−0.9	1.30	−2.3	8.37	0.3	14.69	0.0	16.88	−0.2	21.65	−0.1
avg. (df = 9)	9.12	1.3	18.43	2.4	0.47	−0.1	0.93	−0.8	1.30	−2.4	8.45	1.3	14.88	1.3	17.13	1.2	21.89	1.1
limiting: df = 14	14.00		28.00		0.38		0.76		0.86		13.34		21.06		23.68		29.14	
k = 15/n = 100	14.60	4.3	30.69	9.6	0.38	0.4	0.75	−1.2	0.79	−7.8	13.92	4.4	21.99	4.4	24.74	4.5	30.52	4.7
k = 15/n = 250	14.13	1.0	28.75	2.7	0.38	0.4	0.77	2.1	0.92	7.0	13.45	0.8	21.28	1.0	23.91	1.0	29.53	1.3
k = 15/n = 500	14.07	0.5	27.84	−0.6	0.37	−0.8	0.74	−2.0	0.78	−9.4	13.40	0.5	21.15	0.4	23.73	0.2	29.26	0.4
k = 15/n = 750	14.03	0.2	28.39	1.4	0.38	0.5	0.77	2.3	0.95	10.3	13.36	0.2	21.16	0.5	23.82	0.6	29.26	0.4
k = 15/n = 1000	14.04	0.3	28.44	1.6	0.38	0.5	0.77	2.0	0.92	7.5	13.37	0.2	21.11	0.2	23.79	0.4	29.49	1.2
k = 15/n = 2500	14.02	0.2	28.19	0.7	0.38	0.2	0.75	−0.9	0.78	−9.4	13.35	0.1	21.13	0.3	23.76	0.3	29.25	0.4
k = 15/n = 5000	14.01	0.0	28.06	0.2	0.38	0.1	0.75	−0.4	0.83	−3.3	13.34	0.0	21.09	0.1	23.70	0.1	29.15	0.0
avg. (df = 14)	14.13	0.9	28.62	2.2	0.38	0.2	0.76	0.3	0.85	−0.7	13.46	0.9	21.27	1.0	23.92	1.0	29.49	1.2

Table 5 Descriptive statistics for the free marginals case under the null hypothesis

	mean	%	var	%	cv	%	skew	%	kurt	%	q50	%	q90	%	q95	%	q99	%
limiting: df = 4	4.00		8.00		0.71		1.41		3.00		3.36		7.78		9.49		13.28	
k = 5/n = 100	4.38	9.6	9.11	13.8	0.69	−2.6	1.29	−8.9	2.43	−18.8	3.76	12.0	8.43	8.4	10.18	7.3	14.10	6.2
k = 5/n = 250	4.16	4.1	8.76	9.6	0.71	0.5	1.39	−1.5	2.77	−7.7	3.47	3.4	8.16	4.9	9.98	5.2	13.81	4.0
k = 5/n = 500	4.08	2.0	8.35	4.4	0.71	0.2	1.41	−0.2	2.93	−2.3	3.42	1.9	7.97	2.4	9.71	2.3	13.58	2.3
k = 5/n = 750	4.03	0.8	8.16	2.0	0.71	0.2	1.43	1.4	3.16	5.5	3.39	1.0	7.84	0.8	9.52	0.3	13.46	1.4
k = 5/n = 1000	4.03	0.7	8.09	1.2	0.71	−0.2	1.38	−2.1	2.85	−4.8	3.38	0.7	7.87	1.2	9.55	0.7	13.27	−0.1
k = 5/n = 2500	4.01	0.1	7.95	−0.6	0.70	−0.5	1.39	−1.9	2.84	−5.4	3.36	0.1	7.78	0.0	9.46	−0.3	13.27	−0.0
k = 5/n = 5000	4.00	0.1	8.01	0.1	0.71	−0.1	1.44	2.1	3.28	9.2	3.37	0.4	7.79	0.1	9.47	−0.2	13.29	0.1
avg. (df = 4)	4.10	2.5	8.35	4.4	0.71	−0.4	1.39	−1.6	2.89	−3.5	3.45	2.8	7.98	2.5	9.70	2.2	13.54	2.0
limiting: df = 9	9.00		18.00		0.47		0.94		1.33		8.34		14.68		16.92		21.67	
k = 10/n = 100	9.53	5.9	20.14	11.9	0.47	−0.1	0.92	−2.0	1.27	−5.0	8.86	6.2	15.54	5.8	17.89	5.7	22.82	5.3
k = 10/n = 250	9.14	1.5	18.51	2.8	0.47	−0.1	0.95	0.7	1.36	2.2	8.47	1.5	14.90	1.5	17.16	1.4	22.05	1.8
k = 10/n = 500	9.04	0.4	18.04	0.2	0.47	−0.3	0.93	−1.5	1.29	−3.6	8.40	0.7	14.70	0.1	16.91	−0.1	21.75	0.4
k = 10/n = 750	9.05	0.6	18.13	0.7	0.47	−0.2	0.93	−1.5	1.26	−5.5	8.38	0.4	14.78	0.7	16.97	0.3	21.68	0.1
k = 10/n = 1000	9.00	0.0	18.02	0.1	0.47	0.0	0.95	0.5	1.34	0.7	8.34	−0.0	14.70	0.1	16.94	0.1	21.66	−0.0
k = 10/n = 2500	9.03	0.3	18.23	1.3	0.47	0.4	0.95	0.4	1.36	1.8	8.38	0.4	14.70	0.1	17.02	0.6	21.80	0.6
k = 10/n = 5000	9.02	0.2	18.08	0.5	0.47	−0.0	0.96	1.7	1.39	3.9	8.37	0.3	14.71	0.2	16.98	0.4	21.82	0.7
avg. (df = 9)	9.12	1.3	18.45	2.5	0.47	−0.0	0.94	−0.2	1.32	−0.8	8.46	1.4	14.86	1.2	17.12	1.2	21.94	1.3
limiting: df = 14	14.00		28.00		0.38		0.76		0.86		13.34		21.06		23.68		29.14	
k = 15/n = 100	14.58	4.2	30.67	9.5	0.38	0.5	0.76	0.8	0.85	−0.3	13.88	4.1	21.98	4.3	24.74	4.5	30.43	4.4
k = 15/n = 250	14.15	1.1	28.51	1.8	0.38	−0.1	0.76	0.2	0.85	−0.8	13.49	1.1	21.27	1.0	23.91	1.0	29.56	1.4
k = 15/n = 500	14.05	0.4	28.17	0.6	0.38	−0.1	0.77	2.0	0.94	9.7	13.39	0.4	21.14	0.4	23.76	0.3	29.31	0.6
k = 15/n = 750	14.03	0.2	27.91	−0.3	0.38	−0.4	0.74	−2.6	0.78	−9.4	13.37	0.2	21.11	0.2	23.66	−0.1	29.11	−0.1
k = 15/n = 1000	14.03	0.2	27.86	−0.5	0.38	−0.5	0.75	−0.7	0.84	−2.0	13.38	0.3	21.11	0.2	23.67	−0.1	29.04	−0.3
k = 15/n = 2500	14.01	0.1	28.04	0.1	0.38	−0.0	0.77	1.9	0.88	2.2	13.35	0.1	21.09	0.1	23.75	0.3	29.25	0.4
k = 15/n = 5000	14.02	0.2	28.19	0.7	0.38	0.2	0.77	1.6	0.89	4.4	13.34	0.0	21.09	0.1	23.76	0.3	29.23	0.3
avg. (df = 14)	14.12	0.9	28.48	1.7	0.38	−0.1	0.76	0.5	0.86	0.5	13.46	0.9	21.26	0.9	23.89	0.9	29.42	1.0

Table 6 Descriptive statistics for the normal marginals case under the null hypothesis

	mean	%	var	%	cv	%	skew	%	kurt	%	q50	%	q90	%	q95	%	q99	%
limiting: df = 4	4.00		8.00		0.71		1.41		3.00		3.36		7.78		9.49		13.28	
k = 5/n = 100	4.11	2.7	8.46	5.8	0.71	0.1	1.42	0.7	3.04	1.2	3.45	2.8	7.98	2.6	9.79	3.2	13.66	2.9
k = 5/n = 250	4.04	1.0	8.19	2.4	0.71	0.2	1.43	1.4	3.14	4.7	3.39	1.0	7.86	1.0	9.60	1.2	13.43	1.2
k = 5/n = 500	4.02	0.6	8.09	1.2	0.71	0.0	1.42	0.3	3.04	1.2	3.37	0.4	7.82	0.5	9.53	0.4	13.34	0.5
k = 5/n = 750	4.01	0.3	8.04	0.5	0.71	−0.1	1.41	−0.2	2.98	−0.6	3.37	0.4	7.80	0.3	9.50	0.1	13.35	0.6
k = 5/n = 1000	4.02	0.4	8.05	0.6	0.71	−0.1	1.41	−0.6	2.93	−2.2	3.37	0.4	7.80	0.3	9.51	0.2	13.30	0.2
k = 5/n = 2500	4.00	0.1	8.03	0.4	0.71	0.1	1.44	1.8	3.16	5.2	3.35	−0.2	7.79	0.1	9.48	−0.1	13.33	0.4
k = 5/n = 5000	4.03	0.6	8.12	1.5	0.71	0.1	1.41	−0.2	2.98	−0.8	3.39	1.0	7.83	0.6	9.54	0.6	13.30	0.2
avg. (df = 4)	4.03	0.8	8.14	1.8	0.71	0.0	1.42	0.5	3.04	1.2	3.38	0.8	7.84	0.8	9.56	0.8	13.39	0.9
limiting: df = 9	9.00		18.00		0.47		0.94		1.33		8.34		14.68		16.92		21.67	
k = 10/n = 100	9.20	2.2	18.86	4.8	0.47	0.1	0.95	0.3	1.32	−0.6	8.52	2.1	15.02	2.3	17.30	2.3	22.15	2.2
k = 10/n = 250	9.05	0.6	18.12	0.7	0.47	−0.2	0.92	−2.3	1.25	−5.9	8.41	0.8	14.77	0.6	16.98	0.4	21.64	−0.1
k = 10/n = 500	9.06	0.7	18.25	1.4	0.47	0.0	0.95	0.5	1.32	−1.2	8.38	0.4	14.78	0.7	17.01	0.5	21.86	0.9
k = 10/n = 750	9.02	0.3	18.16	0.9	0.47	0.2	0.95	0.5	1.42	6.8	8.37	0.3	14.72	0.2	16.95	0.2	21.67	0.0
k = 10/n = 1000	8.99	−0.1	17.86	−0.8	0.47	−0.3	0.94	−0.3	1.32	−1.2	8.33	−0.2	14.68	−0.0	16.88	−0.2	21.57	−0.4
k = 10/n = 2500	9.01	0.2	18.05	0.3	0.47	−0.0	0.95	0.8	1.39	4.4	8.38	0.4	14.68	−0.0	16.92	0.0	21.70	0.2
k = 10/n = 5000	9.01	0.1	18.03	0.2	0.47	0.0	0.95	0.9	1.37	3.0	8.34	−0.0	14.66	−0.2	16.96	0.2	21.71	0.2
avg. (df = 9)	9.05	0.6	18.19	1.1	0.47	−0.0	0.94	0.1	1.34	0.8	8.39	0.5	14.76	0.5	17.00	0.5	21.76	0.4
limiting: df = 14	14.00		28.00		0.38		0.76		0.86		13.34		21.06		23.68		29.14	
k = 15/n = 100	14.32	2.3	29.30	4.7	0.38	0.0	0.76	0.2	0.87	1.4	13.65	2.3	21.53	2.2	24.20	2.2	29.88	2.5
k = 15/n = 250	14.09	0.7	28.27	1.0	0.38	−0.2	0.76	0.6	0.90	4.8	13.45	0.8	21.20	0.6	23.82	0.6	29.22	0.3
k = 15/n = 500	14.05	0.4	28.33	1.2	0.38	0.2	0.75	−1.1	0.82	−4.4	13.40	0.5	21.19	0.6	23.80	0.5	29.16	0.1
k = 15/n = 750	14.00	−0.0	27.85	−0.5	0.38	−0.2	0.73	−3.9	0.72	−15.8	13.35	0.1	21.09	0.1	23.65	−0.1	28.88	−0.9
k = 15/n = 1000	14.07	0.5	28.40	1.4	0.38	0.2	0.77	2.1	0.91	6.3	13.42	0.6	21.16	0.5	23.84	0.7	29.50	1.2
k = 15/n = 2500	14.04	0.3	28.30	1.1	0.38	0.3	0.77	2.5	0.99	15.4	13.39	0.4	21.08	0.1	23.77	0.4	29.37	0.8
k = 15/n = 5000	14.02	0.1	28.10	0.4	0.38	0.0	0.75	−0.6	0.83	−3.3	13.37	0.2	21.07	0.0	23.73	0.2	29.20	0.2
avg. (df = 14)	14.08	0.6	28.36	1.3	0.38	0.0	0.76	−0.0	0.86	0.6	13.43	0.7	21.19	0.6	23.83	0.6	29.32	0.6

Table 7 Kolmogorov-Smirnov test for fit to the limiting distribution

		fix		free		nv	
k	*n*	*D*	*p*	*D*	*p*	*D*	*p*
5	100	0.0700	< 0.0001	0.0608	< 0.0001	0.0155	< 0.0001
5	250	0.0256	< 0.0001	0.0212	< 0.0001	0.0063	< 0.0001
5	500	0.0127	< 0.0001	0.0108	< 0.0001	0.0039	0.0052
5	750	0.0087	< 0.0001	0.0071	< 0.0001	0.0033	0.0241
5	1000	0.0063	< 0.0001	0.0062	< 0.0001	0.0040	0.0037
5	2500	0.0036	0.0123	0.0033	0.0288	0.0037	0.0096
5	5000	0.0028	0.0842	0.0038	0.0060	0.0061	< 0.0001
10	100	0.0521	< 0.0001	0.0499	< 0.0001	0.0188	< 0.0001
10	250	0.0135	< 0.0001	0.0139	< 0.0001	0.0072	< 0.0001
10	500	0.0093	< 0.0001	0.0076	< 0.0001	0.0061	< 0.0001
10	750	0.0043	0.0013	0.0061	< 0.0001	0.0046	0.0005
10	1000	0.0037	0.0074	0.0019	0.4893	0.0028	0.0971
10	2500	0.0035	0.0150	0.0061	< 0.0001	0.0044	0.0007
10	5000	0.0044	0.0009	0.0039	0.0044	0.0027	0.1048
15	100	0.0447	< 0.0001	0.0421	< 0.0001	0.0243	< 0.0001
15	250	0.0104	< 0.0001	0.0126	< 0.0001	0.0088	< 0.0001
15	500	0.0079	< 0.0001	0.0060	< 0.0001	0.0060	< 0.0001
15	750	0.0048	0.0002	0.0042	0.0017	0.0026	0.1301
15	1000	0.0049	0.0001	0.0043	0.0011	0.0069	< 0.0001
15	2500	0.0028	0.0926	0.0026	0.1269	0.0054	< 0.0001
15	5000	0.0019	0.4897	0.0022	0.2775	0.0030	0.0593

Table 8 The *p*-values of the theoretical quantiles applied to the bootstrap distributions

		$\alpha = .10$			$\alpha = .05$			$\alpha = .01$		
n	k = 5/df = 4	nv	free	fix	nv	free	fix	nv	free	fix
100	*p*-value	0.11	0.13	0.13	0.06	0.07	0.07	0.01	0.01	0.01
	rel. err.	8 %	28 %	30 %	13 %	32 %	33 %	18 %	41 %	28 %
250	*p*-value	0.10	0.12	0.12	0.05	0.06	0.06	0.01	0.01	0.01
	rel. err.	3 %	16 %	18 %	5 %	21 %	22 %	6 %	25 %	30 %
500	*p*-value	0.10	0.11	0.11	0.05	0.06	0.06	0.01	0.01	0.01
	rel. err.	2 %	7 %	8 %	2 %	10 %	13 %	3 %	13 %	21 %
750	*p*-value	0.10	0.10	0.11	0.05	0.05	0.05	0.01	0.01	0.01
	rel. err.	1 %	3 %	5 %	1 %	2 %	4 %	3 %	9 %	8 %
1000	*p*-value	0.10	0.10	0.10	0.05	0.05	0.05	0.01	0.01	0.01
	rel. err.	1 %	4 %	1 %	1 %	3 %	2 %	1 %	0 %	1 %
2500	*p*-value	0.10	0.10	0.10	0.05	0.05	0.05	0.01	0.01	0.01
	rel. err.	0 %	0 %	2 %	0 %	−1 %	3 %	2 %	0 %	11 %
5000	*p*-value	0.10	0.10	0.10	0.05	0.05	0.05	0.01	0.01	0.01
	rel. err.	3 %	1 %	−1 %	3 %	0 %	−1 %	1 %	1 %	−6 %
	k = 10/df = 9									
100	*p*-value	0.11	0.13	0.13	0.06	0.07	0.07	0.01	0.02	0.02
	rel. err.	10 %	27 %	27 %	13 %	34 %	34 %	20 %	52 %	46 %
250	*p*-value	0.10	0.11	0.11	0.05	0.05	0.06	0.01	0.01	0.01
	rel. err.	3 %	6 %	7 %	3 %	8 %	10 %	−1 %	14 %	15 %
500	*p*-value	0.10	0.10	0.10	0.05	0.05	0.05	0.01	0.01	0.01
	rel. err.	3 %	0 %	4 %	3 %	0 %	5 %	8 %	3 %	6 %
750	*p*-value	0.10	0.10	0.10	0.05	0.05	0.05	0.01	0.01	0.01
	rel. err.	1 %	3 %	2 %	1 %	2 %	2 %	1 %	1 %	6 %
1000	*p*-value	0.10	0.10	0.10	0.05	0.05	0.05	0.01	0.01	0.01
	rel. err.	0 %	1 %	0 %	−1 %	1 %	0 %	−3 %	0 %	−1 %
2500	*p*-value	0.10	0.10	0.10	0.05	0.05	0.05	0.01	0.01	0.01
	rel. err.	0 %	1 %	1 %	0 %	4 %	2 %	1 %	5 %	−5 %
5000	*p*-value	0.10	0.10	0.10	0.05	0.05	0.05	0.01	0.01	0.01
	rel. err.	−1 %	1 %	0 %	1 %	2 %	−1 %	1 %	5 %	0 %
	k = 15/df = 14									
100	*p*-value	0.11	0.12	0.13	0.06	0.07	0.07	0.01	0.02	0.02
	rel. err.	12 %	24 %	25 %	15 %	31 %	32 %	25 %	47 %	49 %
250	*p*-value	0.10	0.11	0.11	0.05	0.05	0.05	0.01	0.01	0.01
	rel. err.	3 %	5 %	6 %	4 %	7 %	6 %	2 %	14 %	12 %
500	*p*-value	0.10	0.10	0.10	0.05	0.05	0.05	0.01	0.01	0.01
	rel. err.	3 %	2 %	2 %	3 %	2 %	1 %	1 %	5 %	4 %
750	*p*-value	0.10	0.10	0.10	0.05	0.05	0.05	0.01	0.01	0.01
	rel. err.	1 %	2 %	2 %	−1 %	−1 %	4 %	−5 %	−1 %	5 %
1000	*p*-value	0.10	0.10	0.10	0.05	0.05	0.05	0.01	0.01	0.01
	rel. err.	2 %	1 %	1 %	4 %	−1 %	4 %	11 %	−2 %	9 %
2500	*p*-value	0.10	0.10	0.10	0.05	0.05	0.05	0.01	0.01	0.01
	rel. err.	0 %	1 %	2 %	2 %	2 %	2 %	7 %	3 %	3 %
5000	*p*-value	0.10	0.10	0.10	0.05	0.05	0.05	0.01	0.01	0.01
	rel. err.	0 %	1 %	1 %	1 %	2 %	0 %	2 %	2 %	0 %

Note: rel. err.: Relative error of the *p*-values compared to the respective value of α given in the top row

Competing interests

The authors declare that they have no competing interests.

Author details

[1]Department of Psychology, Applied Psychology and Methods Research Unit, Universitaetsstr. 65, 9020 Klagenfurt, Austria. [2]The Health and Life Sciences University, EWZ 1, 6060 Hall/Tirol, Austria.

References

Alexandrowicz, RW: Statistical and practical significance of the likelihood ratio test of the linear logistic test model versus the rasch model. Educ. Res. Eval. **17**, 335–350 (2011)

Alexandrowicz, RW: GANZ RASCH: A free software for categorical data analysis. Soc. Sci. Comput. Rev. **30**, 369–379 (2012)

Andersen, EB: Asymptotic properties of conditional maximum likelihood estimators. J. R. Stat. Soc. Ser. B. **32**, 283–301 (1970)

Andersen, EB: The numerical solution of a set of conditional estimation equations. J. R. Stat. Soc. Ser. B. **34**, 42–54 (1972)

Andersen, EB: A goodness of fit test for the rasch model. Psychometrika. **38**, 123–140 (1973)

Andrich, D: A rating formulation for ordered response categories. Psychometrika. **43**, 561–573 (1978)

Baker, FB, Kim, S-H: Item Response Theory. Parameter Estimation Techniques., 2nd revised and expanded. Marcel Dekker, NY (2004)

Cohen, J: Statistical Power Analysis for the Behavioral Sciences. 2nd ed. Lawrence Erlbaum, Hillsdale, NJ (1988)

Davison, AC, Hinkley, DV: Bootstrap Methods and Their Application. University Press, Cambridge (1997)

Draxler, C: Sample size determination for rasch model tests. Psychometrika. **75**, 708–724 (2010)

Draxler, C, Alexandrowicz, RW: Sample size determination within the scope of conditional maximum likelihood estimation with special focus on testing the rasch model. Psychometrika. **80**, 817–919 (2015)

Efron, B, Tibshirani, RJ: An Introduction to the Bootstrap. Chapman & Hall, Boca Raton, FL (1998)

Enders, CK: Applied Missing Data Analysis. Guilford, NY (2010)

Engelhard Jr., G: Invariant Measurement. Using Rasch Models in the Social, Behavioral, and Health Sciences. Routledge, NY (2013)

Fischer, GH: The linear logistic test model as an instrument in educational research. Acta Psychologica. **37**, 359–374 (1973)

Fischer, GH, Molenaar, IW (Eds). Springer, NY (1995)

Fisher, RA: On the mathematical foundations of theoretical statistics. Philos. Trans. R. Soc. A. **222**, 309–368 (1922)

Formann, AK: A note on the computation of the second order derivatives of the elementary symmetric functions in the rasch model. Psychometrika. **51**, 335–339 (1986)

Glas, CAW, Verhelst, ND: Testing the Rasch Model. In: Fischer, GH, Molenaar, IW (eds.), pp. 69–95. Springer, NY, (1995)

Gustafsson, J-E: A solution of the conditional estimation problem for long tests in the rasch model for dichotomous items. Educ. Psychol. Meas. **40**, 377–385 (1980)

Hoijtink, H, Boomsma, A: On Person Parameter Estimation in the Dichotomous Rasch Model. In: Fischer, GH, Molenaar, IW (eds.), pp. 53–68. Springer, NY, (1995)

Kreiner, S, Christensen, KB: Overall Tests of the Rasch Model. In: Christensen, KB, Kreiner, S, Mesbah, M (eds.), pp. 105–109. Wiley, Hoboken, NJ, (2013)

Masters, GN: A rasch model for partial credit scoring. Psychometrika. **47**, 149–174 (1982)

Maxwell, SE: Sample size and multiple regression analysis. Psychol. Methods. **5**, 434–458 (2000)

Molenaar, IW: Estimation of Item Parameters. In: Fischer, GH, Molenaar, IW (eds.), pp. 39–51. Springer, NY, (1995)

Muthén, BO: Mplus Technical Appendices [Manual], Los Angeles (1998)

Neyman, J, Scott, EL: Consistent estimates based on partially consistent observations. Econometrica. **16**, 1–32 (1948)

Pawitan, Y: In All Likelihood: Statistical Modelling and Inference Using Likelihood. Clarendon, Oxford (2001)

Rasch, G: Probabilistic Models for Some Intelligence and Attainment Tests. Danmarks Pædagogiske Institut, Copenhagen (1960)

Rasch, G: An Individualistic Approach to Item Analysis. In: Lazarsfeld, PF, Henry, NW (eds.), pp. 89–107. The M.I.T. Press, Cambridge, MA, (1966)

Rost, J: The Growing Family of Rasch Models. In: Boomsma, A, van Duijn, MAJ, Snijders, TAB (eds.), pp. 25–42. Springer, NY, (2001)

R Core Team, R: A Language and Environment for Statistical Computing [Computer software]. R Foundation for Statistical Computing, Vienna, Austria (2015). Retrieved from http://www.R-project.org/ (Accessed 16 Jan 2016)

Thode, HC: Testing for Normality. Marcel Dekker, NY (2002)

van den Wollenberg, AL: A simple and effective method to test the dimensionality axiom of the rasch model. Appl. Psychol. Meas. **6**, 83–91 (1982)

von Davier, M: Winmira 2001 [computer software] (2001)

Warm, TA: Weighted likelihood estimation of ability in item response theory. Psychometrika. **54**, 427–450 (1989)

2

pTAS distributions with application to risk management

Matthias Fischer[1*] and Kevin Jakob[2]

*Correspondence: Matthias.Fischer@fau.de
[1] Department of Statistics & Econometrics, FAU Erlangen-Nürnberg, Erlangen, Germany
Full list of author information is available at the end of the article

Abstract
The family of positive tempered α-stable (pTAS) or sometimes also tempered one-sided α-stable distributions dates back to Tweedie (1984) and Hougaard (1986) who discussed it in the context of frailty distribution in life table methods for heterogenous populations. The pTAS family generalizes the well-known gamma distribution and allows for heavier tails depending on the parameter α. Because of this property, pTAS distributions appear to be useful in the context of risk management. Against this background, the contribution of his work is three-fold: Firstly, we summarize the properties of the pTAS family. Secondly, we describe its numerical implementation and illustrate the functions by means of R examples in the Appendix. Thirdly, we empirically demonstrate that this family can be successfully applied in risk management. Concretely, applications to credit and operational risk are given.

Keywords: Stable distributions, Positive distributions, Estimation, Operational risk, Credit risk

Mathematics Subject Classification: 44A10; 60E07; 91B28

1 Derivation and properties of the pTAS family

1.1 Evolution of the pTAS family

The distribution of a (non-degenerate) random variable X is called stable if there exist constants $a_n > 0$ and b_n such that, for any $n > 1$, if $X_1, X_2, \ldots$ are i.i.d. copies of X and $S_n \equiv \sum_{i=1}^{n} X_i$, then $S_n \stackrel{d}{=} a_n X + b_n$. The distribution is called strictly stable if $b_n = 0$. The normalizing constants are necessarily of the form $a_n = n^{1/\alpha}$ for some $\alpha \in (0, 2]$, where α is called the index or characteristic exponent of the distribution. We also say that X is α-stable if it is stable with index α and characteristic function

$$\varphi(t;\alpha,\beta,\gamma,\delta) = \begin{cases} \exp\left(-\gamma^\alpha |t|^\alpha (1 - i\beta \tan(\pi\alpha/2)\mathrm{sgn}(t)) + i\delta t\right), & \alpha \neq 1 \\ \exp\left(-\gamma |t| (1 - i\beta \frac{2}{\pi} \log(t)\mathrm{sgn}(t)) + i\delta t\right), & \alpha = 1 \end{cases}$$

where $\beta \in [-1, 1]$ (skewness parameter), $\delta \in \mathbb{R}$ (location parameter) and $\gamma > 0$ (scale parameter), briefly $X \sim S_\alpha(\beta, \gamma, \delta)$. Forthon, the focus is on positive or one-sided α-stable (pAS) variables, i.e. $X > 0$ a.s. which corresponds to these cases (parameter sets) where $0 < \alpha < 1$, $\beta = 1$ and $\delta \geq 0$. In this case and for $\delta = 0$, the Laplace transform is finite for all $s > 0$ and given by (see, e.g. Nolan (2003) or Janson (2011, Theorem 3.12))

$$\mathcal{L}_{\mathrm{pAS}}(s;\alpha,\gamma) = \mathbb{E}\left(e^{-sX}\right) = \exp\left(-\xi s^\alpha\right), \quad \alpha \in (0,1] \text{ and } \xi = \frac{\gamma\alpha}{\cos(\pi\alpha/2)}. \qquad (1.1)$$

The positive or one-sided tempered α-stable distribution has its origin in Tweedie (1984) and Hougaard (1986). The process of tempering (also exponential tilting or Esscher transformation, Esscher (1932)) shortens the right-hand tail of the stable distribution so that, in contrast with the stable case, moments of all orders exist. Technically, a new proper density g is defined by

$$g_Y(y;\alpha,\gamma,\theta) := f_{\text{pAS}}(y;\alpha,\gamma)\frac{\exp(-\theta y)}{\mathcal{L}_{\text{pAS}}(\theta;\alpha,\gamma)}, \quad \theta \geq 0. \tag{1.2}$$

The distribution of the corresponding random variable Y is called positive tempered α-stable (pTAS) distribution. Using (1.1), (1.2), its Laplace transformation derives as

$$\mathcal{L}_Y(s) = \exp\left(-\frac{\gamma^\alpha}{\cos(\pi\alpha/2)}\left((\theta+s)^\alpha - \theta^\alpha\right)\right). \tag{1.3}$$

1.2 Properties of pTAS distributions and some remarks

Hougaard (1986) proved that all moments exist if $\theta > 0$. In particular, using (1.3),

$$\mu = \mathbb{E}(Y) = \frac{\alpha\xi}{\theta^{1-\alpha}} \text{and } \sigma^2 = \text{Var}(Y) = \frac{\alpha(1-\alpha)\xi}{\theta^{2-\alpha}},$$

where $\nu = \sigma/\mu$ denotes the coefficient of variation. Moreover, skewness and kurtosis, defined by the third and fourth standardized moments read as

$$\mathbb{S}(Y) = \frac{\nu(2-\alpha)}{1-\alpha} \quad \text{and} \quad \mathbb{K}(Y) = \frac{\nu^2(2-\alpha)(3-\alpha)}{(1-\alpha)^2} + 3. \tag{1.4}$$

In the relevant literature, different parameterizations appear, summarized in Table 1, below.

Figure 1 illustrates the possible combinations of skewness and kurtosis that can be modeled via a pTAS distribution (gray shaded area). One can see that the area of possible combinations is bounded by the gamma distribution from below whereas the possible combinations of the Lognormal distribution are covered by the pTAS family. Therefore, we think that the pTAS distribution can serve as a good alternative whenever data are assumed to follow a lognormal distribution in general but some more flexibility regarding the tail behavior is required. For comparison reasons, we also depict the Weibull distribution which is often used in operational risk models (see Section 3.2).

Hougaard (1986) showed that all pTAS densities are unimodal and proved the following scaling-property: $cX \sim \mathcal{P}(\alpha, c^\alpha\delta, \theta/c)$ when $X \sim \mathcal{P}(\alpha,\delta,\theta)$. Furthermore, the pTAS family is self-decomposable (i.e. if X follows a pTAS distribution then for each $\lambda \in (0,1)$ there exists a random variable Y independent of X such that $\mathcal{L}(\mathcal{X}) = \mathcal{L}(\lambda\mathcal{X} + \mathcal{Y})$) and

Table 1 Different Parametrizations: $\mathcal{P}(K) \leftrightarrows \mathcal{P}(T)$ with $\beta = \alpha, \alpha = \theta^\alpha \cdot \delta/(1-\alpha), \lambda = (1-\alpha)/\delta$. $\mathcal{P}(H) \leftrightarrows \mathcal{P}(T) \leftrightarrows \mathcal{P}(P)$ using $\delta = \mu\left(\frac{1-\alpha}{\mu\nu^2}\right)^{1-\alpha}$, $\gamma = \left[\frac{\mu\cos(\pi\alpha/2)}{\alpha}\right]^{\frac{1}{\alpha}}\left[\frac{1-\alpha}{\mu\nu^2}\right]^{\frac{1-\alpha}{\alpha}}$ and $\theta = \frac{1-\alpha}{\mu\nu^2}$

Reference	Parameter	Abbreviation
Hougaard (1986)	(α,δ,θ)	$\mathcal{P}(H)$
Tweedie (1984)[a], Haas and Pigorsch (2009)	(α,γ,δ)	$\mathcal{P}(T)$
Palmer et al. (2008)	(α,μ,ν)	$\mathcal{P}(P)$
Küchler and Tappe (2011)	(β,α,λ)	$\mathcal{P}(K)$

[a]Tweedie (1984) considered exponential dispersion models with variance function of the form $V(\mu) \propto \mu^p$ such models with p>2 coincide with the pTAS family with $\alpha = (p-2)/(p-1)$, see Jørgensen (1987), chapter 4, for a detailed account

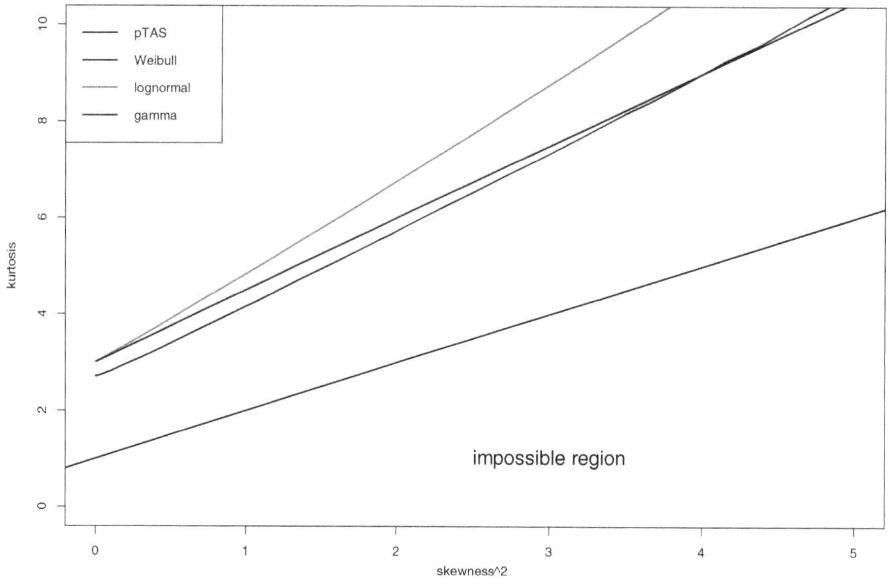

Fig. 1 Skewness-Kurtosis Plot of pTAS family, Weibull and lognormal distribution

infinitely divisible (i.e. for all $n \in N$ exists a sequence of iid random variables $\left(X_j^{1/n}\right)_{j=1}^n$ such that $X \overset{d}{=} X_1^{1/n} + \ldots + X_n^{1/n}$. In this case, the characteristic function reads

$$\varphi_X(u) = \mathbb{E}(e^{iuX}) = \exp\left(i\gamma u - 0.5\sigma^2 u^2 + \int_{-\infty}^{\infty} \left(e^{iux} - 1 - iux1_{|x|<1}\right) \nu(dx)\right),$$

where $\gamma \in R, \sigma \geq 0$ and ν denotes the Lévy measure. The triplet (γ, σ^2, ν) is called the Lévy triplet. If the Lévy measure $\nu(dx)$ has a density $k(x)$, then this density is called a Lévy density. Barndorff-Nielson and Shephard (2001) discuss the Lévy-density of the pTAS distribution. After a suitable re-parametrization, one obtains

$$k(x) = \frac{\gamma 2^\alpha \alpha \exp(-\theta x)}{\Gamma(1-\alpha) x^{-(1+\alpha)}}, \quad x > 0.$$

for $0 < \alpha < 1, \theta > 0, \gamma > 0$. According to Lemma 4 in Küchler and Tappe (2011) it also holds that

$$\varphi_X(u) = \exp\left(\alpha\Gamma(-\beta)\left[(\lambda - iu)^\beta - \lambda^\beta\right]\right).$$

The pTAS density reduces to a simple closed form in three special cases. At first, letting $\alpha \to 0$, the gamma distribution with scale $\alpha_\Gamma = \frac{\theta^\alpha \delta}{1-\alpha}$, shape $\beta_\Gamma = \frac{1-\alpha}{\theta}$ is recovered. Secondly, setting $\alpha = 1/2$, the inverse Gaussian distribution with $\lambda_{IG} = \frac{\delta^2 \theta^{2\alpha-1}}{(1-\alpha)}$ arises. Finally, another special case arises when $\alpha = 1/3$ where the probability density function (pdf) can be expressed in closed form:

$$f_{\alpha=1/3}(y; \delta, \theta) = \frac{\sqrt{3}}{\pi}\left(\frac{\delta}{y}\right)^{1.5} K_{1/3}\left(2\sqrt{\delta^3 y^{-1}}\right) \exp(3\delta\theta^{1/3} - \theta y)\Big)$$

where $K_\lambda(x)$ denotes the modified Bessel function of the third kind with index λ.

The special cases as well as some density functions for further values of α are shown in Fig. 2.

The pTAS family can be considered as an alternative to the generalized inverse Gaussian (GIG) family (see Koudou and Ley (2014) for a review of this family) which itself arises

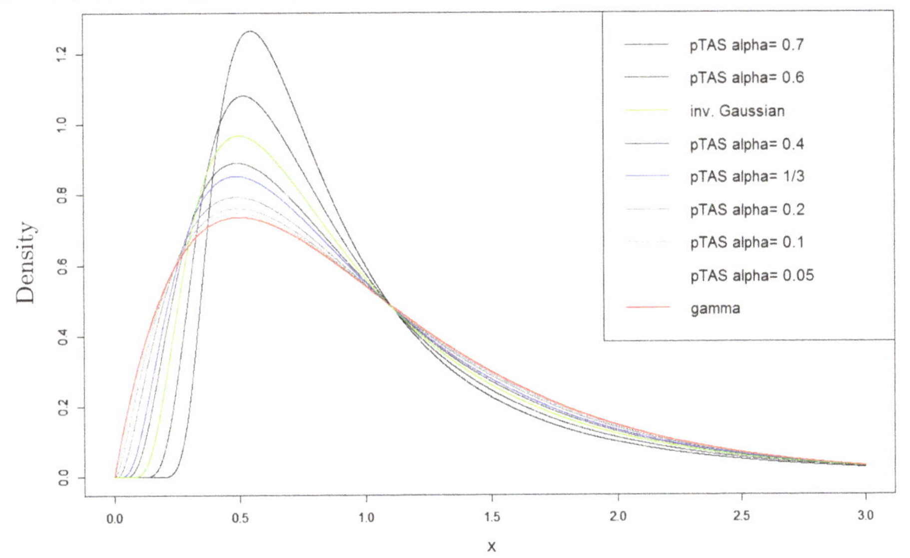

Fig. 2 Density of pTAS distribution for different values of α as well as the special cases of the inverse Gaussian ($\alpha = 0.5$) and the gamma distribution ($\alpha \to 0$)

from the inverse Gaussian (IG) distribution by exponential tilting. Barndorff-Nielson and Shephard (2001) introduce the four-parameter, so-called modified stable (MS) distributions which nests both GIG family and pTAS family (setting $\kappa = 0.5$ and $\nu = -\kappa$ in their notation), where ν denotes the fourth (additional) parameter.

2 Implementation issues

This chapter describes numerical algorithms which can be used for the standard functions (e.g. density and distribution function, random number generation and quantile function) of a pTAS random variable. The algorithms are described independently of a specific programming language. However, we give some examples how to use the pTAS R-package, which contains the described functions, in Appendix 4. The R-package is not yet published but can be obtained from the authors by request.

2.1 Density and distribution function

Since the pdf and cumulative distribution function (cdf) of the pTAS family are in general not available in closed form, but only via the Laplace transform, we have to use numerical inversion techniques to calculate values of the density or distribution function. Therefore, we follow a general approach given by Abate et al. (2000), which will be described briefly in this section. For further details please refer to the mentioned literature.

Starting from the Bromwich integral, the value of the pdf f at $t > 0$ can be recovered from the Laplace transform $\hat{f} := \mathcal{L}(f)$ by [1]

$$f(t) = \frac{1}{2\pi i} \int_{b-i\infty}^{b+i\infty} e^{st}\hat{f}(s)\,\mathrm{d}s = \frac{2e^{bt}}{\pi} \int_{0}^{\infty} \mathrm{Re}\left(\hat{f}(b+iu)\right)\cos(ut)\,\mathrm{d}u, \tag{2.1}$$

where $i = \sqrt{-1}$ and b is a real number greater than all singularities of $\hat{f}$. The second transformation is achieved via a change of variables.

Approximating the integral in (2.1) by the trapezoidal rule with step size $h > 0$, we get

$$f(t) \approx \frac{he^{bt}}{\pi}\left(\hat{f}(b) + 2\sum_{k=1}^{\infty}\operatorname{Re}\left(\hat{f}(b+ikh)\right)\cos(kht)\right). \tag{2.2}$$

This equation gives rise to three kinds of error, a discretization error caused by the step size h, a truncation error and a round-off error because of the numerical operations performed by calculating the series.

The discretization error can be reduced by interpreting (2.1) as a Fourier series of a function $g(t) := e^{-bt}f(t)$, whereby the coefficients of the series itself can be expressed via the values of the Laplace transform. The error then can be reduced by increasing the ratio b/h. The round-off error can be controlled by choosing $h = \frac{\pi}{lt}$ and $b = \frac{A}{2lt}$ for some $l \in \mathbb{N} > 0$ and $A = \frac{2l}{2l+1}m\ln 10$, where in turn $m > 0$ should be chosen such that 10^{-m} is close to the machine precision. In order to improve the truncation error, the so called Euler summation technique can be applied, which gives us another parameter $n \in \mathbb{N}$ determining the number of coefficients being (fully) respected for the summation. Similarly to Abate et al. (2000) we recommend the following default values: $l = 1, n = 38$ and $m = 11$.

Finally, the pdf is approximated via

$$f(t) \approx \frac{\exp\left(\frac{A}{2l}\right)}{2lt}\sum_{k=1}^{m}\binom{m}{k}2^{-m}S_{n+k}, \tag{2.3}$$

with

$$S_0 = \hat{f}\left(\frac{A}{2lt}\right) + 2\sum_{j=1}^{l}\operatorname{Re}\left[\hat{f}\left(\frac{A}{2lt} + \frac{ij\pi}{lt}\right)\exp\left(\frac{ij\pi}{l}\right)\right] \quad \text{and}$$

$$S_k = S_{k-1} + (-1)^k 2\sum_{j=1}^{l}\operatorname{Re}\left[\hat{f}\left(\frac{A}{2lt} + \frac{ij\pi}{lt} + \frac{ik\pi}{t}\right)\exp\left(\frac{ij\pi}{l}\right)\right] \quad k \in \mathbb{N}_{>0}. \tag{2.4}$$

Using the definition of the Laplace transform and partial integration it is easy to see that for $F(t) = \int_0^\infty f(t)\,dt$

$$\hat{F}(s) := \mathcal{L}(F)(s) = \mathcal{L}(f)(s)/s. \tag{2.5}$$

Therefore, with just a minor modification, the same method used to approximate the pdf can be used to approximate the cdf of a pTAS distribution.

2.2 Quantiles and random numbers

To calculate the quantile for a given level $p \in (0, 1)$, Ridout (2008) proposed a modified version of Newton's method to invert the cdf. Because, especially in the upper and lower tail of the distribution, the pdf may be close to zero, which may lead to an iteration step outside a given interval $[t_{\min}, t_{\max}]$. In this case, the Newton step is replaced by a bisection method.

Given a tolerance $\epsilon > 0$, a maximum number of iterations $N_{\max} \in \mathbb{N}$ and a closed interval $[t_{\min}, t_{\max}]$ such that $\exists t \in [t_{\min}, t_{\max}]: |F(t) - p| \leq \epsilon$ the following algorithm can be used to calculate the p-quantile of a pTAS distribution:

In order to determine the initial values of the lower and upper bounds, we propose to store some values for t and $F(t)$ on a grid $t_{i=1,\dots,N}$ depending on mean and standard

Algorithm 1 Inversion of the CDF

Set initial values
  $n \leftarrow 1$
  $t_1 \leftarrow (t_{\min} + t_{\max})/2$
While $n \leq N$ and $|F(t) - p| > \epsilon$
  $t_{n+1} \leftarrow t_n - (F(t_n) - p)/f(t_n)$ # Newton step
  if $t_{n+1} > t_{\max}$ or $t_{n+1} < t_{\min}$ # bisection method
    $t_{n+1} \leftarrow (t_{\min} + t_{\max})/2$
  if $F(t_{n+1}) > p$ # update lower/upper bounds
    $t_{\max} \leftarrow t_{n+1}$
  if $F(t_{n+1}) < p$
    $t_{\min} \leftarrow t_{n+1}$
  $n \leftarrow n + 1$
if $n > N$
  Warning: Iteration not successful.
else
  return t_n

deviation. Using precalculated values for $t_{\min}$ and $t_{\max}$ increases performance especially when quantile transformations should be performed very often.

If multiple quantiles should be calculated at the same time for different probabilities $p_{m=1,\dots,M}$, the probabilities should be sorted in ascending order before the transformation. Assuming that $p_1 \leq \ldots \leq p_M$ we can apply Algorithm 1 on the first element p_1 and use the resulting value $t_1 = F^{-1}(p_1)$ as starting point (and also as lower bound) for the next element p_2.

Algorithm 2 Multiple Inversion of CDF

Choose initial bounds (from precalculated grid):
  $t_{\min} > 0 : F(t_{\min}) < \min_{m=1,\dots,M} p_m$
  $t_{\max} > 0 : F(t_{\max}) > \max_{m=1,\dots,M} p_m$
Calculate first quantile via Algorithm 1:
  $t_1 = F^{-1}(p_1)$
For $m = 2, \ldots, M$
  $t_m \leftarrow F^{-1}(p_m)$ # via Algorithm 1 with starting value t_{m-1}
  $t_{\min} \leftarrow t_m$

For random number generation the inverse probability integral transform method can be applied very easily by using Algorithm 2 based on sorted uniform drawings.

2.3 Estimating parameters

Given a series of N positive observations $x_1, \ldots, x_N$ a pTAS distribution can be fitted via various methods. First of all, a numerical maximum likelihood estimation (MLE) can be used. Within the package (see appendix A) we use the *mle()* function of the R-package *stats4* which in turn uses the *optim()* optimizer of R in order to perform a maximum likelihood estimation. In particular, for observations $t_1, \ldots, t_N$ the parameters are determined by

$$\left(\alpha^*, \mu^*, \sigma^*\right) = \text{argmin}_{\alpha,\mu,\sigma} \; - \sum_{n=1}^{N} \log f_{(\alpha,\mu,\sigma)}\left(t_n\right),$$

whereas for the calculation of the pdf $f_{(\alpha,\mu,\sigma)}(t)$ the algorithm described in 2.1 is used. As starting values we use the empirical mean and variance and a value of 0.5 for α. In order to ensure that the parameters stay in their specific range, we use the L-BFGS-B method with box constraints. Alternatively, distribution parameters can be estimated based on mean and variance, which determine parameters μ and ν regarding the $\mathcal{P}_P$ parametrization and either the skewness, the kurtosis or a quantile to estimate parameter α. Given a skewness or kurtosis value (besides mean and variance), parameter α can be calculated by using Eq. (1.4). Please note, that the minimal skewness for a pTAS distribution equals 2ν. Please also note, that if α should be estimated based on a given kurtosis value the solution may not be unique (see Eq. 1.4).

A similar problem occurs if α should be estimated based on a given quantile t^* for probability p^*. Therefore, we propose to start with a grid $\alpha_{i=1,\ldots,N-1}$ (e.g. $\alpha_i = \frac{i}{N}$ for $i = 1, \ldots, N-1$ for some $N \in \mathbb{N}$) and calculate quantiles $t_i = F^{-1}_{\mathcal{P}_P(\alpha_i,\mu,\nu)}(p^*)$ in order to determine the number of possible solutions as well as lower and upper bounds for each of them. Given a lower and upper bound for each possible solution, a standard root finding method (e.g. bisection method) can be applied. In order to choose a proper value of N, we recommend to visualize the problem at first with the help of a plot like the one given in Fig. 3, which shows an example containing two possible solutions for α. In this case, we fixed the parameters μ and ν based on mean and variance and want to find a solution for α such that $F_{\mathcal{P}_P(\alpha,\mu,\nu)}(14.36) = 0.999$. As Fig. 3 shows, we get the two solutions $\alpha_1 = 0.5$ and $\alpha_2 = 0.98$.

In cases with no unique solution, the resulting parameters may lead to very different distributions (e.g. with different tails) which, in the context of risk management, can cause different risk figures. Therefore, we recommend to choose the estimation method with respect to the proposed area of application and to compare estimation results between

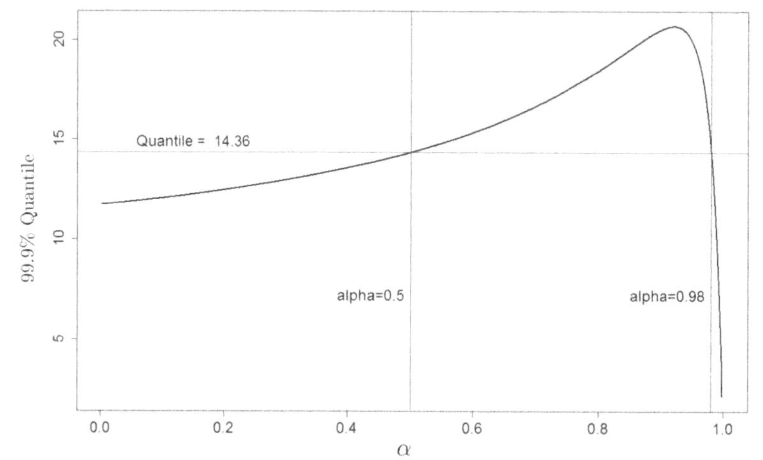

Fig. 3 99.9 % quantile depending on parameter α ($\mu = 1$ and $\nu = 1.5$ are fixed)

different methods if they are not unique. Otherwise, using inappropriate parameter estimators may lead to an increase of model risk.

3 Application to risk management

After showing how the pTAS distribution can be implemented and parameters can be estimated, we provide two extensive risk management applications of the pTAS family.

3.1 Credit risk

Our first application deals with the quantification of credit risk. Given a particular credit portfolio of N counterparties, financial institutions use credit portfolio models to estimate the distribution of the portfolio loss over a fixed time horizon (usually one year) due to counterparties' defaults [2]. For each counterparty we denote the exposure at default by $\mathrm{EAD}_i > 0$, the loss given default (as fraction of EAD_i) by $\mathrm{LGD}_i \in [0, 1]$ and the probability of default by $\mathrm{PD}_i \in (0, 1)$. The portfolio loss L then reads as

$$L = \sum_{i=1}^{N} \mathrm{EAD}_i \cdot \mathrm{LGD}_i \cdot D_i,$$

where D_i is a Bernoulli distributed random number with probability PD_i representing the default ($D_i = 1$) or the survival of counterparty i.

Given these counterparty specific information, a crucial task of a credit portfolio model is to model the changes of PD_i over the specified time horizon by taking into consideration systematic influences due to country or business dependencies as well as idiosyncratic changes. For our example we use the CreditRisk$^+$ model, which is available via the GCPM R-package on CRAN. We give a short introduction on those model parts, which are necessary for this example. For detailed information please refer to Credit Suisse First Boston International (1997) or Gundlach and Lehrbass (2004).

Within the CreditRisk$^+$ model the possible change of counterparty i's PD is modeled by the conditional $\overline{\mathrm{PD}}_i$, which is given by

$$\overline{\mathrm{PD}}_i = \mathrm{PD}_i \left(w_{i,0} + \sum_{k=1}^{K} w_{i,k} S_k \right),$$

where $w_{i,k} \in [0, 1]$ represent the affiliation to one or multiple out of K sectors (business-country combinations). In contrast, $w_{i,0} = 1 - \sum_{k=1}^{K} w_{i,k} \geq 0$ represent the idiosyncratic component. The variables $S_k > 0$ (so-called sector variables) represent the economic situation of sector $k = 1, \ldots, K$. An economic boom is associated with values $S_k < 1$, which yields to a conditional PD below the unconditional one (i.e. $\overline{\mathrm{PD}}_i < \mathrm{PD}_i$), whereas a recession is expressed by values $S_k > 1$, which increases counterparties' PD. Within the original framework of Credit Suisse First Boston International (1997), the sector variables are modeled via a gamma distribution, which insures that (together with some additional assumptions) the portfolio loss distribution can be calculated analytically [3].

As mentioned above, the gamma distribution was chosen for performance issues. From an economic point of view, this assumption may be questionable. Instead of a gamma distribution, one can also use a lognormal distribution, which is more heavy tailed and therefore more conservative compared to a gamma distribution. However, the lognormal distribution also possesses only two parameters. Therefore, using mean and variance [4] to parametrize the distribution of S_k the heaviness of the tail can not be controlled explicitly. If we use a pTAS distribution instead, we can fit the sector distribution to mean and variance of observed PD changes and still have control over the tail via parameter α.

For our example, we use portfolio of 5,000 counterparties belonging to ten different sectors. The data used to estimate the sector distributions are monthly PD values over 10 years, which are estimated via a Merton type model (see Merton (1974)) from marked data (i.e. stock prices and liabilities) for over 20,000 corporations and aggregated on sector level. The portfolio as well as the underlying data are explained in more detail in Fischer and Jakob (2015) and Jakob and Fischer (2014). We estimate pTAS distributions for each sector by using MLE and the empirically observed skewness and compare the results on portfolio risk figures with the original setting (i.e. gamma distributions parametrized via mean and variance) and a framework using the lognormal distribution and the Weibull distribution. Because of the relatively small number of observations (10 years of monthly data), we do not use the kurtosis or quantile estimation methods in this example. For the Monte Carlo simulation of the CreditRisk$^+$ model the GCPM R-package is used. The dependency between sector variables are modeled via a Student-t copula with 3.8 degrees of freedom, estimated using a maximum likelihood approach[5].

Table 2 shows the empirical skewness of each sector as well as the estimated parameter α and the skewness of a Gamma, a lognormal and a Weibull distribution parametrized only by mean and variance. As already indicated by Fig. 1, the lognormal distribution possesses a skewness and kurtosis similar to a pTAS distribution with higher values of α. By looking at Table 2 we can see, that the observed skewness is often underestimated by the lognormal distribution, which is also confirmed by Fig. 4, where the risk figures of the lognormal framework are slightly below those of the pTAS(skewness) framework. However, this may only be the case if skewness is in a suitable range. By using a pTAS distribution one always has greater flexibility to account for semi-heavy (or not so heavy) tailed distributions compared to a lognormal or a gamma distribution. In addition, Fig. 5 shows the empirical observations of the sector variables S_k together with the densities of the fitted distributions exemplarily for Sector 4. The Figure shows that a pTAS distribution estimated via a MLE approach fits the data much better compared to all other distributions. The pTAS distribution estimated via a MLE approach also possesses the heaviest tail of all presented competitors, which in turn causes significant higher risk figures as shown by Fig. 4.

The VaR is considerably higher

Table 2 Skewness values and estimated parameters for different sector distributions

		Sector									
		S_1	S_2	S_3	S_4	S_5	S_6	S_7	S_8	S_9	S_{10}
Skewness	Empirical	2.59	1.74	2.24	2.10	1.74	2.46	2.44	2.16	1.84	2.57
	Weibull	0.98	0.52	0.84	0.63	0.37	0.93	1.05	0.70	0.62	1.54
	Gamma	1.29	0.97	1.19	1.04	0.87	1.25	1.34	1.10	1.04	1.69
	lognormal	2.21	1.56	2.00	1.71	1.38	2.13	2.31	1.81	1.69	3.13
	pTAS (skewness)	2.59	1.74	2.24	2.10	1.74	2.46	2.44	2.16	1.84	2.57
	pTAS (MLE)	4.57	2.92	3.68	3.73	2.74	4.73	5.62	3.43	3.43	5.68
α	pTAS (skewness)	0.67	0.62	0.64	0.67	0.67	0.66	0.62	0.66	0.61	0.51
	pTAS (MLE)	0.84	0.80	0.81	0.84	0.81	0.85	0.86	0.81	0.82	0.83

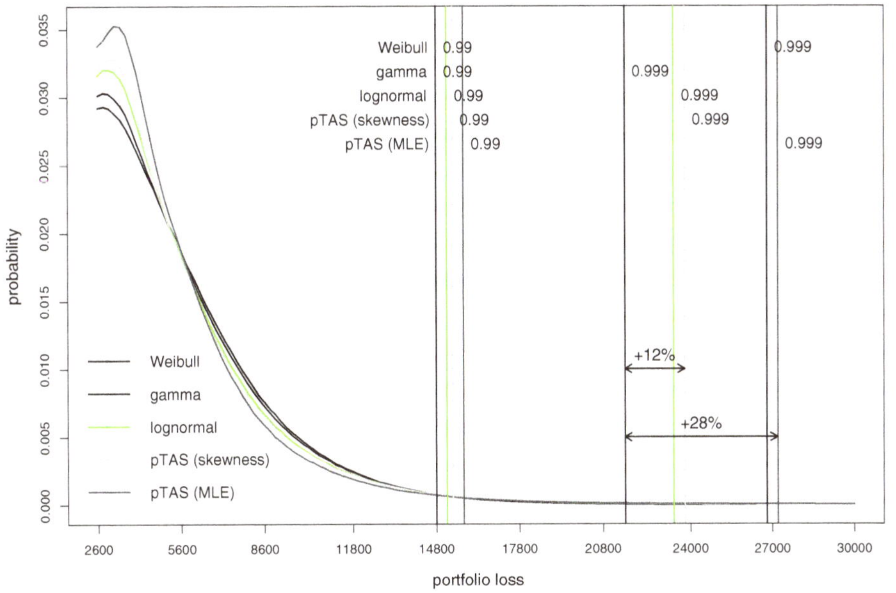

Fig. 4 Portfolio loss distribution for different sector distributions. The *vertical lines* indicate the Value at Risk for the stated loss level

Figure 4 shows the pdf of the portfolio loss distribution together with vertical lines indicating the Value at Risk (VaR) for level τ, which is simply the quantile of the portfolio loss distribution. The VaR is considerably higher if we use a pTAS distribution which considers the skewness or which is estimated via MLE for the sector distribution compared to the standard case of a gamma distribution, which only accounts for mean and variance. Because financial institutions typically use higher values for τ (e.g. $\tau = 0.999$) to calculate the economic capital which is necessary to cover unexpected losses, the use of a simple gamma distribution may imply a significant amount of model risk. In our case, the $VaR_{0.999}$ rises by around 12 % if we use pTAS distributions based on skewness and up to 28 % if we use MLE. Please note, that the effect in general depends on the sectors (i.e.

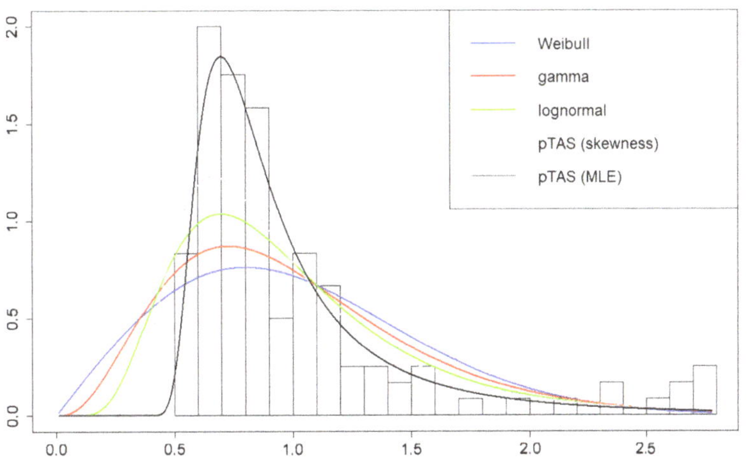

Fig. 5 Sector 4: Histogram of sector realizations together with fitted parametric distrbutions

business lines and countries) of the portfolio as well as the data used for estimating the sector distributions.

3.2 Operational risk

A financial institution is exposed to different risks such as credit risk and market risk, for instance, and is required to put aside a capital buffer against unexpected losses. With the implementation of Basel II recommendations a capital requirement for operational risk (OpRisk) was set under regulation, too. There are many definitions of operational risk and many institutions have adopted their own definitions which better reflects their business model and strategy. The Basel Committee defines operational risk *as the risk of loss resulting from inadequate or failed internal processes, people and systems or from external events.* It becomes obvious that OpRisk is a very broad concept and may include anything from bank robberies, unauthorized trading, failures in the internal systems to terrorist attacks and natural catastrophies. Commonly, the loss data are organized according to seven official Basel II defined event types (Internal Fraud, External Fraud, Employment Practices & Workplace Safety, Clients, Products & Business Practice, Damage to Physical Assets, Business Disruption & System Failures, Execution, Delivery & Process Management) and eight business lines (Corporate Finance, Trading & Sales, Retail Banking, Commercial Banking, Payment & Settlement, Agency Services, Asset Management, Retail Brokerage) which implies 56 categories. In order to derive the OpRisk loss distribution, one typically derives the loss distribution for each relevant category in the first step[6] and aggregates the individual loss distributions to the overall loss distribution, in a second step. For reason of simplicity, we assume in the sequel that the bank operates on one business line and is faced to one event type only. In this case, the loss distribution reduces to a compound sum of the type

$$L = L_1 + \ldots + L_N$$

with random variables N (number of loss in the next year) and corresponding loss severities $L_1, \ldots, L_N$ for each loss. For reasons of simplicity we assume that N and $L_1, \ldots, L_N$ are mutually independent (see, e.g. Neslehova et al. (2006) for an inclusion of dependence). Once these variables are specified and estimated within a parametric setting and for a given loss data set, the loss distribution can be easily simulated within a Monte Carlo framework. Typically, the number of losses are assumed to follow a Poisson distribution with parameter $\lambda > 0$ which corresponds to the expected number of losses within the next period/year. In contrast, specifying the severity distribution is more complex to some extent. To avoid the problems in fitting a single parametric distribution with one or two parameters to the whole data set, one often uses a compound severity distribution, see El Adlouni et al. (2011). This involves dividing the severity distribution into the so-called body and the so-called tail by a threshold, and presuming a different distribution for each part. Both distributions are then combined into a single severity distribution which is commonly termed as a compound distribution. Losses below and above the threshold are referred to as low-severity and high-severity losses. Once, the parametric model(s) for the severity distribution has been selected, the estimation of unknown parameters can be done using the method of moments (MM), Maximum Likelihood (ML, used in the empirical part) or Ordinary Least Square (OLS). In order to compare the goodness-of-fit (GoF), graphical tools like quantile-quantile-plots or suitable GoF-test should be used.

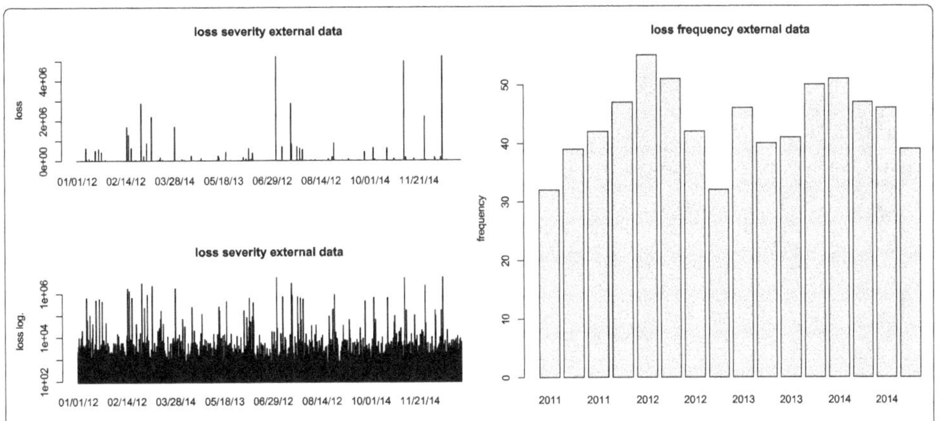

Fig. 6 Empirical loss severities and loss frequencies

The underlying data set is provided in the textbook of Bolancè et al. (2012), chapter 7. It contains 700 loss data from 2011 to 2014 which were multiplied by 1000 for technical reasons. The corresponding loss severities and loss frequencies in the course of time are depicted in Fig. 6.

In order to illustrate the OpRisk calculation, we assume that the number of losses is Poisson distributed and that we have a compound severity distribution, where the body is modelled by the empirical distribution function, whereas the tail is assumed to follow a lognormal, a Weibull, a pTAS, a gamma and a generalized gamma (see, e.g. Stacy (1962)) distribution, respectively. Table 3 summarizes the corresponding OpVaR (i.e. quantile of the loss distribution) for different confidence levels and each of the four distributions. Obviously, the Weibull (which is often applied in banking industry) and the pTAS tail model are close together, whereas the lognormal model implies significant higher OpVaR's. In contrast, both gamma and generalized gamma distribution produce lower OpVaR figures.

In order to compare the goodness-of-fit, both quantile-quantile plots (see Fig. 7) and goodness-of-fit statistics (Kolmogorov-Smirnov and Anderson-Darling, see Table 4 and Chernobay et al. (2015) for a detailed discussion) indicate that the pTAS distribution with $\widehat{\alpha} = 0.3842$ seems to be the preferable choice. Although the generalized gamma distribution provides reasonable AD2UP and KS statistics, too, the user is in danger of being to progressive, taking into account that there are only a few number of observations and that the true parametric model is unknown[7].

Finally, Fig. 7 also depicts the simulated portfolio loss for the OpRisk data set.

4 Conclusion

Within this article, we discussed the family of positive tempered α-stable distributions, which is a flexible distribution family and well suited to model both light and semi-heavy

Table 3 OpVaR for different confidences level and distributions

Percentile →	0.95	0.99	0.995	0.999	0.9995
lognormal	40,937,057	76,509,385	100,584,286	187,533,641	247,986,179
Weibull	37,542,382	47,459,865	51,335,798	60,440,362	64,322,549
pTAS	33,554,021	44,883,975	49,657,499	60,450,462	65,182,740
gamma	32,023,064	41,335,245	45,109,838	53,586,965	57,257,060
gen.gamma	24,836,520	31,636,053	34,333,082	40,596,749	43,033,552

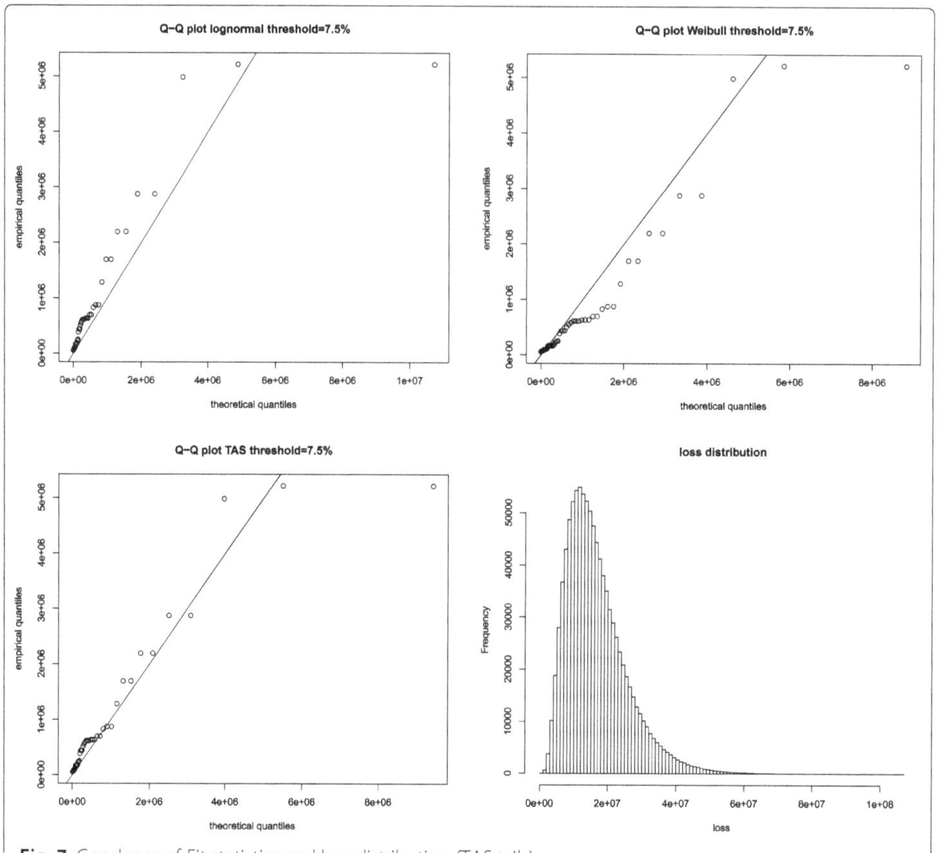

Fig. 7 Goodness-of-Fit statistics and loss distribution (TAS tails)

tailed data on the positive half-axis. Besides the derivation of the family and a summary of certain characteristics, which are relevant for practical applications, we provided an overview of the existing literature on this family. Furthermore, we introduced algorithms that can be used to implement the basic functionalities of the pTAS family (i.e. density and distribution function, calculation of quantiles and random numbers generation), which are also available via the pTAS R-package. By applying the pTAS distribution in the field of credit and operational risk we show that this distribution is more flexible and provides a better fit to empirical data compared to other competitors which are often used. Therefore, as in case of credit risk, the pTAS distribution can help to reduce the amount of model risk or as in case of operational risk, to reflect the risk more adequately which in turn helps banks to allocate economic capital more appropriately. Beyond the two given examples, the pTAS distribution may be also used to model (stock) returns and therefore being beneficial for market risk management as well.

Table 4 Goodness-of-fit statistics

Test statistic	AD2UP	*p*-value	KS	*p*-value
lognormal	2.6582	0.080	0.8235	0.060
Weibull	7.8001	0.024	1.7422	0.005
pTAS	2.7823	0.682	0.6585	0.739
gamma	3.1287	0.077	0.8612	0.197
gen.gamma	2.6755	0.167	0.5674	0.697

Endnotes

[1]Please refer to Schiff (1999, Theorem 4.3) for further details.

[2]Besides counterparties' defaults also changes in a counterparties creditworthiness (so-called rating migrations) cause a portfolio loss. However, for reasons of simplicity we restrict this example to default risk only.

[3]In 1997, when the CreditRisk$^+$ was published, this was a major advantage over simulative models, which use a Monte Carlo simulation to estimate the loss distribution. However, nowadays computers are much more powerful and Monte Carlo simulations are widely used.

[4]The parametrization based on mean and variance is a standard method within the CreditRisk$^+$ model. In order to ensure that $\mathbb{E}\left(\overline{\mathrm{PD}}_i\right) = \mathrm{PD}_i$, which is a common assumption, we have the condition that $\mathbb{E}(S_k) = 1$ for all k.

[5]For readers interested in the topic of copulas within credit portfolio models, we refer to Jakob and Fischer (2014) and Fischer and Jakob (2015).

[6]The Basel Committee have prescribed guidance for three types of methods for the calculation of capital requirement for operational risk. Those are the Basic Indicator Approach, the Standard Approach and the Advanced Measurement Approach. The latter is the most sophisticated of the approaches and this is what this section is about.

[7]Please note that for the GoF-test for left-truncated samples the distribution of the statistic is not parameter-free, and the p-values and the critical values are obtained by means of Monte Carlo simulation. There remains a certain variation of the p-values - a direct comparison is not meaningful.

Appendix

Using the pTAS R-package

With the help of short examples we explain how the functions can be used within an R-session. We will not describe the several functions in all details and parameters. Please refer to the corresponding help pages within the package for more information.

Creating a pTAS distribution

The pTAS package uses an object oriented approach, which means that every distribution (e.g. with different parameters) is represented by a different object of class pTAS. Therefore, one can work with many different distributions at the same time (within the same workspace) without jeopardizing their consistency regarding distributional or numerical parameters.

Hence, the first step is to create a new pTAS distribution object.

```
library(pTAS)
## Loading required package: stats4
# creating pTAS object with parameters according to Palmer(2008)
MYpTAS = pTAS(param.P = c(0.5, 1, 0.75))
```

The MYpTAS object describes a pTAS distribution with parameterization $\mathcal{P}_P(\alpha,\gamma,\theta) = (0.5, 1, 1.5)$. The pTAS function automatically performs some plausibility checks on the given parameters and translates the given parameterization (i.e. according to Palmer et al.

(2008) in this case) to the other ones. In addition, distribution figures such as mean, variance, skewness and kurtosis are calculated.

```
# get parametrization according to Hougaard (1986)
param.H(MYpTAS)
##     alpha     delta     theta
## 0.5000000 0.9428090 0.8888889
# get mean and skewness
mean.pTAS(MYpTAS)
## [1] 1
skew.pTAS(MYpTAS)
## [1] 2.25
```

To obtain a first impression of the distribution, the density (and also the distribution function) can be shown by simply using the PLOT function.

```
# plot pdf (use additional argument cdf=TRUE to plot cdf)
plot(MYpTAS)
```

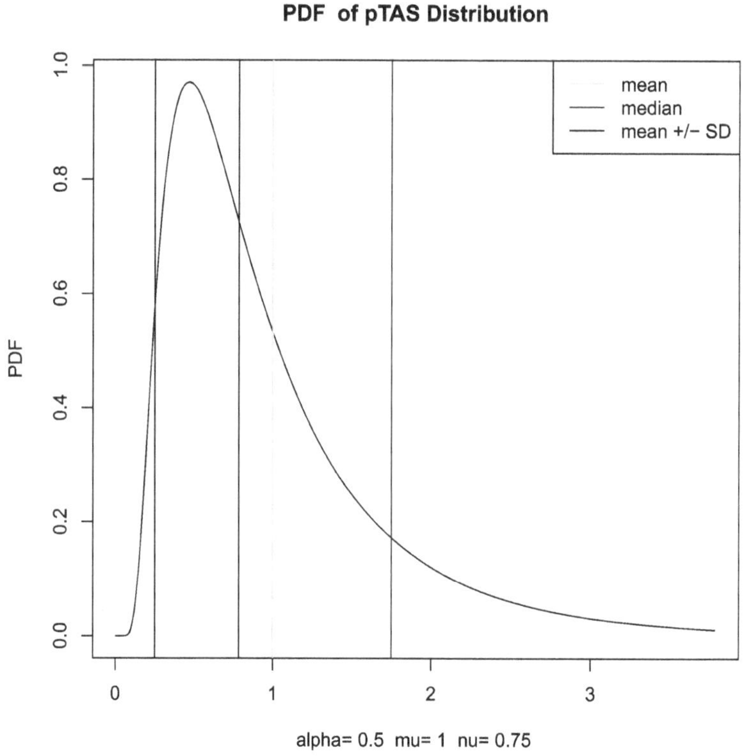

Density, distribution function, quantiles and random numbers

Density and distribution function are available via the standard R-notation (d.../ p...). The numerical parameters described in 2.1 can be set when the object is created via the PTAS function.

```
# calculate pdf and cdf
t = seq(0.5, 2.5, 0.5)
dpTAS(t, MYpTAS)
## [1] 0.96465947 0.53192304 0.24967324 0.12058243 0.06046488
ppTAS(t, MYpTAS)
## [1] 0.2547667 0.6340912 0.8205896 0.9089881 0.9524083
```

Quantiles and random numbers are generated via Algorithm 2.

```
# calculate quantiles
p = c(0.1, 0.25, 0.5, 0.75, 0.9)
qpTAS(p, MYpTAS)
## [1] 0.3355067 0.4950626 0.7855115 1.2644107 1.9291478
# draw random numbers
set.seed(1)  # for reproducibility
rpTAS(5, MYpTAS)
## [1] 1.9935582 0.8937332 0.5111581 0.6264167 0.4452205
```

Estimation methods

For implementation issues, the MLE will be performed on the $\mathcal{P}_P$- parameterization always. For parameters α, μ, ν lower and upper bounds as well as fixed values can be specified. Especially for parameterα an appropriate upper bound may be helpful, because pdf calculations for parametersα close to 1 are numerically challenging. The FIT_MLE function uses the MLE function from the STATS4 package, which in turn uses R's OPTIM optimizer.

```
# draw some random numbers
x = rpTAS(200, MYpTAS)
# perform MLE
MLE_pTAS = fit_mle(x)
# estimation results
param.P(MLE_pTAS)
##     alpha        mu        nu
## 0.5489056 0.9729183 0.6668250
# compared to original values
param.P(MYpTAS)
## alpha    mu    nu
##   0.50  1.00  0.75
```

Detailed information on the optimization results can by obtained via the OPTIM_RES function. This gives a list containing the number of iterations, the convergence result and the Hessian matrix for further calculations (e.g. to calculate confidence intervals). If the optimization did not converge properly, a warning is displayed automatically.

```
optim_res(MLE_pTAS)
## $par
##     alpha        mu        nu
## 0.5489056 0.9729183 0.6668250
##
## $value
## [1] 136.1362
##
## $counts
## function gradient
##       23       23
##
## $convergence
 ## [1] 0
##
## $message
## [1] "CONVERGENCE: REL_REDUCTION_OF_F <= FACTR*EPSMCH"
```

```
##
## $hessian
##            alpha         mu         nu
## alpha   567.3590   198.1225  -475.1069
## mu      198.1225   628.2451  -367.0736
## nu     -475.1069  -367.0736   880.2828
```

Alternatively, distribution parameters can be also estimated based on mean and variance, which determine parameters μ and ν regarding the $\mathcal{P}_P$ parametrization and either the skewness, the kurtosis or a quantile to estimate parameter α. If α should be estimated based on a given kurtosis value, which possible has multiple solutions, a pTAS distribution with either the highest or the lowest value of possible α's or a list containing all distributions will be returned.

If α should be estimated based on a given quantile, an grid search method as described in section 2.3 is applied. For given values $t^* > 0$ and $p^* \in (0, 1)$ the algorithm terminates if $\left|F_{\mathcal{P}_P(\alpha,\mu,\nu)}(t^*) - p^*\right| < \epsilon$ for a prespecified $\epsilon > 0$, whereas values for μ and ν are fixed based on mean and variance. Similar to the estimation based on kurtosis, the solution for α (if one exists) may be not unique. Again, one can determine which distribution should be returned with an additional argument (see example below).

```
# estimation based on skewness/kurtosis
skew_pTAS = fit_moments(mean = 1, var = 0.5, skew = 3)
kurt_pTAS = fit_moments(mean = 1, var = 0.5, kurt = 10)
## alpha1: 0.4757636 alpha2: 1.293467
# estimation based on quantile
quant_pTAS_1 = fit_quantile(mean = 1, var = 1, x = 5, prob = 0.99)
## Warning in fit_quantile(mean = 1, var = 1, x = 5, prob =
0.99):
Solution for alpha is not unique, choosing solution with
highest alpha.
param.P(quant_pTAS_1)
##      alpha        mu        nu
## 0.8898322 1.0000000 1.0000000
quant_pTAS_2 = fit_quantile(mean = 1, var = 1, x = 5, prob = 0.99,

which.alpha = "lowest")  # return lower alpha
## Warning in fit_quantile(mean = 1, var = 1, x = 5, prob =
0.99,...):
Solution for alpha is not unique, choosing solution with
lowest alpha.
param.P(quant_pTAS_2)
##      alpha        mu        nu
## 0.5147693 1.0000000 1.0000000
```

Acknowledgement

The authors thank two anonymous referees and an associate editor for their helpful comments and suggestions which significantly improved the presentation of the paper.

Authors' contributions

MF carried out the derivation and general properties of the pTAS family as well as the application to operational risk. KJ addressed issues regarding the implementation and the credit risk example. The manuscript was drafted together. All authors read and approved the final manuscript.

Competing interests
The authors declare that they have no competing interests.

Author details
[1]Department of Statistics & Econometrics, FAU Erlangen-Nürnberg, Erlangen, Germany. [2]Department of Statistics, University of Augsburg, Augsburg, Germany.

References

Abate, J, Choudhury, GL, Whitt, W: An introduction to numerical transform inversion and its application to probability models. In: Computational Probability, pp. 257–323. Springer Science+Business Media, New York (2000)

Barndorff-Nielson, OE, Shephard, N: Normal modified stable processes. Economics papers 2001-w6. University of Oxford, Oxford (2001)

Bolancè, C, Guillèn, M, Gustafsson, J, Nielson, JP: Quantitative Models for Operational Risk: Extremes, Dependence and Aggregation. Chapman & Hall/CRC Finance Series, New York (2012)

Chernobay, A, Rachev, S, Fabozzi, F: Composites goodness-of-fit tests for left-truncated loss samples. In: Lee, CF, Lee, J (eds.) Handbook of Financial Econometrics and Statistics, pp. 575–596. Springer, (2015)

Credit Suisse First Boston International: CreditRisk$^+$ A Credit Risk Management Framework (1997). http://www.csfb.com/institutional/research/assets/creditrisk.pdf. Accessed 27 Jan 2014

El Adlouni, S, Ezzahid, E, Moutazzim, J: Mixed distributions for loss severity modelling with zeros in the operational risk losses. Int. J. Appl. Math. Stat. **11**(21), 96–109 (2011)

Esscher, F: On the probability function in the collective theory of risk. Scand. Actuarial J. **1932**(3), 175–195 (1932). doi:10.1080/03461238.1932.10405883

Fischer, M, Jakob, K: Copula-Specific Credit Portfolio Modeling. In: Glau, K, Scherer, M, Zagst, R (eds.) Innovations in Quantitative Risk Management. Springer Proceedings in Mathematics & Statistics, pp. 129–145. Springer International Publishing, (2015)

Gundlach, M, Lehrbass, F: CreditRisk+ in the Banking Industry. Springer Finance. Springer, Berlin Heidelberg (2004)

Haas, M, Pigorsch, C: Financial economics: Fat-tailed distributions. In: Encyclopedia of Complexity and Systems Science vol. 4, pp. 3404–3435. Springer, New York (2009)

Hougaard, P: Survival models for heterogeneous populations derived from stable distributions. Biometrika. **73**(2), 387–396 (1986)

Jakob, K, Fischer, M: Quantifying the impact of different copulas in a generalized CreditRisk+ framework An empirical study. Depend. Model. **2**, 1–21 (2014)

Janson, S: Stable distributions. arXiv preprint arXiv:1112.0220 (2011). Accessed 2015-08-16

Jørgensen, B: Exponential dispersion models (with discussion). J. R. Stat. Soc. Series B. **49**(2), 127–162 (1987)

Koudou, AE, Ley, C: Characterizations of gig laws: A survey. Probab. Surv. **11**, 161–176 (2014)

Küchler, U, Tappe, S: Tempered stable distributions and applications to financial mathematics (2011)

Merton, RC: On the pricing of corporate debt: The risk structure of interest rates. J. Finance. **29**(2), 449–470 (1974)

Neslehova, J, Embrechts, P, Chavez-Demoulin, V: Infinite mean models and the LDA for operational risk. Journal of Operational Risk. **1**(1), 3–25 (2006)

Nolan, J: Stable Distributions: Models for Heavy-tailed Data. Birkhauser, Boston (2003). In progress available via: http://fs2.american.edu/jpnolan/www/stable/chap1.pdf

Palmer, KJ, Ridout, MS, Morgan, BJ: Modelling cell generation times by using the tempered stable distribution. J. R. Stat. Soc. Series C (Appl. Stat.) **57**(4), 379–397 (2008)

Ridout, MS: Generating random numbers from a distribution specified by its Laplace transform. Stat. Comput. **19**(4), 439–450 (2008). doi:10.1007/s11222-008-9103-x

Schiff, JL: The Laplace Transform: Theory and Applications. Springer-Verlag, New York (1999)

Stacy, EW: A generalization of the gamma distribution. Ann. Math. Stat. **3**(33), 1187–1192 (1962)

Tweedie, MCK: An index which distinguishes between some important exponential families. In: Statistics: Applications and New Directions: Proc, Indian Statistical Institute Golden Jubilee International conference, pp. 579–604, (1984)

3

Exact inference on scaling parameters in norm and antinorm contoured sample distributions

Wolf-Dieter Richter

Correspondence:
wolf-dieter.richter@uni-rostock.de
Department of Mathematics,
University of Rostock, Rostock,
Germany

Abstract

Exact distributions of generalized Chi-square and Fisher statistics are used to derive confidence intervals and significance tests for inferring on one or two scaling parameters, respectively, under non-standard assumptions w.r.t. the multivariate sample distribution. The latter may have convex or radially concave density level sets and heavy or light distribution centers and tails. Independent and $l_{n,p}$-dependent sample variables are considered.

Keywords: Estimating and testing scaling parameters, Non-standard model assumptions, Generalized Chi-square and Fisher distributions, Heavy and light distribution tails, Heavy and light distribution centers, Independence sampling, Dependence sampling

MSC2010: Primary: 62F03, 62F25, 62C07 secondary: 6E15

1 Introduction

The well known Chi-square and Fisher distributions were introduced already a long time ago in (Fisher 1924a, b;1925; Hermert 1876; Pearson 1900), respectively. Since then, methods of statistical inference on variances are extensively studied under various aspects in numerous papers, see e.g. (Mirvaliev 1987; Plackett 1983; Statlect 2016) and the recent survey in (Bolboaca et al. 2011), for only to mention a few of them.

It is often natural to assume that expectations are unknown and thus to additionally be estimated when studying variances. In some cases, however, it may be reasonable to assume that expectations are known. Assume, for example, a company produces tubes of a certain norm length. The production dependent deviations from this length may according to (Bomsdorf 1989), p. 162 be considered as normally distributed with expectation zero and unknown variance. If the company indicates to customers a certain variance of tube length then one might be interested in proving this.

The successful fulfillment of manufacturing orders requires according to (Haidenhain 2011) machine tools with high thermal stability. Rapidly changing operating conditions for machine tools, however, may lead to unknown variances of technical features having a known expectation and make it difficult to constantly ensure accuracy in production processes.

Moreover, comparing in another situation instrumental accuracy by measuring a certain standard several times with two gages may generate random samples having the same known mean and unknown variances which one might be interested then to compare with each other.

If in such situations independent random variables $X_1, \ldots, X_n$ and $Y_1, \ldots, Y_m$ follow Gaussian laws with known expectations μ_1, μ_2 and unknown standard deviations σ_1, σ_2, respectively, then, with $\hat{\sigma}_1^2 = \frac{1}{n}\sum_1^n (X_i - \mu_1)^2$ being the maximum likelihood estimator of the variance σ_1^2, $n\left(\frac{\hat{\sigma}_1}{\sigma_1}\right)^2$ follows a Chi-square density with n d.f., and $\frac{(\hat{\sigma}_1/\sigma_1)^2}{(\hat{\sigma}_2/\sigma_2)^2}$ follows Fisher's F-distribution with (n, m) d.f.

The aim of the present paper is to allow more general distributions of the sample vector, and to study exact distributions of generalized Chi-square and Fisher statistics being based upon the maximum likelihood estimators of the model's scaling parameters. The multivariate sample distributions considered to this end are convex or radially concave contoured. These two distribution classes are subclasses of the family of star-shaped distributions.

The paper is organized as follows. Star-shaped distributions are introduced in Section 2. A one sample non-standard scaling model for i.i.d. power exponentially distributed random variables and a certain extension of it are introduced in Section 3.1. Confidence estimation of and significance testing of hypotheses on the scaling parameter are discussed in Sections 3.2 and 3.4 for p-power exponential and more general sample distributions, respectively. Preparations for the latter case are given in Section 3.3. A two sample non-standard scaling parameter model ist studied then in Section 3.5 where estimation and testing are dealt with. A brief comparison of norms and antinorms, generating convex or radially concave level sets of the multivariate sample density, respectively, is given in the Appendix.

Finally, we remark that an introduction to non-standard models and exact distribution theory is given in (Richter 2012, 2014). For closely related studies on expectations and variances based upon $l_{n,p}$-norms and goodness of fitting using $l_{n,p}$-norms for data sets that follow a power exponential distribution, we refer to (Livadiotis 2007, 2012, 2014). For a comparison of the variances of two dependent random variables, see (Wilcox 2015).

2 Star-shaped sample distributions

2.1 The general class

The most general non-standard distributions of the multivariate sample vector we are considering in the present paper are star-shaped distributions. More specific distributions are $l_{n,p}$-symmetric and p-generalized elliptically contoured distributions considered, e.g., in (Richter 2012, 2014), respectively. Let $K \subset \mathbb{R}^n$ denote a star body with the origin being an inner point, and let S be its topological boundary which will be called a star sphere. The Minkowski functional of K is defined by $h_K(x) = \inf\{\lambda > 0 : x \in \lambda K\}$, $x \in \mathbb{R}^n$ and satisfies $h_K(x) = 1, x \in S$. Assume that a random vector X defined on a probability space $(\Omega, \mathfrak{A}, P)$ allows the stochastic representation

$$X \stackrel{d}{=} R \cdot U_S$$

where R is a non-negative random variable and U_S follows independently of R the S-generalized uniform distribution ω_S on S. Then X is called origin-including star-shaped distributed. Note that $X \stackrel{d}{=} Y$ means that X and Y follow the same distribution law, and

$$\omega_S(A) = \frac{O_S(A)}{O_S(S)}, A \in \mathfrak{B}_S = \mathfrak{B}^n \cap S$$

where $\mathfrak{B}^n$ denotes the Borel σ-field in $\mathbb{R}^n$ and O_S the star generalized surface content measure on $\mathfrak{B}_S$. If μ stands for the Lebesgue measure on $\mathfrak{B}^n$ and $CPC(A) = \{x \in \mathbb{R}^n : x/h_K(x) \in A\}$ is the central projection cone generated by $A \in \mathfrak{B}_S$ then the star generalized surface content of A is formally defined for a particular class of star spheres S in (Richter 2007) and under the weak technical Assumption 1 for general star spheres in (Richter 2014) by

$$O_S(A) = f_A'(1) \tag{1}$$

where

$$f_A'(r) = \frac{d}{dr} f_A(r), \quad f_A(r) = \mu(sector(A,r)), \quad sector(A,r) = CPC(A) \cap K(r)$$

and $K(r) = r \cdot K = \{(rx_1, \ldots, rx_n)^T \in \mathbb{R}^n : (x_1, \ldots, x_n)^T \in K\}$. Integral, or differential geometric, characterizations of O_S which can equivalently be taken as a definition of O_S are given for particular classes of star bodies in (Richter 2009, 2013) and for arbitrary two-dimensional convex star bodies in (Richter 2015b). Note that, e.g., the well known (Richter 2007) $l_{n,p}$-generalized (non-Euclidean) surface content of the $l_{n,p}$-unit sphere,

$$\omega_{n,p} = \frac{2^n \Gamma\left(\frac{1}{p}\right)^n}{p^n \Gamma\left(\frac{n}{p}\right)}, \; p > 0,$$

is equal to the Euclidean surface content, if and only if $p = 2$.

There are widespread miss-understandings w.r.t. this circumstance in the literature. The reader should be aware that there is some literature where formulas are derived which are called surface content and uniform distribution, respectively, however neither defining in advance the notion of surface content nor referring to literature where such definition is given.

The star generalized surface measure on the boundary of n-dimensional convex or radially concave star bodies is dealt with under different aspects in (Richter 2016, c, 2016). A general result in (Richter 2016) says that

$$O_S(A) = \int\limits_{G(A)} \hat{h}_{\hat{K}}(N(x))dx$$

where

$$\hat{K} = \{\lambda x : x \in Grad(S), 0 \leq \lambda \leq 1\}, \quad Grad(S) = \{\nabla h_K(x), x \in S\}$$

and

$$\hat{h}_{\hat{K}}(t) = \inf\left\{\lambda > 0 : t \in \lambda \hat{K}\right\}, \; t \in \operatorname{pos} \hat{K}$$

is a certain modification of the Minkowski functional of $\hat{K}$. Here, $\operatorname{pos} x = \{\lambda x : \lambda \geq 0\}, x \in \mathbb{R}^n$ and $\operatorname{pos} M = \bigcup\limits_{x \in M} \operatorname{pos} x$.

If X is origin-including star-shaped distributed and $\nu \in st(K)$ where $st(K)$ denotes the set of all points w.r.t. which K is star-shaped then $X - \nu$ is also origin-including star-shaped distributed. If, however, ν does not belong to $st(K)$ then $X - \nu$ will be called just star-shaped distributed. The first of these two cases is dealt with in (Richter 2014), Section 5.1 where non-concentric elliptically contoured distributions are introduced. Both cases are considered when studying non-central Chi-square distributions in (Ittrich et al. 2000).

For a density having non-concentric circular contours with centers on a straight line, see (Arnold et al. 2008).

Example. Let $a = (a_1, \ldots, a_n)^T \in \mathbb{R}^n$ be a vector having positive components, $p \geq 1$, and $K = \left\{x \in \mathbb{R}^n : \left(\sum_{i-1}^{n} |\frac{x_i}{a_i}|^p\right)^{1/p} \leq 1\right\}$. Then, according to (Richter 2014), the star-generalized surface content measure on the boundary S of K is

$$O_S(A) = a_n \int_{G(A)} \frac{d(x_1, \ldots, x_{n-1})}{\left(1 - \sum_{i=1}^{n} |\frac{x_i}{a_i}|^p\right)^{1-1/p}}, \; A \in \mathfrak{B}_S. \tag{2}$$

The special case $p = 2$ has been dealt with in (Richter 2013).

2.2 Norm contoured distributions

Note that a symmetric convex star body having the origin as an inner point can always be generated by a norm, $h_K(.) = ||.||$, and that according to (Richter 2015c)

$$O_S(A) = \int_{G(A)} ||N(x)||^* dx, \; A \in \mathfrak{B}_S \tag{3}$$

where $G(A) = \left\{\vartheta \in \mathbb{R}^{n-1} : \exists \eta = \eta(\vartheta) \, s.t. \, \left(\vartheta^T, \eta(\vartheta)\right)^T \in A\right\}$ and $||.||^*$ denotes the norm dual to $||.||$ and $N(\vartheta) = (\nabla\eta(\vartheta)^T, -1)^T$ is the outer normal vector to S at $(\vartheta^T, \eta(\vartheta))^T$. If K is the star body from the Example in Section 2.1 then (3) takes the specific form (2).

Let us emphasize that the star generalized surface content of an arbitrary measurable subset A of the star sphere S is a sum of integrals, in general. If S^+ and S^-, however, denote upper and lower half spheres of the norm sphere S, and $K^* = \{y \in \mathbb{R}^n : ||y||^* \leq 1\}$, then

$$O_S(A) = \int_{G(A \cap S^+)} h_{K^*}(N(x))dx + \int_{G(A \cap S^-)} h_{K^*}(N(x))dx, \; A \in \mathfrak{B}_S$$

where

$$G\left(A \cap S^{+(-)}\right) = \left\{\vartheta \in \mathbb{R}^{n-1} : \exists \eta = \eta(\vartheta) \, s.t. \, \left(\vartheta^T, \eta\right)^T \in A \cap S^{+(-)}\right\}.$$

For simplicity, and in accordance with Definition 1, Lemma 1 and Corollary 1 in (Richter 2014), however, $O_S(A)$ will be represented throughout this paper symbolically by a single integral.

Finally note that formula (1) applies whenever it is possible to evaluate the volume of a convex body. W.r.t. this research area we refer, e.g., to (Pisier 1989) and a discussion on page 11 in (Richter 2011).

Note that K^* is always a symmetric convex star body and defines therefore a metric geometry in the sense of (Burago et al. 2001).

2.3 Antinorm contoured distributions

If K is radially concave w.r.t. a fan $\mathcal{F} = \{C_1, C_2, \ldots\}$ and belongs to the particular class $AN1$ of star bodies introduced in (Richter 2015c) then its Minkowski functional is an antinorm, $h_K(.) = \nmid . \nmid$. For the notion of antinorm, we refer to (Moszyńska and Richter 2012) and the Appendix. Let

$$h_K^{\mathcal{F}}(u) = \sum_i I_{C_i}(u) \inf\left\{u^T y : y \in S \cap C_i\right\}, u \in \mathbb{R}^n$$

be the anti-support function of K w.r.t. $\mathcal{F}$ and

$$K^o = \left\{\lambda(u)u : 0 \leq \lambda(u)h_K^{\mathcal{F}}(u) \leq 1, u \in S_E^{(n-1)}\right\}$$

the anti-polar set of K where $S_E^{(n-1)}$ denotes the Euclidean unit sphere in $\mathbb{R}^n$ then according to (Richter 2015c)

$$O_S(A) = \int_{G(A)} h_{K^o}(N(x))dx \tag{4}$$

where $N(x)$ is the inner normal vector to S at $x \in S$.

Example. Let the antinorm ball and sphere K and S, respectively, be formally defined as in the preceding example but with $p \in (0, 1]$, and let a still be a vector with positive components. Then the S-generalized surface content measure $\mathfrak{O}_S$ satisfies representation (2). Notice, however, that we are evaluating star generalized surface contents of subsets of antinorm spheres, here, while star generalized surface contents of subsets of norm spheres are considered in the preceding example. Notice that $1 - 1/p > 0$ if $p > 1$ and $1 - 1/p < 0$ if $0 < p < 1$.

2.4 Continuous distributions

Let a function $g : [0, \infty) \to [0, \infty)$ satisfy the assumptions $0 < I(g) < \infty$ where $I(g) = \int_0^\infty r^{n-1} g(r)dr$. Such function is called a density generating function (dgf). A dgf can be used modeling light or heavy distribution centers and tails. Moreover, it allows even modeling complete level distributions. Well known examples are power-t exponential dgf $g_{PE}(r) = \exp\left\{-\frac{r^p}{p}\right\}$, Kotz type dgf $g_K(r) = r^{M-1}e^{-\beta r^\gamma}, \beta, \gamma > 0, 2M + n > 2$ and Pearson-VII-type dgf $g_{PT7}(r) = \left(1 + \frac{r}{m}\right)^{-M}, M > n/2, m > 0$ where $I(g_{PE}) = p^{n/p-1}\Gamma(n/p), I(g_K) = \frac{\Gamma((n+M-1)/\gamma)}{\gamma\beta^{(n+M-1)/\gamma}}$ and $I(g_{PT7}) = m^n B(M-n, n)$, respectively. Student- and Cauchy-type dgf's are special cases of g_{PT7}.

If the star-shaped distributed random vector X has a probability density (p.d.), $\varphi_{g,K,\nu}$ say, then

$$\varphi_{g,K,\nu}(x) = C(g, K)g(h_K(x - \nu)), x \in \mathbb{R}^n.$$

Densities of this type have been considered in (Arnold et al. 2008; Fernández et al. 1995). Here,

$$C(g, K) = 1/(I(g)O_S(S)), \tag{5}$$

and the p.d. of the generating variate R is

$$f(r) = I_{(0,\infty)}(r)r^{n-1}g(r).$$

The corresponding probability distribution allows the geometric measure representation

$$\Phi_{g,K,\nu}(B) = \frac{1}{I(g)} \int_0^\infty r^{n-1} g(r) \mathfrak{F}_S(B,r) dr, \; B \in \mathfrak{B}^n \tag{6}$$

where

$$\mathfrak{F}_S(B,r) = \omega_S\left(\left[\frac{1}{r}(B-\nu)\right] \cap S\right), \; r > 0$$

denotes the intersection proportion function (ipf) of the set B.

If K is the Euclidean unit ball then $K^* = K$ and O_S is the Euclidean surface content measure. In this case, representation (6) has been considered already in (Richter 1985, 1991). For specific non-Euclidean cases, (6) has been considered in (Richter 2007, 2009, 2013, 2014).

Surveys of different types of applications of formula (6) are given for $n = 2$ in (Richter 2015b) and for arbitrary n in (Richter 2012, 2014). For two elementary applications, see Section 2.6.

2.5 Building subclasses and class extensions

2.5.1 Building subclasses of distributions by modeling R

Let us denote the probability law of the generating variate R by P^R, and let ν be a σ-finite measure on the Borel σ-field $\mathfrak{B}^{(1)}$ (or its restriction to the positive real line) such that P^R is absolutely continuous w.r.t. ν. If ν is the Lebesgue measure on $\mathfrak{B}^{(1)}$ then $\frac{dP^R}{d\nu}(r) = f(r)$ is the p.d. considered in Section 2.4. In Sections 2.2 and 2.3, P^R can be chosen as the distribution of any non-negative random variable thus $\frac{dP^R}{d\nu}$ being a mixture of an absolutely continuous, a discrete and a singular component, in general.

2.5.2 Extending distribution classes by restricting U

We emphasize for the rest of this section that S circumscribes the origin as in inner point. Let $\mathcal{M}$ be a measurable subset of the star sphere S. We call a random vector $U_\mathcal{M}$ star-generalized uniformly distributed on $\mathcal{M}$ (or the $\mathcal{M}$-restriction of U) if $U_\mathcal{M}$ follows the distribution law

$$P^{U_\mathcal{M}}(A) = \frac{O_S(A)}{O_S(\mathcal{M})}, \; A \in \mathcal{B}_S \cap \mathcal{M}.$$

Note that $U_S \overset{d}{=} U$ and that pos$\mathcal{M}$ is not an absorbing set unless for $\mathcal{M} = S$.

For an example of an $\mathcal{M}$-restriction of U occuring in depth-based classification for distributions with non-convex support, we refer to (Hlubinka and Vencalek 2013).

We call

$$\mathcal{D}_S(\mathcal{M}) = \{\mathcal{L}\{R \cdot U_\mathcal{M}\}, \; R \text{ is a non-negative random variable}\}$$

where $\mathfrak{L}(Y)$ means the distribution law of Y the class of directionally $\mathcal{M}$-out-of-S restricted distributions and

$$\mathcal{D}_S = \{\mathcal{D}_S(\mathcal{M}) : \mathcal{M} \text{ is a measurable subset of the star sphere } S\}$$

the class of directionally out-of-S restricted distributions. Finally,

$$\mathcal{D} = \{\mathcal{D}_S : S \text{ is a star sphere}\}$$

is called the class of directionally out-of a star sphere restricted distributions in $\mathbb{R}^n$.

2.6 Two elementary types of applications of formula (6)

2.6.1 The ipf is the indicator function of an interval

Assume that the random vector X follows the continuous star-shaped distribution law $\Phi_{g,K,\nu}$, and let the statistic $T : \mathbb{R}^n \to [0, \infty)$ be defined by $T(x) = h_K(x - \nu)$. The ipf of the set $B(t) = \{x \in \mathbb{R}^n : T(x) < t\}$ allows then the representation

$$\mathfrak{F}_S(B(t), r) = \frac{O_S\left(\left[\frac{1}{r}(B(t) - \nu)\right] \cap S\right)}{O_S(S)} = \begin{cases} 1 & r \in I(t) \\ 0 & \text{elsewhere} \end{cases}$$

where S is still the boundary of K and $I(t)$ is the particular interval $[0, t)$. It follows from (6) that

$$P(T(X) < t) = \Phi_{g,K,\nu}\left(\{x \in \mathbb{R}^n : h_K(x - \nu) < t\}\right) = \frac{1}{I(g)O_S(S)} \int_0^\infty r^{n-1} g(r) I_{[0,t)} dr.$$

Now, a density formula can be derived from the representation

$$P(T(X) < t) = \frac{1}{I(g)O_S(S)} \int_{I(t)} r^{n-1} g(r) dr.$$

Similar results hold for arbitrary intervals $I(t)$. For a specific application of the present type we refer to the calculation of generalized Chi-square densities.

2.6.2 The ipf is constant

Let $\mathfrak{F}_{B(t)}(r) = C(t)$ where $C(t) \in (0, 1)$ does not depend on r. Then it follows from (6) that $P(T < t) = C(t)$, and T is called a g-robust statistic. For specific such applications we refer to the geometric derivation of generalized Student and Fisher distributions. The notion of robustness considered here was studied already in (Fang et al. 1990) for elliptically contoured distributions.

Several other types of sets for which the ipf has been calculated already in the past can be found in some of author's papers surveyed in (Richter 2012, 2014).

3 Inference on scaling parameters

3.1 The power exponential sample distribution

3.1.1 A basic scaling parameter model

Let the i.i.d. random variables $X_1, \ldots, X_n$ follow the power exponential (or p-generalized Gaussian or p-generalized Laplace) density $f_p(.; \mu, \sigma)$, where $p > 0$ and $\mathbb{E}X_1 = \mu$ are known, and

$$f_p(x; \mu, \sigma) = \frac{C_p}{\sigma} \exp\left\{-\frac{|x - \mu|^p}{p\sigma^p}\right\}, -\infty < x < \infty. \tag{7}$$

Note that $C_p = p^{1-1/p}/(2\Gamma(1/p))$ and $V(X_1) = \sigma^2\Gamma\left(\frac{3}{p}\right)/\Gamma\left(\frac{1}{p}\right)$. Let X_0 follow the p.d. of the standardized p-generalized normal distribution, $f_p(.; 0, 1)$. Then X_1 satisfies the representation $X_1 \stackrel{d}{=} \mu + \sigma X_0$. The quantity σ is called a scaling parameter.

3.1.2 Norm and antinorm contoured sample distributions

The density of the sample vector $X = (X_1, \ldots, X_n)$ is

$$\prod_{i=1}^n f_p(x_i; \mu, \sigma) = \frac{C_p^n}{\sigma^n} \exp\left\{-\sum_{i=1}^n \frac{|x_i - \mu|^p}{p\sigma^p}\right\}, x = (x_1, \ldots, x_n) \in \mathbb{R}^n.$$

The density super level sets $\left\{x \in \mathbb{R}^n : \prod_{i=1}^{n} f_p(x_i; \mu, \sigma) \geq t\right\}$, $t > 0$ are generated by the functional

$$x \to |x|_p = \left(\sum_{i=1}^{n} |x_i|^p\right)^{1/p}$$

which is a norm if $p \geq 1$ and an antinorm if $0 < p \leq 1$. Here, $K = B_p = \{x \in \mathbb{R}^n : |x|_p \leq 1\}$.

For getting a more concrete idea of the shape of this density let us recall that the boundaries of the two-dimensional density super level sets are the star circles

$$\left\{(x_1, x_2)^T \in \mathbb{R}^2 : |x_1|^p + |x_2|^p = r^p\right\}, \; r > 0.$$

The star discs circumscribed by these p-circles are convex if $p \geq 1$ and radially concave if $0 < p \leq 1$. For the latter notion, we refer to (Moszyńska and Richter 2012). Similar properties hold in the case of arbitrary finite dimension.

In the geometric measure representation (6), the star generalized surface content measure allows for every $p > 0$ and every measurable subset A of the upper (or lower) half of the star sphere S the representation

$$O_S(A) = \int_{G(A)} \left(1 - \sum_{1}^{n-1} |x_i|^p\right)^{(1-p)/p} dx \tag{8}$$

for $p \geq 1$ and $p \in (0, 1]$ being each time a special case of (3) and (4), respectively.

3.1.3 Extended scaling parameter model

Let now $X_1, \ldots, X_n$ be i.i.d. random variables following the densities

$$f_p\left(x; \mu_i, \sigma^2 a_i^2\right) = \frac{C_p}{\sigma a_i} \exp\left\{-\frac{|x - \mu_i|^p}{p(\sigma a_i)^p}\right\}, \; x \in \mathbb{R}, \; i = 1, \ldots, n \tag{9}$$

where $\mathbb{E}X_i = \mu_i \in \mathbb{R}, p > 0, a_i > 0$ are known, and $\sigma > 0$ is unknown. Note that

$$\mathbb{V}(X_i) = \sigma^2 a_i^2 \frac{\Gamma(3/p)}{\Gamma(1/p)}$$

and that X_i allows the representation

$$X_i \overset{d}{=} \mu_i + \sigma a_i X_0$$

where X_0 has the p.d. $f_p(.; 0, 1)$ and σ is a common scaling parameter.

3.2 Inference on a single scaling parameter

3.2.1 Basic scaling parameter model.

The maximum likelihood estimator of σ in the model (7) is the sufficient statistic

$$\hat{\sigma} = mle(\sigma) = \left(\frac{1}{n} \sum_{i=1}^{n} |X_i - \mu|^p\right)^{1/p}.$$

Note that

$$\mathbb{E}(\hat{\sigma}^p) = \sigma^p \text{ and } \mathbb{V}(\hat{\sigma}^p) = \frac{p}{n} \sigma^{2p},$$

$\hat{\sigma}^p$ thus being an unbiased and consistent estimator for σ^p. The distribution of $\hat{\sigma}$ can be derived, e.g., on using formula (6) with the help of formula (8). The result follows below from the consideration for the extended scaling parameter model (9).

In the sense of Section 2.6, it is an elementary problem to evaluate the S-generalized surface content of subsets of S according to (8) in the present case because we are dealing here with a set the ipf of which is just the indicator function of an interval. This way it is proved (within a more general framework) in (Richter 2009) (and with a slight modification already in (Richter 2007)) that $n(\hat{\sigma}/\sigma)^p$ follows the χ^p-(or p-generalized χ^2-) distribution with n d.f., $n(\hat{\sigma}/\sigma)^p \sim \chi^p(n)$ where $Y \sim Q$ means that the random variable Y follows the probability distribution Q. The p.d. of the $\chi^p(n)$ distribution is according to (Richter 2007, 2009)

$$f_{n,p}(t) = \frac{t^{n/p-1}e^{-t/p}}{p^{n/p}\Gamma(n/p)} I_{(0,\infty)}(t). \tag{10}$$

Although assigning in its derivation the number $\omega_{n,p}$ the Euclidean interpretation instead of the right non-Euclidean one, density (10) was derived later again in (Livadiotis 2012). For a considerable generalization of this density, see (14) below and (Richter 2009). We denote the quantile of order $q \in (0,1)$ of the $\chi^p(n)$ distribution by $\chi^p_q(n)$. Thus, by definition $P\left(n\left(\hat{\sigma}/\sigma\right)^p < \chi^p_q(n)\right) = q$. Note that the $\chi^p(n)$ distribution is a particular Gamma distribution having parameters $\frac{n}{p}$ and $\frac{1}{p}$. The quantiles of this distribution can therefore be computed using the inverse Gamma function being available in standard software.

3.2.2 *Confidence intervals for σ^p*

The random intervals

$$\left[\frac{|X-\mu 1_n|_p^p}{\chi^p_{1-\alpha_1}(n)}, \frac{|X-\mu 1_n|_p^p}{\chi^p_{\alpha_2}(n)}\right], \left[\frac{|X-\mu 1_n|_p^p}{\chi^p_{1-\alpha}(n)}, \infty\right) \text{ and } \left(-\infty, \frac{|X-\mu 1_n|_p^p}{\chi^p_{\alpha}(n)}\right] \tag{11}$$

are level $1-\alpha$ two-sided, and one-sided upper and lower confidence intervals for σ^p, respectively, with $\alpha = \alpha_1 + \alpha_2$ in the two-sided case, and where $1_n = (1, \ldots, 1)^T \in \mathbb{R}^n$. If $p = 2$ then $\chi^p_q(n)$ is the common quantile of order q of the Chi-square distribution with n d.f., $\chi^2_q(n)$. The case $p = 1$ is dealt with in (Kotz et al. 2001), formula (2.6.112), where, however, $2/\chi^2_{1-\alpha/2}(2n)$ is used instead of $1/\chi^1_{1-\alpha/2}(n)$. Tables 2, 3, 4 and 5 present quantiles $\chi^p_q(n)$ for several values of n, p and q.

3.2.3 *Significance testing*

If one wants to test the hypothesis $H_0 : \sigma = \sigma_0$ versus the two-sided alternative $H_A : \sigma \neq \sigma_0$ then the decision rule of a level-α significance test is defined as to reject H_0 if

$$\sum_{i=1}^{n}\left|\frac{X_i-\mu}{\sigma_0}\right|^p \in \left(-\infty, \chi^p_{\alpha_2}(n)\right) \bigcup \left(\chi^p_{1-\alpha_1}(n), \infty\right), \tag{12}$$

where $\alpha = \alpha_1 + \alpha_2, 0 < \alpha_i < 1/2, i = 1, 2$. Tests for one-sided hypotheses are analogously constructed.

3.2.4 *Extended scaling parameter model*

In the model (9), the maximum likelihood estimator of the common scaling parameter σ is

$$mle(\sigma) = \widehat{\sigma} = \frac{1}{n^{1/p}}|X-\nu|_{a,p}$$

where $\nu = (\mu_1, \ldots, \mu_n)^T$ and

$$|x|_{a,p} = \left(\sum_{i=1}^{n}\left|\frac{x_i}{a_i}\right|^p\right)^{1/p,} \quad x \in \mathbb{R}^n.$$

The p.d. of X is called a p-generalized axes aligned elliptically contoured density having the contour defining star body $K = B_{a,p} = \{x \in \mathbb{R}^n : |x|_{a,p} \leq 1\}$, see (Richter 2014). This density, as well as $\hat{\sigma}$ allow functional representations

$$f_X(x) = \frac{C_p^n}{\sigma^n a_1 \cdot \ldots \cdot a_n} g_{PE}(h_{\sigma B_{a,p}}(x - \nu)), x \in \mathbb{R}^n$$

and

$$\widehat{\sigma} = \frac{1}{n^{1/p}} h_{B_{a,p}}(X - \nu).$$

For the derivation of the density of $n\left(\frac{\hat{\sigma}}{\sigma}\right)^p$, we start from the equation

$$P\left(n\left(\frac{\widehat{\sigma}}{\sigma}\right)^p < t\right) = \Phi_{g_{PE}, B_{a,p}, 0_n}\left(t^{1/p} B_{a,p}\right).$$

We recall that the intersections $B \cap S_{a,p}(r), r > 0$, where $S_{a,p}(r) = rS_{a,p}$ and $S = S_{a,p}$ is the boundary of $K = B_{a,p}$ are called sometimes the 'indivisibles' of a set B.

The geometric disintegration method behind formula (6) divides integration into first an integration on a part of a generalized sphere and second an integration w.r.t. r:

$$P\left(n\left(\frac{\widehat{\sigma}}{\sigma}\right)^p < t\right) = \frac{1}{I(g_{PE})} \int_0^\infty r^{n-1} g_{PE}(r) \mathfrak{F}_{S_{a,p}}\left(t^{1/p} B_{a,p}, r\right) dr$$

where according to Section 2.6

$$\mathfrak{F}_{S_{a,p}}\left(t^{1/p} B_{a,p}, r\right) = I_{[0, t^{1/p}]}(r).$$

Thus

$$P\left(n\left(\frac{\widehat{\sigma}}{\sigma}\right)^p < t\right) = \frac{1}{I(g_{PE})} \int_0^{t^{1/p}} r^{n-1} g_{PE}(r) dr, \ t > 0.$$

Note that this result does not depend on $a = (a_1, \ldots, a_n)^T$. Finally,

$$n\left(\frac{\widehat{\sigma}}{\sigma}\right)^p \sim \chi^p(n).$$

If one replaces the functional $|.|_p^p$ by $|.|_{a,p}^p$ in the definitions of the intervals (11) then the new intervals are confidence intervals for σ^p in the extended scaling parameter model (9).

3.3 The p-generalized Student's sample distribution and beyond

We study here the distribution of the statistic $n\left(\frac{\hat{\sigma}}{\sigma}\right)^p$ if the distribution of the sample vector $X = (X_1, \ldots, X_n)^T$ is not the n-dimensional power exponential one but any other element of the family of $l_{n,p}$-symmetric distributions. Before doing this, let us recall that one of the most famous non-normal spherically contoured distributions is the Student's t-distribution because of its heavy tails and the remarkable property that all its marginal distributions have a density generating function of the same known type. Moreover, note that the components of a spherically contoured vector are uncorrelated but not independent unless in the Gaussian case. All these properties hold also true for the p-generalized Student's t-distribution with f degrees of freedom which was studied for dimension one in (Richter 2007) and for arbitrary finite dimension n in Arellano-Valle and Richter (2012).

In the latter case, the dgf may be chosen as

$$g_{St(n,p,f)}(r) = \left(1 + \frac{r^p}{f}\right)^{-\frac{f+n}{p}}, r > 0,$$

and the density of X is

$$t_{n,p}(x;f) = D_{n,p,f}\left(1 + \frac{|x|_p^p}{f}\right)^{-\frac{f+n}{p}}, x \in \mathbb{R}^n,$$

where $D_{n,p,f} = \frac{\left(\frac{p}{2}\right)^n \Gamma\left(\frac{n+f}{p}\right)}{\Gamma\left(\frac{f}{p}\right)\left(\Gamma\left(\frac{1}{p}\right)\right)^n f^{n/p}}$. It follows that the density of the statistic $n\left(\frac{\hat{\sigma}}{\sigma}\right)^p$ is then

$$f^{\chi}_{n,g_{St},p}(t) = \frac{t^{\frac{n}{p}-1}}{f^{\frac{n}{p}} B\left(\frac{n}{p}, \frac{f}{p}\right)\left(1 + \frac{t^p}{f}\right)^{\frac{f+n}{p}}}, t \in \mathbb{R}. \quad (13)$$

If, however, the density of the sample vector X is an arbitrary $l_{n,p}$-symmetric one having dgf g,

$$f(x) = C(g,p)g(|x|_p), \ x \in \mathbb{R}^n,$$

then, according to Richter (2009),

$$f^{\chi}_{n,g,p}(t) = \frac{t^{\frac{n}{p}-1} g(t)}{p I_{n,g,p}}, \ t \in \mathbb{R}, \quad (14)$$

where $I_{n,g,p} = \int_0^\infty r^{n-1} g(r^p) dr$.

This density can be considered as a (p,g)-generalization of the classical Chi-square density with n d.f., or as a g-generalization of the p-generalized Chi-square density $f_{n,p}$ in (10). We denote the quantile of order q of the $\chi^p_g(n)$-distribution by $\chi^p_{g;q}(n), 0 < q < 1,$ that is $\chi^p_g(n)\left(\chi^p_{g;q}(n)\right) = q$. Note that $\chi^p_{g_{PE};q}(n) = \chi^p_q(n)$.

Let a random variable following density (14) be denoted by R^p. Note that tables of quantiles of the distribution of R are recently provided in Müller and Richter (2016) for $n \in \{1,2,3\}, p \in \left\{\frac{1}{2}, 1, 3\right\}$ and different dgfs g of power-exponential and Student's type as well of Pearson type 2.

3.4 Inference for generalized sample distributions

If the sample vector X is distributed according to an $l_{n,p}$-symmetric distribution, see e.g. (Richter 2009), with a dgf g being different from the power exponential one, $g \neq g_{PE}$, then $X_1, \ldots, X_n$ are not further independent. In this case of $l_{n,p}$-dependent observation variables, it is possible to study decision rules for exact inference on scaling parameters as before. Confidence and significance levels of the analogously defined estimators and tests, however, change in dependence of the actual shape of g. It is an immediate consequence from the equation

$$P\left(\chi^p_{g,\alpha_2}(n) < n\left(\frac{\hat{\sigma}}{\sigma}\right)^p < \chi^p_{g,1-\alpha_1}(n)\right) = 1 - \alpha_1 - \alpha_2$$

that the level-α confidence intervals in (11) extend to the present g-generalized case if all quantiles $\chi^p_q(n)$ are replaced with $\chi^p_{g;q}(n)$. Similarly, the same replacement of quantiles g-generalizes the level-α tests considered in (12).

3.5 The ratio of two scaling parameters

3.5.1 The p-generalized Fisher statistic and Fisher distribution

Let $X_1, \ldots, X_n$ and $Y_1, \ldots, Y_m$ be completely independent random variables with

$$X_i \text{ having p.d. } f_p(.;\mu_1,\sigma_1),\ i = 1,\ldots,n \text{ and } Y_i \text{ having p.d. } f_p(.;\mu_2,\sigma_2),\ i = 1,\ldots,m$$

where $p > 0$ and the expectations μ_1, μ_2 are known and the scaling parameters σ_1, σ_2 are unknown. With the notation $X = (X_1, \ldots, X_n)^T, Y = (Y_1, \ldots, Y_m)^T$, the $n+m$-dimensional random vector $\left(X^T Y^T\right)^T$ follows the distribution $\Phi_{g,K,\nu}$ where the dgf is the power exponential one, $g(r) = g_{PE}(r) = I_{(0,\infty)}(r) \exp\left\{-\frac{r^p}{p}\right\}$, the star body K is the p-generalized ellipsoid

$$K = B_{p,\sigma_1,\sigma_2} = \left\{\left(x^T y^T\right)^T \in \mathbb{R}^{n+m} : |\left(x^T y^T\right)^T|_{p,\sigma_1,\sigma_2} \le 1\right\},$$

and the shift vector ν is $\left(\mu_1 1_n^T \mu_2 1_m^T\right)^T \in \mathbb{R}^{n+m}$. Here,

$$|\left(x^T y^T\right)^T|_{p,\sigma_1,\sigma_2} = \left(\frac{|x|_p^p}{\sigma_1^p} + \frac{|y|_p^p}{\sigma_2^p}\right)^{1/p}, \left(x^T y^T\right)^T \in \mathbb{R}^{n+m}.$$

Let, in accordance with Section 3.2,

$$\hat{\sigma}_1^p = \frac{1}{n}|X - \mu_1 1_n|_p^p, \quad \hat{\sigma}_2^p = \frac{1}{m}|Y - \mu_2 1_m|_p^p, \tag{15}$$

and put

$$T = \frac{(\hat{\sigma}_1/\sigma_1)^p}{(\hat{\sigma}_2/\sigma_2)^p}. \tag{16}$$

Note that, for $t > 0$,

$$\begin{aligned} P(T < t) &= \Phi_{g_{PE},B_{p;\sigma_1,\sigma_2},\nu}\left(\left\{\left(x^T y^T\right)^T \in \mathbb{R}^{n+m} : \frac{|\frac{x-\mu_1 1_n}{\sigma_1}|_p^p}{|\frac{y-\mu_2 1_m}{\sigma_2}|_p^p} < \frac{n}{m}t\right\}\right) \\ &= \Phi_{g_{PE},B_{p;1,1},0}\left(C_{n,m;p}(t)\right) \end{aligned}$$

where

$$C_{n,m;p}(t) = \left\{\left(x^T y^T\right)^T \in \mathbb{R}^{n+m} : \frac{|x|_p^p}{|y|_p^p} < \frac{n}{m}t\right\}$$

denotes a cone in $\mathbb{R}^{n+m}$. The ipf of this cone is studied in Richter (2009). It is shown there that $C_{n,m;p}(t)$ belongs to the second elementary type of sets considered in Section 2.6, thus

$$P(T < t) = \mathfrak{F}S_{a,\sigma_1,\sigma_2}(C_{n,m;p}(t), 1),\ t > 0$$

does not depend on the dgf g. To be specific,

$$\mathfrak{F}S_{a,\sigma_1,\sigma_2}(C_{n,m;p}(t), 1) = \frac{p}{B\left(\frac{n}{p}, \frac{m}{p}\right)} \int_{\operatorname{arccot}\left(\left(\frac{n}{m}t\right)^{1/p}\right)}^{\pi/2} \frac{(\cos\varphi)^{n-1}(\sin\varphi)^{m-1}}{((\cos\varphi)^p + (\sin\varphi)^p)^{n/p}} d\varphi.$$

Hence, the p.d. of T is

$$f_{n,m;p}(t) = \frac{(n/m)^{n/p} t^{n/p-1}}{B(n/p, m/p)(1 + nt/m)^{(n+m)/p}} I_{(0,\infty)}(t).$$

The distribution law of T, denoted by $F_{n,m}(p)$, is called in Richter (2009) the p-generalized Fisher distribution with (n, m) d.f. Finally, it turns out that

$$T = \frac{m\sigma_2^p |X - \mu_1 1_n|_p^p}{n\sigma_1^p |Y - \mu_2 1_m|_p^p} \sim F_{n,m}(p).$$

Let us denote the q'th order quantile of the p-generalized Fisher distribution having (n, m) d.f. by $F_{n,m;q}(p), q \in (0, 1)$.

3.5.2 Confidence intervals

Two-sided, possibly asymmetric, and one-sided upper and lower confidence intervals for $\frac{\sigma_1^p}{\sigma_2^p}$ are, correspondingly,

$$\xi\left[\frac{1}{F_{n,m;1-\alpha_1}(p)}, \frac{1}{F_{n,m;\alpha_2}(p)}\right], \ \xi\left[\frac{1}{F_{n,m;1-\alpha}(p)}, \infty\right) \text{ and } \xi\left(-\infty, \frac{1}{F_{n,m;\alpha}(p)}\right]$$

where $\xi[e, f] = \{\xi \cdot s : e \le s \le f\}$ and

$$\xi = \frac{m|X - \mu_1 1_n|_p^p}{n|Y - \mu_2 1_m|_p^p}.$$

3.5.3 Significance test

For deciding

$$H_0 : \sigma_1 = \sigma_2 \text{ vs. } H_A : \sigma_1 \neq \sigma_2 \tag{17}$$

we use the distribution of T if H_0 is true to construct a level α significance test. In this case, $T = \frac{m}{n}\frac{|X-\mu_1 1_n|_p^p}{|Y-\mu_2 1_m|_p^p}$. Reject H_0 if $T < F_{n,m;\alpha_2}(p)$ or if $T > F_{n,m;1-\alpha_1}(p)$, respectively, where $\alpha_1 > 0, \alpha_2 > 0, \alpha_1 + \alpha_2 = \alpha < 1$.

Proposition 1. *If sample sizes n and m are equal then this rule is equivalent to rejecting H_0 if the likelihood ratio statistic is smaller than a constant $c \in (0, 1)$ chosen s.t.*

$$P(\textit{reject } H_0 \textit{ although } H_0 \textit{ is true}) = \alpha.$$

To show this, let us consider the likelihood function

$$L(\sigma_1, \sigma_2) = (n+m)\ln C_p - n\ln\sigma_1 - m\ln\sigma_2 - \frac{1}{p\sigma_1^p}\sum_{i=1}^{n}|x_i - \mu_1|^p - \frac{1}{p\sigma_2^p}\sum_{i=1}^{m}|y_i - \mu_2|^p.$$

The maximum likelihood estimators of the scaling parameters are given in (15). If $\sigma_1 = \sigma_2 = \sigma$, say, then $L^*(\sigma) = L(\sigma, \sigma)$ is the likelihood function for estimating σ, and

$$\hat{\sigma} = mle(\sigma) = \left(\frac{\sum_{i=1}^{n}|x_i - \mu_1|^p + \sum_{i=1}^{m}|y_i - \mu_2|^p}{n+m}\right)^{1/p}.$$

The likelihood ratio is

$$Q = \frac{\sup_{\sigma_1=\sigma_2} L(\sigma_1, \sigma_2)}{\sup_{\sigma_1>0,\sigma_2>0} L(\sigma_1, \sigma_2)} = \frac{L^*(\hat{\sigma})}{L(\hat{\sigma}_1, \hat{\sigma}_2)}$$

$$= \left(\frac{(n+m)^{n+m}}{n^n m^m}\right)^{1/p} \frac{|x - \mu_1 1_n|_p{}^n |y - \mu_2 1_m|_p{}^m}{\left(|x - \mu_1 1_n|_p{}^p + |y - \mu_2 1_m|_p{}^p\right)^{(n+m)/p}}.$$

Let $\xi = |x - \mu_1 1_n|_p{}^{p/2}$ and $\eta = |y - \mu_2 1_m|_p{}^{p/2}$. Because $n = m$ it follows that $Q \leq t, t \in (0,1)$ iff $l(\xi) \geq 0$ where $l(\xi) = \xi^2 + p\xi + q, p = -2\eta/t^{p/(2m)}$ and $q = \eta^2$. The two positive solutions of the equation $l(\xi) = 0$ are

$$\xi_1 = -p/2 + (p^2/4 - q)^{1/2} = \frac{\eta}{t^{p/(2m)}}\left(1 + (1 - t^{p/m})^{1/2}\right),$$

and

$$\xi_2 = -p/2 - \left(p^2/4 - q\right)^{1/2} = \frac{\eta}{t^{p/(2m)}}\left(1 - \left(1 - t^{p/m}\right)^{1/2}\right)$$

where $\eta > 0$. Because $l(.)$ is a convex function, the rejection area of H_o is the set of all ξ satisfying $\xi \geq \xi_1$ or $\xi \leq \xi_2$. The latter inequalities are equivalent to

$$\frac{|x - \mu_1 1_n|_p^p}{|y - \mu_2 1_m|_p^p} \geq \frac{2\left(1 + (1 - t^{p/m})^{1/2}\right)}{t^{p/m}} - 1$$

and

$$\frac{|x - \mu_1 1_n|_p^p}{|y - \mu_2 1_m|_p^p} \leq \frac{2\left(1 - (1 - t^{p/m})^{1/2}\right)}{t^{p/m}} - 1.$$

This proves the equivalence of the considered decision rules.

Remark 1. *As we have seen in this section, inferring on the scaling parameters ratio in $l_{n,p}$-symmetric sample distributions can be done in a g-robust way. For a discussion of this property see Richter (2009), and note that it goes back in the case $p = 2$ to Fang et al. (1990).*

4 Conclusions

As to shortly summarize the results of this paper, we derived exact confidence intervals and significance tests for single scaling parameters and ratios of two such parameters if the n-dimensional sample vector follows a norm or antinorm contoured $l_{n,p}$-symmetric distribution. This assumption includes modeling heavy and light tails of the sample vector's distribution as well as independence and $l_{n,p}$-dependence sampling. Certain classical results are (g,p)-generalized in the present work. The additional methodological benefit coming from the present paper may be estimated having a look at Richter (2012) where a rather general frame of future work is sketched.

Appendix

Table 1 gives a brief comparison of norms and antinorms at the hand of the functional $|.|_q, q \in R$. Note that $N_q = \{x \in \mathbb{R}^n : |x|_q = 0\}$ and $H_i = \left\{(x_1, \ldots, x_n)^T \in \mathbb{R}^n : x_i = 0\right\}, i = 1, \ldots, n$ are hyper planes in $\mathbb{R}^n$.

Table 1 Norms and antinorms

Parameter	q	$\in [1,\infty)$	$\in (0,1]$	$\in (-\infty,0)$
Functional	$\|.\|_q$	Norm	Antinorm	Semi-antinorm, $z \neq 0$
Unit ball	B_q	Convex	Radially concave w.r.t. standard fan in $\mathbb{R}^n$	Radially concave w.r.t. standard fan in $\mathbb{R}^n$
Inverse image of 0	N_q	$\{0\}$	$\{0\}$	$\cup_1^d H_i$

Table 2 Quantiles $\chi_q^p(n)$, $p = 3.3$

q \ n	1	2	3	4	5	10	15	20	25	30
0.005	0.00	0.00	0.01	0.05	0.12	1.14	2.92	5.17	7.73	10.51
0.01	0.00	0.00	0.02	0.08	0.20	1.47	3.51	6.00	8.77	11.76
0.025	0.00	0.01	0.06	0.18	0.37	2.08	4.54	7.39	10.49	13.78
0.05	0.00	0.02	0.12	0.32	0.60	2.75	5.58	8.75	12.15	15.71
0.1	0.00	0.06	0.26	0.58	0.99	3.70	6.98	10.55	14.29	18.16
0.9	2.95	5.19	7.07	8.78	10.39	17.70	24.42	30.86	37.12	43.25
0.95	4.56	7.17	9.30	11.20	12.98	20.93	28.13	34.96	41.57	48.02
0.975	6.31	9.21	11.53	13.60	15.52	24.00	31.61	38.79	45.70	52.42
0.99	8.75	11.96	14.50	16.75	18.82	27.91	35.98	43.55	50.81	57.85
0.995	10.67	14.07	16.75	19.11	21.29	30.78	39.16	47.00	54.49	61.74

Table 3 Quantiles $\chi_q^p(n)$, $p = 2.5$

q \ n	1	2	3	4	5	10	15	20	25	30
0.005	0.00	0.00	0.03	0.12	0.26	1.68	3.84	6.43	9.29	12.36
0.01	0.00	0.01	0.06	0.18	0.37	2.06	4.46	7.27	10.33	13.57
0.025	0.00	0.02	0.13	0.33	0.61	2.72	5.50	8.63	11.99	15.50
0.05	0.00	0.05	0.23	0.52	0.89	3.42	6.53	9.95	13.56	17.31
0.1	0.01	0.13	0.43	0.84	1.33	4.36	7.88	11.64	15.55	19.57
0.9	2.83	4.86	6.60	8.20	9.72	16.70	23.19	29.43	35.51	41.50
0.95	4.15	6.49	8.43	10.20	11.86	19.38	26.28	32.87	39.26	45.52
0.975	5.56	8.13	10.24	12.15	13.93	21.92	29.17	36.06	42.71	49.21
0.99	7.50	10.32	12.62	14.68	16.60	25.11	32.77	40.00	46.96	53.72
0.995	9.01	12.00	14.41	16.57	18.58	27.44	35.37	42.83	50.00	56.95

Table 4 Quantiles $\chi_q^p(n)$, $p = 0.75$

q \ n	1	2	3	4	5	10	15	20	25	30
0.005	0.02	0.19	0.50	0.92	1.40	4.35	7.76	11.43	15.25	19.19
0.01	0.03	0.25	0.62	1.08	1.61	4.74	8.31	12.10	16.04	20.08
0.025	0.06	0.36	0.82	1.36	1.96	5.37	9.16	13.14	17.24	21.43
0.05	0.10	0.49	1.02	1.64	2.29	5.96	9.94	14.08	18.33	22.65
0.1	0.17	0.68	1.31	2.00	2.73	6.69	10.89	15.22	19.64	24.10
0.9	2.14	3.64	5.01	6.32	7.59	13.63	19.43	25.10	30.68	36.22
0.95	2.71	4.34	5.82	7.21	8.55	14.89	20.91	26.77	32.52	38.20
0.975	3.27	5.02	6.58	8.04	9.45	16.04	22.25	28.27	34.17	39.99
0.99	4.00	5.88	7.53	9.08	10.57	17.45	23.88	30.09	36.16	42.12
0.995	4.55	6.51	8.23	9.84	11.37	18.45	25.04	31.38	37.55	43.62

Table 5 Quantiles $\chi_q^p(n)$, $p = 0.5$

q \ n	1	2	3	4	5	10	15	20	25	30
0.005	0.05	0.34	0.77	1.29	1.86	5.18	8.88	12.79	16.83	20.96
0.01	0.07	0.41	0.89	1.45	2.07	5.54	9.37	13.39	17.52	21.73
0.025	0.12	0.54	1.10	1.73	2.40	6.11	10.12	14.29	18.56	22.89
0.05	0.18	0.68	1.31	1.99	2.71	6.63	10.80	15.10	19.48	23.93
0.1	0.27	0.87	1.58	2.33	3.11	7.26	11.61	16.07	20.59	25.16
0.9	1.94	3.34	4.64	5.89	7.10	12.95	18.60	24.14	29.62	35.06
0.95	2.37	3.88	5.26	6.57	7.85	13.94	19.77	25.47	31.09	36.64
0.975	2.79	4.38	5.83	7.21	8.54	14.84	20.82	26.66	32.39	38.05
0.99	3.32	5.02	6.55	8.0	9.39	15.92	22.09	28.08	33.95	39.74
0.995	3.72	5.49	7.07	8.57	10.0	16.69	22.99	29.08	35.04	40.91

Competing interests
The author declares that he has no competing interests.

Acknowledgement
The author is thankful to the reviewers for their valuable hints leading to an improvement of the paper.

References
Arellano-Valle, RB, Richter, W-D: On skewed continuous $l_{n,p}$-symmetric distributions. Chil. J. Stat. **3**(2), 193–212 (2012)
Arnold, BC, Castillo, E, Sarabia, JM: Multivariate distributions defined in terms of contours. J. Stat. Planning Inf. **138**, 4158–4171 (2008)
Bolboaca, SD, Jäntschi, L, Sestras, AF, Sestras, RE, Pamfil, DC: Rearson-Fisher Chi-Square revisited. Information. **2**, 528–545 (2011)
Bomsdorf, E: Induktive Statistik. Verlag Josef Eul, Köln (1989)
Burago, D, Burago, Y, Ivanov, S: A course in metric geometry. Graduate studies in mathematics, vol. 33. A.M.S., Providence RI (2001)
Fang, K-T, Kotz, S, Ng, K-W: Symmetric multivariate and related distributions. Chapman and Hall, London (1990)
Fernández, C, Osiewalski, J, Steel, MFJ: Modeling and inference with v-spherical distributions. JASA. **90**(432), 1331–1340 (1995)
Fisher, RA: The conditions under which x^2 measures the discrepancy between observation and hypothesis. J. R. Stat. Soc. **87**, 442–444 (1924a)
Fisher, RA: On a distribution yielding the error functions of several well known statistics. Proc. Int. Cong. Math. Toronto. **2**, 805–813 (1924b)
Fisher, RA: Statistical methods for research workers. In: Classics in the History of Psychology. Internet Resource by C.D. Green, York University, Toronto, (1925). ISBN 0-05-002170-2, http://psychclassics.yorku.ca/Fisher/Methods. Accessed 23 Mar 2016
Haidenhain, J: Machining accuracy of machine tools. Technical information 635 399-22.5 (2011). www.heidenhain.de/de_EN/php/documentation-information/brochures/popup/media/media/file/view/file-0209/file.pdf. Accessed 23 Mar 2016
Helmert, PR: Über die Wahrscheinlichkeit von Potenzsummen der Beobachtungsfehler etc. Z.f. Math. u. Phys. **21**, 102–219 (1876)
Hlubinka, D, Vencalek, O: Depth-based classification for distributions with nonconvex support. J. Probab. Statist. **2013**(629184), 7 (2013)
Ittrich, C, Krause, D, Richter, W-D: Probabilities and large quantiles of noncentral generalized Chi-square distributions. Statistics. **34**, 53–101 (2000)
Kotz, S, Kozubowski, TJ, Podgórski, K: The Laplace distribution and generalizations. Birkhäuser, Boston (2001)
Livadiotis, G: Approach to general methods for fitting and their sensitivity. Physica A. **375**, 518–536 (2007)
Livadiotis, G: Expectation values and variance based on L^p norms. Entropy. **14**, 2375–2396 (2012)
Livadiotis, G: Chi-p distribution: characterization of the goodness of the fitting using L^p norms. J. Statist. Distr. Appl. **1**, 4 (2014)
Mirvaliev, M: The components of chi-squared statistics for goodness of fit tests. J. Sov Math. **38**, 2357–2363 (1987)
Moszyńska, M, Richter, W-D: Reverse triangle inequality. Antinorms and semi-antinorms. Studia Scientiarum Mathematicarum Hungarica. **49**(1), 120–138 (2012)
Müller, K, Richter, W-D: Extreme value distributions for dependent jointly $l_{n,p}$-symmetrically distributed random variables. Depend. Model. **4**, 30–62 (2016)
Pearson, K: On the criterion that a given system of deviations from the probable in the cases of a correlated system of variables is such that it can be reasonably supposed to have arisen from random sampling. Phil. Mag. **50**, 157 (1900)
Pisier, G: The volume of convex bodies and Banach space geometry. University Press, Cambridge (1989)
Plackett, RL: Karl Pearson and the Chi-squared test. Int. Statst. Rev. **51**, 59–72 (1983)
Richter, W-D: Laplace-Gauß integrals, Gaussian measure asymptotic behaviour and probabilities of moderate deviations. Z. Anal. Anw. **4**(3), 257–267 (1985)
Richter, W-D: A geometric method in stochastics (in German). Rostock. Math. Kolloqu. **44**, 63–72 (1991)
Richter, W-D: Generalized spherical and simplicial coordinates. J. Math. Anal. Appl. **336**, 1187–1202 (2007)
Richter, W-D: Continuous $l_{n,p}$-symmetric distributions. Lith. Math. J. **49**(1), 93–108 (2009)
Richter, W-D: Circle numbers for star discs. ISRN Geometry. **16**(Article ID 479262), 16 (2011)
Richter, W-D: Exact distributions under non-standard model assumptions. AIP Conf. Proc. **1479**, 442–445 (2012). doi:10.1063/1.4756160
Richter, W-D: Geometric and stochastic representations for elliptically contoured distributions. Comm. Stat. Theory Methods. **42**, 579–602 (2013)
Richter, W-D: Geometric disintegration and star-shaped distributions. J. Stat. Distributions Appl. **1**, 20 (2014). doi:10.1186/s40488-014-0020-6
Richter, W-D: Star-shaped distributions: Euclidean and non-Euclidean representations. In: Proceedings of the 60th World Statistics Congress 2015. Rio de Janeiro, (2016). Internat. Statist. Inst., https://www.isi-web.org. Accessed 23 Mar 2016
Richter, W-D: Norm contoured distributions in $\mathbb{R}^2$. Lecture Notes of Seminario Interdisciplinare di Mathematica University di Basilicata Italy. **12**, 179–199 (2015b)
Richter, W-D: Convex and radially concave contoured distributions. Probab. Stat. **12**(Article ID 165468), 12 (2015c)
Richter, W-D: Representing continuous star-shaped probability measures in spaces with suitably constructed geometries. AIP Conf. Proc (2016). Accepted for print
Statlect: The digital textbook (2016). http://www.statlect.com/hypothesis_testing_variance.htm. Accessed 23 Mar 2016
Wilcox, R: Comparing the variances of two dependent variables. J. Stat. Distr. Appl. **2**, 7 (2015)

4

A class of continuous bivariate distributions with linear sum of hazard gradient components

Jayme Pinto* and Nikolai Kolev

*Correspondence:
jaymeaugusto@gmail.com
Department of Statistics, University of São Paulo, São Paulo, Brazil

Abstract
The main purpose of this article is to characterize a class of bivariate continuous non-negative distributions such that the sum of the components of underlying hazard gradient vector is a linear function of its arguments. It happens that this class is a stronger version of the Sibuya-type bivariate lack of memory property. Such a class is allowed to have only certain marginal distributions and the corresponding restrictions are given in terms of marginal densities and hazard rates. We illustrate the methodology developed by examples, obtaining two extended versions of the bivariate Gumbel's law.

Keywords: Bivariate hazard gradient, Bivariate lack of memory property, Characterization, Gumbel's bivariate exponential, Marshall-Olkin model, Sibuya'a dependence function

MSC code: Primary: 62H05, 60E05

1 Introduction and motivation

Marshall and Olkin (1967) introduced the classical bivariate lack of memory property (to be denoted by $BLMP_1$) via relation

$$S_{X_1,X_2}(x_1+t, x_2+t) = S_{X_1,X_2}(x_1,x_2)S_{X_1,X_2}(t,t) \tag{1}$$

for all $x_1, x_2 \geq 0$ and $t > 0$, where $S_{X_1,X_2}(x_1,x_2) = P(X_1 > x_1, X_2 > x_2)$ is the joint survival function of non-negative continuous random vector (X_1, X_2). The functional Eq. (1) tells us that, independently of t, the $BLMP_1$ preserves the distribution of both (X_1, X_2) and its residual lifetime vector $\mathbf{X}_t = [(X_1 - t, X_2 - t) \mid X_1 > t, X_2 > t]$.

The only bivariate distribution with exponential marginals which possesses $BLMP_1$ has a joint survival function given by

$$S_{X_1,X_2}(x_1,x_2) = \exp\{-\lambda_1 x_1 - \lambda_2 x_2 - \lambda_3 \max(x_1,x_2)\}, \quad x_1, x_2 \geq 0, \tag{2}$$

where $\lambda_i \geq 0$, $i = 1,2,3$. The Marshall-Olkin (MO) bivariate exponential distribution (2) is generated by the stochastic representation

$$(X_1, X_2) = [\min(T_1, T_3), \min(T_2, T_3)], \tag{3}$$

where T_i are independent and exponentially distributed random variables with parameters $\lambda_i > 0$, $i = 1,2,3$. Unless $\lambda_3 = 0$, the distribution (2) exhibits singularity along the line $x_1 = x_2$ and its contribution to $S_{X_1,X_2}(x_1,x_2)$ is $\alpha = P(X_1 = X_2) = \frac{\lambda_3}{\lambda_1+\lambda_2+\lambda_3}$, e.g. Marshall and Olkin (1967). Hence, the MO bivariate exponential distribution is not absolutely continuous and does not have a probability density with respect to the two-dimensional Lebesgue measure.

Apart from MO bivariate exponential distribution, other known solutions of (1) are the bivariate distributions obtained by Freund (1961), Block and Basu (1974), Proschan and Sullo (1974), Friday and Patil (1977) and all distributions considered by Kulkarni (2006). Consult Chapter 10 in Balakrishnan and Lai (2009) for a related discussion as well.

Johnson and Kotz (1975) introduced another version of the bivariate lack of memory property (the local bivariate lack of memory property), which was rediscovered by Roy (2002) and named $BLMP_2$. The authors require conditional distributions $\{X_1 \mid X_2 > x_2\}$ and $\{X_2 \mid X_1 > x_1\}$ to possess (preserve) the univariate lack of memory property. The only absolutely continuous distribution with such a property is the Gumbel's type I bivariate exponential distribution given by

$$S_{X_1,X_2}(x_1,x_2) = \exp\{-\lambda_1 x_1 - \lambda_2 x_2 - \theta\lambda_1\lambda_2 x_1 x_2\}, \quad x_1, x_2 \geq 0 \tag{4}$$

where $\lambda_i > 0, i = 1,2$ and $\theta \in [0,1]$, see Gumbel (1960). It is negative quadrant dependent since $S_{X_1,X_2}(x_1,x_2) \leq S_{X_1}(x_1)S_{X_2}(x_2)$ for all $x_1, x_2 \geq 0$.

If the first partial derivatives of $S_{X_1,X_2}(x_1,x_2)$ exist, one can define $BLMP_1$ and $BLMP_2$ alternatively. We will explore the conditional failure (hazard) rates defined by

$$r_i(x_1,x_2) = \frac{\partial}{\partial x_i}[-\ln S_{X_1,X_2}(x_1,x_2)]\,,\ i = 1,2.$$

Marshall (1975) and Johnson and Kotz (1975) interpreted $r_i(x_1,x_2)$ as the conditional failure rate of X_i evaluated at x_i under condition that $X_j > x_j$ for $i = 1,2,\ i \neq j$. Equivalently, the conditional hazard rates are the univariate hazard rates of conditional distributions of each variate, given certain inequality of the remainder.

Marshall (1975) named the vector $\mathbf{R}(x_1,x_2) = (r_1(x_1,x_2), r_2(x_1,x_2))$ the hazard gradient of the distribution (X_1,X_2). The hazard gradient vector $\mathbf{R}(x_1,x_2)$ (if exists) uniquely determines the bivariate distribution by means of line integral

$$S_{X_1,X_2}(x_1,x_2) = \exp\left\{-\int_C \mathbf{R}(\mathbf{z})\mathrm{d}\mathbf{z}\right\}, \tag{5}$$

where C is any sufficiently smooth continuous path beginning at $(0,0)$ and terminating at (x_1,x_2). Equation (5) holds provided that along the path of integration $S_{X_1,X_2}(x_1,x_2)$ is absolutely continuous and $\mathbf{R}(x_1,x_2)$ exists for almost all x_1 and x_2, see Marshall (1975). The term "almost all" means that the set where the partial derivative does not exist has insignificant measure in the first quadrant, i.e. has a two-dimensional Lebesgue measure zero. See details and a multivariate version in Marshall (1975).

All bivariate distributions which have first partial derivatives and possess $BLMP_1$ are characterized by the equation

$$r_1(x_1,x_2) + r_2(x_1,x_2) = a_0$$

for almost all $x_1, x_2 \geq 0$, where a_0 is a non-negative constant, see Theorem 2 in Kulkarni (2006). In particular, the last relation is valid for the MO bivariate exponential distribution (2) with $a_0 = \lambda_1 + \lambda_2 + \lambda_3$ for all $x_1 \neq x_2$, since

$$\mathbf{R}(x_1,x_2) = (r_1(x_1,x_2), r_2(x_1,x_2)) = \begin{cases} (\lambda_1+\lambda_3, \lambda_2), & \text{if } x_1 > x_2, \\ \text{does not exist}, & \text{if } x_1 = x_2, \\ (\lambda_1, \lambda_2+\lambda_3), & \text{if } x_1 < x_2. \end{cases}$$

Thus, the sum $r_1(x_1, x_2) + r_2(x_1, x_2)$ is not defined along the line $x_1 = x_2$ having a zero two-dimensional Lebesgue measure. The reason is that the conditional probability $P(X_i > x_i \mid X_j > x_j)$ $i, j = 1, 2,\ i \neq j$, experiences a cusp at $x_1 = x_2$, i.e. it is continuous but not differentiable at $x_1 = x_2$ and the corresponding failure rate $r_i(x_1, x_2)$ is not defined, see Singpurwalla (2006), page 99.

In fact, Johnson and Kotz (1975) identify $BLMP_2$ by a local "constancy" of the failure rates $r_i(x_1, x_2),\ i = 1, 2$, see their sections 3(iv) and 5.4. For distribution (4) one gets $r_i(x_1, x_2) = \lambda_i + a_1 x_{3-i},\ i = 1, 2$, where a_1 is a non-negative constant. Thus, substituting $a_0 = \lambda_1 + \lambda_2$ and $a_1 = \theta\lambda_1\lambda_2$, the sum of conditional hazard rates in $BLMP_2$ case is specified by

$$r_1(x_1, x_2) + r_2(x_1, x_2) = a_0 + a_1 x_1 + a_1 x_2$$

for all $x_1, x_2 \geq 0$.

It would be challenging to link $BLMP_1$ and $BLMP_2$ in a new class of bivariate continuous distributions, to be denoted by $\mathcal{L}(\mathbf{x}; \mathbf{a})$, satisfying relation

$$r(x_1, x_2) = r_1(x_1, x_2) + r_2(x_1, x_2) = a_0 + a_1 x_1 + a_2 x_2 \tag{6}$$

for almost all $x_1, x_2 \geq 0$, where $\mathbf{x} = (x_1, x_2)$ and $\mathbf{a} = (a_0, a_1, a_2)$ is a parameter vector with non-negative elements, including the possibility $a_1 \neq a_2$.

To show one more member of the class $\mathcal{L}(\mathbf{x}; \mathbf{a})$, let us consider joint survival function

$$S_{X_1,X_2}(x_1, x_2) = \exp\left\{-\lambda_1 x_1^2 - \lambda_2 x_2^2 - \lambda_3 \max(x_1, x_2)\right\}, \quad x_1, x_2 \geq 0, \tag{7}$$

which has a singular component along the line $x_1 = x_2$ and meets (3). One can verify that relation (6) is satisfied by setting $a_0 = \lambda_3 > 0,\ a_1 = 2\lambda_1 > 0$ and $a_2 = 2\lambda_2 > 0$. The distribution given by (7) belongs to the generalized Marshall-Olkin (GMO) distributions introduced by Li and Pellerey (2011). The random variables T_1, T_2 and T_3 in (3) are assumed to be independent in the class of GMO distributions, relaxing Marshall-Olkin assumption of exponential marginality. It is direct to check that $S_{X_1,X_2}(x_1, x_2) \geq S_{X_1}(x_1) S_{X_2}(x_2)$ for all $x_1, x_2 \geq 0$, i.e. the bivariate distribution in (7) is positive quadrant dependent.

Thus, one can find many examples of bivariate continuous distributions possessing $BLMP_1$ and $BLMP_2$ represented by (1) and (4), respectively, as well as those exhibiting positive or negative quadrant dependence that belong to the class $\mathcal{L}(\mathbf{x}; \mathbf{a})$ specified by (6). The class $\mathcal{L}(\mathbf{x}; \mathbf{a})$ is composed of non-negative bivariate distributions which are absolutely continuous or continuous with singularity along the line $L = \{x_1 = x_2 \geq 0\}$ such that the sum of the components of underlying hazard gradient is a linear function of both arguments x_1 and x_2.

It happens that the class $\mathcal{L}(\mathbf{x}; \mathbf{a})$ (therefore both $BLMP_1$ and $BLMP_2$) is a particular case of the Sibuya-type bivariate lack of memory property recently introduced by Pinto and Kolev (2015c) as follows.

Definition 1. The non-negative continuous bivariate distribution (X_1, X_2) with marginal survival functions $S_{X_1}(x_1)$ and $S_{X_2}(x_2)$ possesses *Sibuya-type bivariate lack of memory property* (to be abbreviated S-BLMP), if and only if

$$\frac{S_{\mathbf{X}_t}(x_1, x_2)}{S_{X_{1t}}(x_1) S_{X_{2t}}(x_2)} = \frac{S_{X_1,X_2}(x_1, x_2)}{S_{X_1}(x_1) S_{X_2}(x_2)} \tag{8}$$

for all $x_1, x_2, t \geq 0$, where $S_{\mathbf{X}_t}(x_1, x_2)$ is the joint survival function of residual lifetime vector $\mathbf{X}_t$ and $S_{X_{it}}(x_i)$ are its marginal survival functions, $i = 1, 2$.

The S-BLMP is a new concept. Definition 1 tells us that the random vector (X_1, X_2) and its residual lifetime vector $\mathbf{X}_t$ should share, for all $t \geq 0$, the same dependence function $D_{X_1,X_2}(x_1, x_2)$ introduced by Sibuya (1960) as follows

$$D_{X_1,X_2}(x_1,x_2) = \ln \frac{S_{X_1,X_2}(x_1,x_2)}{S_{X_1}(x_1)S_{X_2}(x_2)}.$$

We will refer to $D_{X_1,X_2}(x_1, x_2)$ as "Sibuya'a dependence function" hereafter. It exhibits interesting connections with important cases of dependencies, see Kolev (2016) for related facts.

The article is organized as follows. In Section 2 we first discuss the S-BLMP and justify our attention to its stronger version: the linear Sibuya-type BLMP introduced by Pinto (2014), see Definition 2. We show its equivalence with the class $\mathcal{L}(\mathbf{x}; \mathbf{a})$ and present characterizations. Naturally, Theorems 1 and 2 are consequences of corresponding statements in Pinto and Kolev (2015c). It will be recognized in Section 3 that only certain marginal distributions of the class $\mathcal{L}(\mathbf{x}; \mathbf{a})$ are allowed. The corresponding conditions in terms of marginal densities and hazard rates are presented. Restrictions of parameters a_0, a_1 and a_2 are given in Theorem 3 and Proposition 1. As a result, one would be able to generate members of the class $\mathcal{L}(\mathbf{x}; \mathbf{a})$, some of them being extensions of Gumbel's law (4), see examples 1A and 3. A discussion and conclusions finalize the paper.

2 A stronger version of S-BLMP

The marginal survival functions of residual lifetime vector $\mathbf{X}_t$ are given by

$$S_{X_{1t}}(x_1) = \frac{S_{X_1,X_2}(x_1+t,x_2)}{S_{X_1,X_2}(t,t)} \quad \text{and} \quad S_{X_{2t}}(x_2) = \frac{S_{X_1,X_2}(x_1,x_2+t)}{S_{X_1,X_2}(t,t)}.$$

Therefore, the relation (8) can be rewritten as

$$S_{X_1,X_2}(x_1+t,x_2+t) = S_{X_1,X_2}(x_1,x_2)S_{X_1,X_2}(t,t)B(x_1,x_2;t), \tag{9}$$

for all $x_1, x_2, t \geq 0$, where the continuous function $B(x_1, x_2; t) = \frac{S_{X_{1t}}(x_1)S_{X_{2t}}(x_2)}{S_{X_1}(x_1)S_{X_2}(x_2)}$ is such that $B(x_1, x_2; 0) = B(0, 0; t) = 1$. The function $B(x_1, x_2; t)$ is named "aging factor", see page 457 in Balakrishnan and Lai (2009).

Without some simplifying assumptions on $B(x_1, x_2; t)$ the class of bivariate distributions possessing S-BLMP is too cumbersome to be of use. General characterizations of the S-BLMP in terms of functional equations involving Sibuya's dependence function are presented by Pinto and Kolev (2015c), see their Lemma 1 for example.

In order to get useful models, one should investigate a stronger version of S-BLMP. In this paper we will perform a detailed analysis for a particular aging function of the form $B(x_1, x_2; t) = \exp\{-a_1x_1t - a_2x_2t\}$, where a_1 and a_2 are given nonnegative constants. One can immediately recognize that in addition to relation (8) in Definition 1 one must assume that

$$S_{X_{it}}(x_i) = S_{X_i}(x_i)\exp\{-a_ix_it\} \quad \text{for} \quad a_i \geq 0,\ i = 1, 2. \tag{10}$$

Thus, we arrive to the stronger version of the S-BLMP, i.e. the linear Sibuya-type BLMP introduced by Pinto (2014) as follows.

Definition 2. The non-negative continuous bivariate distribution (X_1, X_2) possesses *linear Sibuya BLMP* (to be abbreviated LS-BLMP), if and only if (8) and (10) are satisfied for all $x_1, x_2, t \geq 0$ and $a_i \geq 0,\ i = 1, 2$.

Therefore, the bivariate continuous distributions possessing LS-BLMP can be equivalently represented by relation

$$S_{X_1,X_2}(x_1 + t, x_2 + t) = S_{X_1,X_2}(x_1, x_2) S_{X_1,X_2}(t, t) \exp\{-a_1 x_1 t - a_2 x_2 t\} \tag{11}$$

for all $x_1, x_2 \geq 0$ and $t > 0$. Really, let (11) be true. Put $x_i = 0$ in (11) to get relations (10), $i = 1, 2$. Substitute the exponent from (10) into (11) to obtain (8). Conversely, let (8) and (10) be fulfilled. Use expression (10) in the left hand side of (8) to restore (11).

To justify the choice of specific form $\exp\{-a_1 x_1 t - a_2 x_2 t\}$ of the "aging factor" in (11), let us consider a system composed by two elements with lifetimes represented by nonnegative continuous random variables X_1 and X_2. Suppose that during the first t units of time the system is protected by breakdowns (by warranty or insurance, say). It is reasonable to assume that $t < \min(x_1, x_2)$. After those first t units of time the system can be affected by two independent "fatal shocks" governed by homogeneous Poisson processes. Assume finally that the i-th unit is damaged with intensity $a_i t$, i.e. the corresponding shock arrival times are exponentially distributed, to be denoted $T_i \sim \text{Exp}(a_i t),\ i = 1, 2$, see Fig. 1.

Therefore, the probability of the system survival $S_{X_1,X_2}(x_1 + t, x_2 + t)$ is given by (11). Really, the right hand side in (11) is the probability $S_{X_1,X_2}(t, t)$ that both elements survive the protected initial t units of time multiplied by $S_{X_1,X_2}(x_1, x_2) \exp\{-a_1 x_1 t - a_2 x_2 t\}$, being the probability of absence of shocks during the following x_i units of time for the i-th element, $i = 1, 2$.

The next characterization theorem gives the joint survival function of bivariate continuous distributions possessing LS-BLMP.

Theorem 1. *The continuous bivariate distribution (X_1, X_2) has LS-BLMP defined by (11) for all $x_1, x_2 \geq 0$ and $t > 0$ if and only if $S_{X_1,X_2}(x_1, x_2)$ is a non-degenerate bivariate survival function given by*

$$S_{X_1,X_2}(x_1, x_2) = \begin{cases} S_{X_1}(x_1 - x_2) \exp\left\{-a_0 x_2 - a_1 x_1 x_2 - \frac{a_2 - a_1}{2} x_2^2\right\}, & \text{if } x_1 > x_2 \geq 0, \\ \exp\left\{-a_0 x_1 - \frac{a_1 + a_2}{2} x_1^2\right\}, & \text{if } x_1 = x_2 \geq 0, \\ S_{X_2}(x_2 - x_1) \exp\left\{-a_0 x_1 - a_2 x_1 x_2 - \frac{a_1 - a_2}{2} x_1^2\right\}, & \text{if } x_2 > x_1 \geq 0, \end{cases} \tag{12}$$

where $a_0, a_1, a_2 \geq 0$ and $S_{X_i}(x_i)$ are the marginal survival functions, $i = 1, 2$.

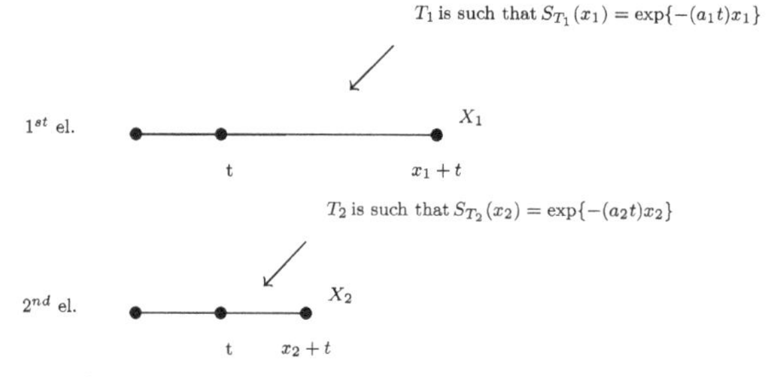

Fig. 1 An interpretation of relation (11)

Proof. Follows step by step the proof of Theorem 4 in Pinto and Kolev (2015c) with $A_i(x) = a_i x,\ i = 1, 2.$ □

Theorem 1 tells us that (12) is the solution of the functional Eq. (11).

In Pinto and Kolev (2015c) is also shown that S-BLMP characterizes the distributions belonging to the class defined by relation

$$r_1(x_1, x_2) + r_2(x_1, x_2) = a_0 + A_1(x_1) + A_2(x_2), \tag{13}$$

where $a_0 > 0$ and the continuous integrable functions $A_i(x_i)$ are such that $A_i(0) = 0$ and $A_i(x_i) > -a_0$ for all $x_i > 0,\ i = 1, 2$.

One can deduce that when $B(x_1, x_2; t) = \exp\{-a_1 x_1 t - a_2 x_2 t\}$, then relation (13) transforms into (6), i.e. $A_i(x_i) = a_i x_i,\ i = 1, 2$ and we obtain the $\mathcal{L}(\mathbf{x}; \mathbf{a})$ class.

When $S_{X_1,X_2}(x_1, x_2)$ is absolutely continuous, its vector hazard gradient $\mathbf{R}(x_1, x_2)$ exists everywhere in the interior of the set

$$\mathcal{A} = \left\{(x_1, x_2) \in \mathbb{R}_+^2 \mid S_{X_1,X_2}(x_1, x_2) > 0\right\},$$

where $\mathbb{R}_+^2$ is the first quadrant. In other words, relation (6) is well defined for all $x_1, x_2 \geq 0$. If it happens that $S_{X_1,X_2}(x_1, x_2)$ is continuous, its vector hazard gradient $\mathbf{R}(x_1, x_2)$ is useful even when it does not exist everywhere in the interior of the set $\mathcal{A}$, see Marshall (1975). Because of possible singularity of the class $\mathcal{L}(\mathbf{x}; \mathbf{a})$ along the line L having zero two-dimensional Lebesgue measure, we will assume hereafter that first partial derivatives of $S_{X_1,X_2}(x_1, x_2)$ exist and are continuous in $\mathcal{A}\backslash L$.

The next characterization theorem holds for bivariate continuous distributions belonging to the class $\mathcal{L}(\mathbf{x}; \mathbf{a})$ whose survival functions possess continuous first partial derivatives, and hence continuous hazard gradient vector $\mathbf{R}(x_1, x_2)$, in $\mathcal{A}\backslash L$.

Theorem 2. *If the first partial derivatives of $S_{X_1,X_2}(x_1, x_2)$ exist and are continuous in $\mathcal{A}\backslash L$, then relation (6) is fulfilled if and only if the joint survival function can be represented by (12).*

Proof. Follows step by step the proof of Theorem 2 in Pinto and Kolev (2015c) with $A_i(x) = a_i x,\ i = 1, 2.$ □

Observe that if our base relation is (11), then the LS-BLMP can be characterized by the joint survival function specified by (12), without the assumption of existence of hazard gradient vector $\mathbf{R}(x_1, x_2)$ as in Theorem 2.

Applying the Sklar's theorem to (12), one can obtain the survival copula corresponding to the class $\mathcal{L}(\mathbf{x}; \mathbf{a})$, see Pinto and Kolev (2015a).

Remark 1. (hazard vector elements in singularity and absolutely continuous cases). The bivariate survival functions considered in Theorem 2 are not necessarily absolutely continuous (and therefore not differentiable in $\mathcal{A}$). In fact, we do not exclude the possibility of existence of a singular component along the line $L = \{x_1 = x_2 \geq 0\}$, i.e. it may happen that $P(X_1 = X_2) > 0$. In such a case, when (x_1, x_2) belongs to the set $\left\{(x_1, x_2) \in \mathbb{R}_+^2 \,|\, x_1 = x_2 = x\right\}$ the function $S_{X_1,X_2}(x_1, x_2)$ is not differentiable as well as the hazard gradient vector $\mathbf{R}(x_1, x_2)$ does not exist. But if in Theorem 2 the continuity of the first partial derivatives of $S_{X_1,X_2}(x_1,x_2)$ holds true in $\mathcal{A}$, then $S_{X_1,X_2}(x_1,x_2)$ is absolutely continuous, see page 357 in Apostol (1974).

If the joint survival function $S_{X_1,X_2}(x_1,x_2)$ is *degenerate* (i.e. has degenerate marginal distributions), from (12) we get

$$S^{de}_{X_1,X_2}(x_1,x_2) = \begin{cases} \exp\left\{-a_0x_2 - a_1x_1x_2 - \frac{a_2-a_1}{2}x_2^2\right\}, & \text{if } x_2 < x_1 = const, \\ \exp\left\{-a_0x - \frac{a_1+a_2}{2}x^2\right\}, & \text{if } x_2 = x_1 = x = const, \\ \exp\left\{-a_0x_1 - a_2x_1x_2 - \frac{a_1-a_2}{2}x_1^2\right\}, & \text{if } x_1 < x_2 = const. \end{cases}$$

Obviously, $S^{de}_{X_1,X_2}(x_1,x_2)$ does not have differentiable failure rates.

Now we will identify members of the class $\mathcal{L}(\mathbf{x};\mathbf{a})$ with independent marginals. Hence, when $(X_1,X_2) \in \mathcal{L}(\mathbf{x};\mathbf{a})$, we will find solutions of functional equation

$$S_{X_1,X_2}(x_1,x_2) = S_{X_1}(x_1)S_{X_2}(x_2) \quad \text{for all} \quad x_1,x_2 \geq 0.$$

Let $r_{X_i}(x_i)$ be hazard rates of random variables $X_i,\ i = 1,2$. The independence between X_1 and X_2 implies that $r_i(x_1,x_2) = r_{X_i}(x_i),\ i = 1,2$. Therefore, relation (6) transforms into

$$r(x_1,x_2) = r_{X_1}(x_1) + r_{X_2}(x_2) = a_0 + a_1x_1 + a_2x_2.$$

The last equation is equivalent to both

$$r_{X_1}(x_1) = \alpha_1 + a_1x_1 \quad \text{and} \quad r_{X_2}(x_2) = \alpha_2 + a_2x_2,$$

where $\alpha_1 \in [0,a_0]$ and $\alpha_2 = a_0 - \alpha_1$. Thus, we obtain the following result.

Corollary 1. *The vector (X_1,X_2) with independent marginals belongs the class $\mathcal{L}(\mathbf{x};\mathbf{a})$ if and only if the marginal survival functions $S_{X_i}(x_i)$ have one of the following three possible analytic forms*

$$\exp\{-\alpha_ix_i\}, \quad \exp\{-0.5a_ix_i^2\} \quad or \quad \exp\{-\alpha_ix_i - 0.5a_ix_i^2\}, \quad i = 1,2,$$

where $\alpha_1 + \alpha_2 = a_0$ with $\alpha_1 \in [0,a_0]$ and $a_i \geq 0,\ i = 1,2$.

Therefore, the class $\mathcal{L}(\mathbf{x};\mathbf{a})$ has only 9 members with independent marginals. It is interesting to note that if $S_X(x) = \exp\{-\alpha x_i - 0.5ax^2\}$, (see the third option in Corollary 1), then X has a linear failure rate $r_X(x) = \alpha_1 + ax$. These type of univariate distributions have been introduced by Kodlin (1967), see Sen (2006) as well. Additionally, since the two first analytic forms in Corollary 1 can be included in the third one (taking $a_i = 0$ or $\alpha_i = 0$), the members of the class $\mathcal{L}(\mathbf{x};\mathbf{a})$ with independent marginals can also be seen as an extension of the (univariate) linear hazard rate model to the bivariate setup.

Let us note that the bivariate linear failure distribution introduced by Hangal and Ahmadi (2011) belongs to the class $\mathcal{L}(\mathbf{x};\mathbf{a})$.

Finally, one can analyze relation (6) from a completely different point of view. Define the function

$$\psi(x_1,x_2) = r(x_1,x_2) - a_0 \quad \text{and set} \quad G_1(x_1) = \psi(x_1,0),\ G_2(x_2) = \psi(0,x_2).$$

The linear functions $G_i(x_i) = a_ix_i$ satisfy the functional equation

$$G_i(x_i + y_i) = G_i(x_i) + G_i(y_i) \quad \text{for all} x_i,y_i \geq 0,\ i = 1,2.$$

Therefore, $G_i(x)$, $i = 1,2$, are additive functionals and only continuous solutions of Cauchy functional equation $f(x+y) = f(x) + f(y)$, see Theorem 1.1 in Sahoo and Kannappan (2011). In fact, we proved the following statement.

Lemma 1. *The class $\mathcal{L}(\mathbf{x};\mathbf{a})$ of non-negative bivariate continuous distributions specified by relation (6) can be equivalently defined by linear (additive) functionals $G_1(x_1) = r(x_1,0) - a_0$ and $G_2(x_2) = r(0,x_2) - a_0$, being the only continuous solutions of the Cauchy functional equation $f(x+y) = f(x) + f(y)$.*

3 Restrictions on the marginal densities or failure rates

Theorem 2 characterizes bivariate continuous distributions belonging to the class $\mathcal{L}(\mathbf{x};\mathbf{a})$, i.e. having joint survival function $S_{X_1,X_2}(x_1,x_2)$ specified by relation (12), which imply some more restrictions on the margins of these distributions. In other words, the corresponding joint survival function is valid only for certain marginal distributions of X_1 and X_2.

Here we will obtain the associated constraints in terms of marginal densities and hazard rates. As a result, we will get the admissible values of the parameter vector $\mathbf{a} = (a_0,a_1,a_2)$. The methodology will be illustrated by several typical examples.

3.1 Marginal density restrictions

The next statement shows the corresponding parameter constraints in terms of marginal densities when $a_1 + a_2 > 0$. The case $a_1 = a_2 = 0$, e.g. the $BLMP_1$, is detailed studied by Kulkarni (2006).

Theorem 3. *Let X_i be a random variable with absolutely continuous density $f_{X_i}(x_i)$, $i = 1,2$. Let $a_0,a_1,a_2 \geq 0$ and $a_1 + a_2 > 0$. Then $S_{X_1,X_2}(x_1,x_2)$ in (12) is a proper bivariate survival function if and only if*

$$A(x_i,x_j) - a_i x_j + \frac{d}{dx_i}\log f_{X_i}(x_i - x_j) + a_i[x_j A(x_i,x_j) - 1]\frac{S_{X_i}(x_i - x_j)}{f_{X_i}(x_i - x_j)} \geq 0 \tag{14}$$

where $A(x_i,x_j) = a_0 + a_i x_i + (a_j - a_i)x_j$ for all $x_i \geq x_j \geq 0$, $i \neq j$, $i,j = 1,2$. The singularity contribution into the joint survival function is given by

$$\begin{aligned}\alpha = P(X_1 = X_2) &= \left[f_{X_1}(0) + f_{X_2}(0) - a_0\right]\sqrt{\frac{\pi}{2(a_1+a_2)}} \\ &\quad \times \exp\left\{\frac{a_0^2}{2(a_1+a_2)}\right\}\left[1 - Erf\left(\frac{a_0}{\sqrt{2(a_1+a_2)}}\right)\right],\end{aligned} \tag{15}$$

where $Erf(x) = \frac{2}{\sqrt{\pi}}\int_0^x \exp\{-t^2\}dt$. In addition, the survival function specified by (12) is absolutely continuous if and only if $f_{X_1}(0) + f_{X_2}(0) = a_0$.

Proof. Let $S_{X_1,X_2}(x_1,x_2)$ be given by (12). Then, it is a proper if

$$\frac{\partial^2}{\partial x_1 \partial x_2} S_{X_1,X_2}(x_1,x_2) \geq 0,$$

After some algebra the last condition transforms into inequality (14).

Since $S_{X_1,X_2}(x_1,x_2)$ may have a singular component along the line $x_1 = x_2$, then $\alpha = P(X_1 = X_2) \in [0,1]$. The bivariate survival function $S_{X_1,X_2}(x_1,x_2)$ in (12) will be proper if and only if both the absolutely continuous part $S^{ac}_{X_1,X_2}(x_1, x_2)$ and the singular part $S^{si}_{X_1,X_2}(x_1, x_2)$ are survival functions and

$$S_{X_1,X_2}(x_1, x_2) = (1-\alpha)S^{ac}_{X_1,X_2}(x_1, x_2) + \alpha S^{si}_{X_1,X_2}(\max\{x_1, x_2\})$$

for $\alpha \in [0, 1]$. An equivalent expression in terms of joint densities is given by

$$f_{X_1,X_2}(x_1, x_2) = (1-\alpha)f^{ac}_{X_1,X_2}(x_1, x_2) + \alpha f^{si}_{X_1,X_2}(\max\{x_1, x_2\}),$$

where

$$(1-\alpha)f^{ac}_{X_1,X_2}(x_1, x_2) = \frac{\partial^2}{\partial x_1 \partial x_2} S_{X_1,X_2}(x_1, x_2).$$

To ensure existence of $S_{X_1,X_2}(x_1, x_2)$ we should evaluate the probability α which is equivalent to impose that $1-\alpha = P(X_1 > X_2) + P(X_2 > X_1) \in [0, 1]$, so one has to calculate the probabilities in the last sum. We have

$$P(X_1 > X_2) = \int_0^\infty \int_0^u (1-\alpha)f^{ac}_{X_1,X_2}(u, v)\, dv\, du.$$

Computing the inner integral $I(u) = \int_0^u (1-\alpha)f^{ac}_{X_1,X_2}(u, v)\, dv$ we get

$$I(u) = -f_{X_1}(0) \exp\left\{-a_0 u - \frac{a_2 + a_1}{2} u^2\right\} - a_1 u \exp\left\{-a_0 u - \frac{a_2 + a_1}{2} u^2\right\} + f_{X_1}(u).$$

Therefore,

$$P(X_1 > X_2) = \int_0^\infty I(u)\, du$$

$$= -f_{X_1}(0) \int_0^\infty \exp\left\{-a_0 u - \tfrac{a_2+a_1}{2} u^2\right\} du - a_1 \int_0^\infty u \exp\left\{-a_0 u - \tfrac{a_2+a_1}{2} u^2\right\} du + 1. \quad (16)$$

In order to solve integrals in (16), we use the following two expressions taken from Gradshteyn and Ryzhik (2007) for positive constants c and d:

$$\int_0^\infty \exp\{-cu^2 - du\}\, du = \frac{1}{2}\sqrt{\frac{\pi}{c}} \exp\left\{\frac{d^2}{4c}\right\}\left[1 - Erf\left(\frac{d}{2\sqrt{c}}\right)\right],$$

(see equation 3.322.2 on page 336), and

$$\int_0^\infty u \exp\{-cu^2 - du\}\, du = \frac{1}{2c} - \frac{d}{4c}\sqrt{\frac{\pi}{c}} \exp\left\{\frac{d^2}{4c}\right\}\left[1 - Erf\left(\frac{d}{2\sqrt{c}}\right)\right],$$

(see equation 3.462.5 on page 365).

Substituting $c = \frac{a_1+a_2}{2}$ and $d = a_0$ we obtain from (16)

$$P(X_1 > X_2) = \frac{a_2}{a_1 + a_2} - \left[f_{X_1}(0) - \frac{a_0 a_1}{a_1 + a_2}\right]\sqrt{\frac{\pi}{2(a_1 + a_2)}} \exp\left\{\frac{a_0^2}{2(a_1 + a_2)}\right\} \times \left[1 - Erf\left(\frac{a_0}{\sqrt{2(a_1 + a_2)}}\right)\right].$$

In a similar way we get

$$P(X_2 > X_1) = \frac{a_1}{a_1 + a_2} - \left[f_{X_2}(0) - \frac{a_0 a_2}{a_1 + a_2}\right]\sqrt{\frac{\pi}{2(a_1 + a_2)}} \exp\left\{\frac{a_0^2}{2(a_1 + a_2)}\right\} \times \left[1 - Erf\left(\frac{a_0}{\sqrt{2(a_1 + a_2)}}\right)\right].$$

From the last expressions and (16) we arrive to (15). To conclude the proof, observe that $S_{X_1,X_2}(x_1,x_2)$ in (12) is absolutely continuous if and only if $\alpha = 0$, which is equivalent to required condition $f_{X_1}(0) + f_{X_2}(0) = a_0$. □

Notice that if inequality (14) is fulfilled for some $a_0 = u > 0$, then it is also satisfied for all $a_0 \geq u$. Denote by

$$\tau = \text{the greatest lower bound of the set of possible values of } a_0 \text{ satisfying(14).}$$

Depending on the marginal densities, it may happen in some special cases that $\tau > f_{X_1}(0) + f_{X_2}(0)$, which contradicts (15) since always $\alpha = P(X_1 = X_2) \geq 0$, i.e. $a_0 \leq f_{X_1}(0) + f_{X_2}(0)$. Such high τ-values would be outside of the parameter space $\mathcal{A}$ of the class $\mathcal{L}(\mathbf{x}; \mathbf{a})$.

The range of possible values of a_0 are shown in Proposition 1. It is crucial for the construction of proper bivariate survival functions belonging to the class $\mathcal{L}(\mathbf{x}; \mathbf{a})$ from pre-specified marginal densities as we will see later.

Proposition 1. *Suppose* $\tau \leq f_{X_1}(0) + f_{X_2}(0)$. *If*

$$a_0 \in \left[\max\{\tau, \max(f_{X_1}(0), f_{X_2}(0))\}, f_{X_1}(0) + f_{X_2}(0)\right], \tag{17}$$

then Theorem 3 is fulfilled.

Proof. In the absence of singularity (whenever $\alpha = P(X_1 = X_2) = 0$), one concludes from (15) that $f_{X_1}(0) + f_{X_2}(0) - a_0 = 0$. Therefore, always $a_0 \leq f_{X_1}(0) + f_{X_2}(0)$, which is the upper bound for a_0 in (17).

The increase of the singular contribution into $S_{X_1,X_2}(x_1,x_2)$ implies increasing of the probability $\alpha = P(X_1 = X_2)$ up to 1. Let us denote by

$$E(a_0, a_1, a_2) = a_0 \sqrt{\frac{\pi}{2(a_1 + a_2)}} \exp\left\{\frac{a_0^2}{2(a_1 + a_2)}\right\} \left[1 - Erf\left(\frac{a_0}{\sqrt{2(a_1 + a_2)}}\right)\right].$$

It is direct to check that $0 \leq E(a_0, a_1, a_2) \leq 1$. We may represent (16) as

$$P(X_1 > X_2) = 1 - \frac{f_{X_1}(0)}{a_0} E(a_0, a_1, a_2) - \frac{a_1}{a_1 + a_2}[1 - E(a_0, a_1, a_2)].$$

The right hand side of the last equation is non-negative if $f_{X_1}(0) \leq a_0$. By analogy, from the expression for $P(X_2 > X_1)$ we obtain $f_{X_2}(0) \leq a_0$.

Notice that if $a_0 \in [\max\{f_{X_1}(0), f_{X_2}(0)\}, f_{X_1}(0) + f_{X_2}(0)]$ then $\alpha \in [0, 1]$. Finally, the lower bound in (17) can be obtained by taking into account the restriction on a_0 imposed by inequality (14) and possible related $\tau-$values. □

Remark 2. (absolutely continuous rule). Observe that whenever the upper bound a_U for a_0 given by (17) is attainable, i.e. if $a_U = a_0 = f_{X_1}(0) + f_{X_2}(0)$, one obtains an absolutely continuous bivariate distribution for which (6) is valid. Therefore, Eq. (15), besides representing the constraint $a_1 + a_2 > 0$, also offers a way to identify the presence of singularity.

For absolutely continuous distributions belonging to the class $\mathcal{L}(\mathbf{x}; \mathbf{a})$, it may happen that the lower and upper bound in (17) coincide, being even zero when $f_{X_1}(0) = f_{X_2}(0) =$

0, or equivalently, when $r_{X_1}(0) = r_{X_2}(0) = 0$, indicating that $a_0 = 0$. For example, consider a joint distribution given by $S_{X_1,X_2}(x_1, x_2) = \exp\{-0.5a_1x_1^2 - 0.5a_2x_2^2\}$. Observe that both $f_{X_1}(0) = f_{X_2}(0) = 0$ implying $a_0 = 0$ and $\alpha = P(X_1 = X_2) = 0$. This special absolutely continuous bivariate distribution with independent marginals case may be treated as an "exception", compare with Corollary 1.

Finally, note that all members of the class $\mathcal{L}(\mathbf{x}; \mathbf{a})$ with independent marginals are absolutely continuous.

Remark 3. (singularity of the BLMP$_1$ models). In the particular case $a_1 = a_2 = 0$, the condition (14) transforms into inequality (ii) of Theorem 5.1 in Marshall and Olkin (1967). In addition, the possible interval values of a_0 in (17) are compatible with those given by Kulkarni (2006) in her Remark 1.

Remark 4. (singularity of the GMO models). Let us consider the joint survival function of (Y_1, Y_2) belonging to the class of GMO distributions and represented by (7), with $a_i = 2\lambda_i$, $i = 1, 2$, and $\lambda_3 = a_0$. It is direct to check that

$$P(Y_1 = Y_2) = a_0\sqrt{\frac{\pi}{2(a_1 + a_2)}} \exp\left\{\frac{a_0^2}{2(a_1 + a_2)}\right\}\left[1 - Erf\left(\frac{a_0}{\sqrt{2(a_1 + a_2)}}\right)\right].$$

Observe that the right hand side in the last equation is just the function $E(a_0, a_1, a_2)$ used in the proof of Proposition 1.

We noted in the proof of Theorem 3 that if the survival function $S_{X_1,X_2}(x_1, x_2)$ given by (12) is proper then $\frac{\partial^2}{\partial x_1 \partial x_2} S_{X_1,X_2}(x_1, x_2)$ should be non-negative. This condition is equivalent to the requirement

$$S_{X_1,X_2}(x_1, y_1) + S_{X_1,X_2}(x_2, y_2) - S_{X_1,X_2}(x_1, y_2) - S_{X_1,X_2}(x_2, y_1) \geq 0$$

for any two points (x_1, y_1) and (x_2, y_2) in $\mathbb{R}_+^2$ such that $x_1 \leq x_2$ and $y_1 \leq y_2$. For example, if $x_1 \leq x_2 \leq y_1 \leq y_2$ and $a_2 = 0$ we conclude from (12) that the last inequality regarding the joint survival function is equivalent to

$$\frac{S_{X_2}(y_1 - x_1) - S_{X_2}(y_2 - x_1)}{S_{X_2}(y_1 - x_2) - S_{X_2}(y_2 - x_2)} \leq \exp\left\{-(x_2 - x_1)\left[a_0 + \frac{a_1}{2}(x_2 + x_1)\right]\right\}.$$

In general, such constraints between marginal survival functions are not easily verified. Relations (14) and (15) in Theorem 3 give alternative conditions in terms of absolutely continuous marginal densities $f_{X_i}(x), i = 1, 2$. However, depending on the complexity of the analytical form of the densities involved, it may be difficult to check these restrictions.

3.2 Marginal failure rates restrictions

The next result offers another set of equivalent constraints for parameters of $\mathcal{L}(\mathbf{x}; \mathbf{a})$, but in terms of marginal failure rates $r_{X_i}(x),\ i = 1, 2$.

Theorem 4. *Let the marginal failure rates $r_{X_i}(x)$, $i = 1, 2$, be differentiable functions and for some nonnegative constants a_0, a_1 and a_2, with $a_1 + a_2 > 0$, the following relations hold*

$$0 \le r_{X_i}(x_i) \le a_0 + a_i x_i; \tag{18a}$$

$$r_{X_i}(x_i - x_j)[A(x_i, x_j) - a_i x_j - r_{X_i}(x_i - x_j)] + \frac{d}{dx_i} r_{X_i}(x_i - x_j) + a_i[x_j A(x_i, x_j) - 1] \ge 0, \tag{18b}$$

with $A(x_i, x_j) = a_0 + a_i x_i + (a_j - a_i)x_j$, *for* $x_i \ge x_j \ge 0,\ i, j = 1, 2,\ i \ne j$;

$$a_0 \in \left[\max\{\tau, \max(r_{X_1}(0), r_{X_2}(0))\}, r_{X_1}(0) + r_{X_2}(0)\right], \tag{18c}$$

where τ *denotes the greatest lower bound of the set of values of* a_0 *for which the inequality (18b) is satisfied.*

Then the joint survival function $S_{X_1,X_2}(x_1, x_2)$ *given by (12) is proper with marginals* $S_{X_i}(x_i) = \exp\left(-\int_0^{x_i} r_{X_i}(u)du\right)$, $x_i \ge 0$, $i = 1, 2$. *The joint distribution is absolutely continuous if and only if* $a_0 = r_{X_1}(0) + r_{X_2}(0)$, *otherwise it possesses a singular component.*

Proof. Let $S_{X_1,X_2}(x_1, x_2)$ be given by (12) and let the univariate failure rates $r_{X_i}(x_i)$ be differentiable functions, $i = 1, 2$. Suppose that $x_2 \ge x_1 \ge 0$. Then substituting $x_1 = 0$ in

$$r_1(x_1, x_2) = \begin{cases} r_{X_1}(x_1 - x_2) + a_1 x_2, & \text{if } x_1 > x_2 \ge 0, \\ a_0 + (a_1 - a_2)x_1 + a_2 x_2 - r_{X_2}(x_2 - x_1), & \text{if } x_2 > x_1 \ge 0 \end{cases}$$

one gets $r_1(0, x_2) = a_0 + a_2 x_2 - r_{X_2}(x_2) \ge 0$ and therefore $r_{X_2}(x_2) \le a_0 + a_2 x_2$. By analogy, we conclude that $r_{X_1}(x_1) \le a_0 + a_1 x_1$ if $x_1 \ge x_2 \ge 0$ and (18a) is established.

All other relations are consequence of Theorem 3. □

Conditions (18a), (18b) and (18c) in Theorem 4 imply several simple practical steps that help to fix the permissible parameter space of coefficients a_0, a_1 and a_2 in $\mathcal{L}(\mathbf{x}; \mathbf{a})$. We discuss them in the next remark.

Remark 5. (parameter space of the class $\mathcal{L}(\mathbf{x}; \mathbf{a})$*).* Inequality (18a) says that the bivariate distributions from the class $\mathcal{L}(\mathbf{x}; \mathbf{a})$ satisfying (6) cannot have marginal distributions with failure rates $r_{X_i}(x_i)$ above the line $a_0 + a_i x_i$ for $i = 1, 2$. For example, distributions with univariate failure rate of the form $a x_i^2$, for $a > 0$ are unable to meet (6).

In addition, since $r_{X_i}(.)$ is a failure rate, then $\int_0^\infty r_{X_i}(u)du = \infty$ and because of (18a) the support of X_i, $i = 1, 2$, cannot be bounded from above, i.e. has to be the entire half line $[0, \infty)$.

To summarize, the parameter space for the coefficients a_0, a_1 and a_2 satisfying (6) in terms of marginal failure rates is given by

$$\{a_0 \in \left[\max\{\tau, \max(r_{X_1}(0), r_{X_2}(0))\}, r_{X_1}(0) + r_{X_2}(0)\right],\ a_1, a_2 \ge 0,\ a_1 + a_2 > 0\},$$

because $r_{X_i}(0) = f_{X_i}(0)$ for $i = 1, 2$. Finally, note that admissible values for coefficients a_1 and a_2 may be further limited as a consequence of inequality (18b) from Theorem 4, see Example 2.

The converse of Theorem 4 also holds for non-degenerate distributions and the statement is given below.

Proposition 2. *If $S_{X_1,X_2}(x_1,x_2)$ is non-degenerate bivariate survival function given by Eq. (12) and has differentiable marginal failure rates then it must satisfy conditions (18a) to (18c) in Theorem 4.*

Proof. Follows step by step the proof of Proposition 1 in Kulkarni (2006). □

The rules established in Theorem 4 may serve as a useful guide for constructing bivariate distributions possessing property (6). The building scheme may be relaxed under additional available information regarding monotone behavior of marginal failure rates. In fact, the class of bivariate distributions $\mathcal{L}(\mathbf{x};\mathbf{a})$ may have arbitrary combination of marginal failure rates: increasing, decreasing, constant, bathtub, etc., implying corresponding restrictions for the parameter space, of course.

3.3 Examples

The next two examples illustrate how relations in Theorem 4 can be applied to construct bivariate distributions from $\mathcal{L}(\mathbf{x};\mathbf{a})$ with given marginal failure rates.

Example 1. (constant failure rate marginals). Assume that

$$S_{X_i}(x)=\exp\{-\lambda_i x\},\quad \text{i.e.}\quad X_i\sim Exp(\lambda_i)\quad \lambda_i>0,\ i=1,2.$$

Then $f_{X_i}(x)=\lambda_i\exp\{-\lambda_i x\}$, $r_{X_i}(x)=\lambda_i$ and $f_{X_i}(0)=r_{X_i}(0)=\lambda_i,\ i=1,2$. From (18c) we obtain the first restriction: $\max(\lambda_1,\lambda_2)\le a_0\le\lambda_1+\lambda_2$.

Let $x_1\ge x_2$. Since $a_1\ge 0$, inequality (18b) transforms into

$$0\le a_1\le(\lambda_1+a_1x_2)[a_0-\lambda_1+a_1(x_1-x_2)+a_2x_2],$$

for all $x_1\ge x_2\ge 0$. The function $(\lambda_1+a_1x_2)[a_0-\lambda_1+a_1(x_1-x_2)+a_2x_2]$ is non-decreasing and its minimum is equal to $\lambda_1(a_0-\lambda_1)$ when $x_1=x_2=0$. Therefore, $0\le a_1\le\lambda_1(a_0-\lambda_1)$.

To find the greatest lower bound τ for which condition (18c) in Theorem 4 is true, is equivalent to verify when $\lambda_2(a_0-\lambda_2)+\lambda_2a_0a_1x_2+\lambda_2a_0a_1x_2^2\ge 0$. The last inequality is satisfied when $a_0\ge\lambda_2$. But we got this lower bound for a_0 already.

Analogously, when $x_2\ge x_1$ we obtain $0\le a_2\le\lambda_2(a_0-\lambda_2)$.

Summarizing, the parameter space is

$$\max(\lambda_1,\lambda_2)\le a_0\le\lambda_1+\lambda_2,\ a_1+a_2>0 \text{ and } 0\le a_i\le\lambda_i(a_0-\lambda_i),\ i=1,2.$$

We will consider two possible cases:

1A. The bivariate survival function will be absolutely continuous if $a_0=\lambda_1+\lambda_2$. Hence, $a_i=\theta_i\lambda_1\lambda_2$ for $\theta_i\in(0,1]$, $i=1,2$. With these specific parameters we get from (12) the representation

$$S_{X_1,X_2}(x_1,x_2)=\begin{cases}\exp\left\{-\left[\lambda_1x_1+\lambda_2x_2+\lambda_1\lambda_2x_2(\theta_1x_1+\frac{\theta_2-\theta_1}{2}x_2)\right]\right\}, & \text{if } x_1\ge x_2,\\ \exp\left\{-\left[\lambda_1x_1+\lambda_2x_2+\lambda_1\lambda_2x_1(\theta_2x_2+\frac{\theta_1-\theta_2}{2}x_1)\right]\right\}, & \text{if } x_2\ge x_1,\end{cases}\tag{19}$$

which may be named *Generalized Gumbel's bivariate exponential distribution.*

Observe, that fixing $\theta_1 = \theta_2 = \theta$ in the last relation we obtain as a particular case the Gumbel's type I bivariate exponential distribution (4).

1B. A bivariate survival function with absolutely continuous and singular components can also be constructed when $a_0 < \lambda_1 + \lambda_2$. Suppose $\lambda_1 > \lambda_2$ and let $a_0 = \lambda_1$. Notice that with this parameter choice the restrictions in (18c) are fulfilled. Hence we obtain $a_1 = 0$ and $a_2 = \theta\lambda_2(\lambda_1 - \lambda_2)$, where $\theta \in (0, 1]$. Substituting these parameter values in (12) we get

$$S_{X_1,X_2}(x_1,x_2) = \begin{cases} \exp\left\{-\left[\lambda_1 x_1 + \frac{\theta\lambda_2(\lambda_1-\lambda_2)}{2}x_2^2\right]\right\}, & \text{if } x_1 \geq x_2, \\ \exp\left\{-[(\lambda_1 - \lambda_2)x_1 + \lambda_2 x_2]\right\} & \\ \times \exp\left\{-\left[\theta\lambda_2(\lambda_1 - \lambda_2)x_1(x_2 - \frac{x_1}{2})\right]\right\}, & \text{if } x_2 \geq x_1. \end{cases}$$

In what follows, we will build a bivariate distribution with increasing marginal failure rates.

Example 2. (increasing failure rate marginals). Consider

$$S_{X_i}(x) = \exp\left\{-\lambda_i x^2 - \lambda_3 x\right\} \quad \text{for} \quad x \geq 0, \lambda_i > 0, \lambda_3 > 0,\ i = 1, 2.$$

Since $f_{X_i}(x) = (2\lambda_i x + \lambda_3)\exp\left\{-\lambda_i x^2 - \lambda_3 x\right\}$, then $r_{X_i}(x) = 2\lambda_i x + \lambda_3$, $i = 1, 2$, and the marginals have increasing failure rate. First limitations on the parameter space come from inequalities (18a) and (18c), i.e. $a_0 \geq \lambda_3$ and $a_i \geq 2\lambda_i$, $i = 1, 2$.

When $x_1 \geq x_2 \geq 0$, from (18b) we get a nonnegative increasing in x_1 and x_2 function

$$2\lambda_1 - a_1 - [2\lambda_1(x_1 - x_2) + \lambda_3 + a_1 x_2]\,[(2\lambda_1 - a_1)(x_1 - x_2) + \lambda_3 - a_0 - a_2 x_2] \geq 0$$

with a minimum at the point $(0, 0)$. Hence we obtain $a_1 \leq 2\lambda_1 + \lambda_3(a_0 - \lambda_3)$.

Analogously, for $x_1 \geq x_2 \geq 0$, we get $2\lambda_2 \leq a_2 \leq 2\lambda_2 + \lambda_3(a_0 - \lambda_3)$.

Summarizing, we have the constraints

$$\lambda_3 \leq a_0 \leq 2\lambda_3 \text{ and } 2\lambda_i \leq a_i \leq 2\lambda_i + \lambda_3(a_0 - \lambda_3),\ i = 1, 2.$$

2A. If $a_0 = 2\lambda_3$ we obtain an absolutely continuous bivariate survival function. In this case $a_i = 2\lambda_i + \theta_i\lambda_3^2, \theta_i \in [0, 1], i = 1, 2$ and letting these values in (12) one gets

$$S_{X_1,X_2}(x_1,x_2) = \begin{cases} \exp\left\{-\left[\lambda_1 x_1^2 + \lambda_3 x_1 + \lambda_2 x_2^2\right]\right\} & \\ \times \exp\left\{-\left[\lambda_3 x_2 + \lambda_3^2 x_2(\theta_1 x_1 + \frac{\theta_2 - \theta_1}{2}x_2)\right]\right\}, & \text{if } x_1 \geq x_2, \\ \exp\left\{-\left[\lambda_2 x_2^2 + \lambda_3 x_2 + \lambda_1 x_1^2\right]\right\} & \\ \times \exp\left\{-\left[\lambda_3 x_1 + \lambda_3^2 x_1(\theta_2 x_2 + \frac{\theta_1 - \theta_2}{2}x_1)\right]\right\}, & \text{if } x_2 \geq x_1. \end{cases}$$

Observe that the expression of the joint survival function involves a complete second degree polynomial in the exponent. In addition, notice that $\theta_1 = \theta_2 = 0$ implies independence between X_1 and X_2.

2B. A bivariate survival function having absolutely continuous and singular component can also be captured substituting

$$a_0 = (1 + \theta_0)\lambda_3, \quad \theta_0 \in [0, 1), \quad a_i = 2\lambda_i + \theta_0\theta_i\lambda_3^2 \quad \text{and} \quad \theta_i \in [0, 1],\ i = 1, 2$$

in (12). Let $\theta_0 = 0$ in the corresponding expression to get relation (7), i.e the distribution from the class of GMO distributions, see Li and Pellerey (2011).

Example 3. ("min" operation based construction). Let the survival function of the bivariate random vector (Y_1, Y_2) follow Gumbel's type I bivariate exponential distribution given by (4) which is a member of the class $\mathcal{L}(\mathbf{x};\mathbf{a})$. Assume that $Y_3 \sim Exp(\lambda_3)$ is independent of (Y_1, Y_2). Therefore, $(X_1, X_2) = [\min(Y_1, Y_3), \min(Y_2, Y_3)]$ belongs to $\mathcal{L}(\mathbf{x};\mathbf{a})$ as well and its survival function is given by

$$S_{X_1,X_2}(x_1,x_2) = \begin{cases} \exp\{-(\lambda_1+\lambda_3)x_1 - \lambda_2 x_2 - \theta\lambda_1\lambda_2 x_1 x_2\}, & \text{if } x_1 \geq x_2, \\ \exp\{-\lambda_1 x_1 - (\lambda_2+\lambda_3)x_2 - \theta\lambda_1\lambda_2 x_1 x_2\}, & \text{if } x_2 \geq x_1. \end{cases} \tag{20}$$

Notice that $S_{Y_1,Y_2}(x_1,x_2)$ is absolutely continuous, but $S_{X_1,X_2}(x_1,x_2)$ displays a singular component along the line $x_1 = x_2$. The expression (20) is given by Pinto and Kolev (2015c) in their Example 1.

It is worth noting that $S_{X_1,X_2}(x_1,x_2)$ given by (20), despite being continuous (but not absolutely continuous), preserves the local constancy of the failure rates $r_1(x_1,x_2)$ and $r_2(x_1,x_2)$ in a very similar fashion as Gumbel's bivariate distribution (4) does, i.e. $r_i(x_1,x_2) = \lambda_i + a_1 x_{3-i},\ i = 1, 2,$. The hazard components of (20) are given by

$$r_i(x_1,x_2) = \begin{cases} \lambda_i + \lambda_3 + \theta\lambda_1\lambda_2 x_{3-i}, & \text{if } x_i > x_{3-i}, \\ \text{does not exist} & \text{if } x_1 = x_2, \\ \lambda_i + \theta\lambda_1\lambda_2 x_{3-i}, & \text{if } x_i < x_{3-i} \end{cases}$$

for $i = 1, 2$. Therefore, we may consider (20) as a *Gumbel's extended bivariate exponential distribution with a singularity* along the line $x_1 = x_2$. Substituting $\lambda_3 = 0$ in (20), one will get absolutely continuous Gumbel's version (4).

Remark 6. (expanding Gumbel's bivariate law). The construction in Example 3 incorporates a singular component into the resulting distribution, which belongs to a wider class of Extended Marshall Olkin (EMO) bivariate distributions introduced by Pinto and Kolev (2015b). In fact, we assume that T_1 and T_2 are dependent random variables, but independent of T_3 in stochastic representation (3). Particular members of the EMO-class are the MO bivariate exponential distribution satisfying (3) and the GMO distributions (remind that the joint distribution given by (7) is member of the GMO-class).

4 Discussion and conclusions

In this paper we study a stronger version of S-BLMP introduced by in Pinto and Kolev (2015c), see Definition 1. We define the class $\mathcal{L}(\mathbf{x};\mathbf{a})$ and characterize it by the following equivalent relations $\mathcal{L}(\mathbf{x};\mathbf{a}) \Leftrightarrow (6) \Leftrightarrow (12)$. In addition,

$$\text{Lemma 1} \Leftrightarrow \mathcal{L}(\mathbf{x};\mathbf{a}) \Leftrightarrow \text{LS} - \text{BLMP} \Leftrightarrow \mathbf{(11)}.$$

Thus, the class $\mathcal{L}(\mathbf{x};\mathbf{a})$ might be treated as a key tool to deepen the BLMP-notion giving possibility to model the aging phenomena in the complement to the "non-aging" one, which fixes the world on Eqs. (1) or (4), via $BLMP_1$ and $BLMP_2$ correspondingly. Our new base equations are (6) or (11). We are convinced that the class introduced is promising in modeling dynamic aging dependence, being much more realistic than the virtual "non-aging world". The class $\mathcal{L}(\mathbf{x};\mathbf{a})$ includes symmetric and asymmetric continuous distributions with possible singularity; those which are positive or negative quadrant dependent; distributions from the GMO and EMO classes, etc. This huge variety of bivariate distributions would help to choose the "right" model consistent with the physical nature of the observations. The selection of bivariate distribution to be used should depend on considerations involving both the physical scenario at hand, and the properties of chosen distribution.

The seminal Gumbel's type I bivariate exponential distribution given by (4) has a central place in probability theory, being object of many characterization results. To count several of them, it is: the only absolutely continuous bivariate distribution possessing $BLMP_2$; belongs to the class $\mathcal{L}(\mathbf{x}; \mathbf{a})$; a key bivariate extreme value distribution, etc. We got two extensions of the Gumbel's type I distribution: one being absolutely continuous and the other having a singular component, represented by (19) and (20) respectively, consult Remark 6 as well. We do believe that further characterizations based on those relations will elevate the generalized Gumbel's laws as a starting point and a base for obtaining new bivariate models, with higher flexibility and chances to better model the genuine dependence structure.

We did not consider in this article inference procedures related to the model (6), nor its application for real data set. But, in Pinto and Kolev (2015b) we perform a Bayesian analysis using the EMO distribution (20) (being a member of the class $\mathcal{L}(\mathbf{x}; \mathbf{a})$) for a soccer data with ties studied by Meintanis (2007) and many other authors. According to the Deviance Information Criterion, our model (20) presented better fit than the Marshall-Olkin bivariate Weibull distribution, recently introduced by Kundu and Gupta (2013), who analyzed the same data set.

In general, one may use techniques for estimation of bivariate density function with partially differentiable kernels, e.g. Scott (1992). Another option is to apply the Kaplan-Meier estimate of bivariate survival function, even in the case of censoring following Dabrowska (1988), for example. Once the model is selected, goodness of the fit can be tested with conventional methods.

Acknowledgments
The authors are grateful to the Editors and the referees for their suggestions which helped to improve this article. The first author is grateful for the support of the Central Bank of Brazil. The second author is partially supported by FAPESP (2013/07375-0) and TUBITAK grants.

References

Apostol, TM: Mathematical Analysis. 2nd Edition. Addison-Wesley, Reading (1974)
Balakrishnan, N, Lai, C-D: Continuous Bivariate Distributions. 2nd Edition. Springer, New York (2009)
Block, HW, Basu, AP: A continuous bivariate exponential extension. J. Am. Stat. Assoc. **69**, 1031–1037 (1974)
Dabrowska, D: Kaplan-Meier estimate on the plane. Ann. Stat. **16**, 1475–1489 (1988)
Freund, E: A bivariate extension of the exponential distribution. J. Am. Stat. Assoc. **56**, 971–977 (1961)
Friday, DS, Patil, GP: A bivariate exponential model with applications to reliability and computer generation of random variables. In: Tsokos, CP, Shimi, IN (eds.) The Theory and applications of Reliability, vol. 1, pp. 527–549. Academic Press, New York, (1977)
Gradshteyn, IS, Ryzhik, IM: Tables of Integrals, Series and Products. 7th Edition. Academic Press, Boston (2007)
Gumbel, E: Bivariate exponential distributions. J. Am. Stat. Assoc. **55**, 698–707 (1960)
Hangal, D, Ahmadi, K: Bivariate linear failure distribution. Int. J. Stat. Manag. Sci. **6**, 73–84 (2011)
Johnson, NL, Kotz, S: A vector multivariate hazard rate. J. Multivariate Anal. **5**, 53–66 (1975)
Kodlin, D: A new response time distribution. Biometrics. **23**, 227–239 (1967)
Kolev, N: Characterizations of the class of bivariate Gompertz distributions. J. Multivariate Anal. **148**, 173–179 (2016)
Kulkarni, HV: Characterizations and modelling of multivariate lack of memory property. Metrika. **64**, 167–180 (2006)
Kundu, D, Gupta, A: Bayes estimation for the Marshall-Olkin bivariate Weibull distribution. Comput. Stat. Data Anal. **57**, 271–281 (2013)
Li, X, Pellerey, F: Generalized Marshall-Olkin distributions and related bivariate aging properties. J. Multivariate Anal. **102**, 1399–1409 (2011)
Marshall, AW: Some comments on the hazard gradient. Stoch. Process. Appl. **3**, 293–300 (1975)
Marshall, AW, Olkin, I: A multivariate exponential distribution. J. Am. Stat. Assoc. **62**, 30–41 (1967)
Meintanis, S: Test of fit for Marshall-Olkin distributions with applications. J. Stat. Plann. Infer. **137**, 3954–3963 (2007)

Pinto, J: Deepening the Notions of Dependence and Aging in Bivariate Probability Distributions. PhD Thesis. Sao Paulo University Press, Sao Paulo (2014)
Pinto, J, Kolev, N: Copula representations for invariant dependence functions. In: Glau, K, Scherer, M, Zagst, R (eds.) Innovations in Quantitative Risk Management, vol.99, pp. 411–421. Springer Series in Mathematics & Statistics, Springer Heidelberg, (2015a)
Pinto, J, Kolev, N: Extended Marshall-Olkin model and its dual version. In: Cherubini, U, Durante, F, Mulinacci, S (eds.) Marshall-Olkin Distributions - Advances in Theory and Applications, vol. 141, pp. 87–113. Springer Series in Mathematics & Statistics, Springer Heidelberg, (2015b)
Pinto, J, Kolev, N: Sibuya-type bivariate lack of memory property. J. Multivariate Anal. **134**, 119–128 (2015c)
Proschan, F, Sullo, P: Estimating the parameters of a bivariate exponential distribution in several sampling situations. In: Proschan, F, Serfling, RJ (eds.) Reliability and Biometry: Statistical Analysis of Life Lengths, pp. 423–440. Society of Industrial and Applied Mathematics, Philadelphia, (1974)
Roy, D: On bivariate lack of memory property and a new definition. Ann. Inst. Stat. Math. **54**, 404–410 (2002)
Sahoo, PK, Kannappan, P: Introduction to Functional Equations. CRC Press, Boca Raton (2011)
Sen, A: Linear failure rate distributions. In: Kotz, S, Read, C, Balakrishnana, N, Vidakovic, B (eds.) Encyclopedia of Statistical Sciences. 2nd ed, pp. 4212–4217. Wiley, New Jersey, (2006)
Scott, D: Multivariate Density Estimation: Theory, Practice and Visualization. Wiley, New York (1992)
Sibuya, M: Bivariate extreme statistics I. Ann. I.st. Stat. Math. **11**, 195–210 (1960)
Singpurwalla, N: Reliability and risk: a Bayesian perspective. Wiley, Chichester (2006)

5

The gamma extended Weibull distribution

Gauss M. Cordeiro[1], Maria do Carmo S. Lima[1], Antonio E. Gomes[2*], Cibele Q. da-Silva[2] and Edwin M. M. Ortega[3]

*Correspondence: aegomes@unb.br
[2]Departamento de Estatística, Universidade de Brasília, 70910-900 Brasília, DF, Brazil
Full list of author information is available at the end of the article

Abstract

The two-parameter Weibull has been the most popular distribution for modeling lifetime data. We propose a four-parameter gamma extended Weibull model, which generalizes the Weibull and extended Weibull distributions, among several other models. We obtain explicit expressions for the ordinary and incomplete moments, generating and quantile functions and mean deviations. We employ the method of maximum likelihood for estimating the model parameters. We propose a log-gamma extended Weibull regression model with censored data. The applicability of the new models is well justified by means of two real data sets.

Keywords: Extended Weibull distribution, Gamma extended Weibull distribution, Generating function, Maximum likelihood estimation, Mean deviation, Quantile function

MSC: 62E10, 62E20, 62F10

1 Introduction

There are hundreds of continuous univariate distributions and recent developments focus on constructing general distributions from classic ones. Many distributions have been used as models to make inferences about a population based on a set of empirical data from that population. Determining an adequate model to make inferences is a very important problem. The Weibull distribution is commonly used for modeling systems with monotone failure rates. However, the data sets in reliability analysis, especially over the life-cycle of the product, can involve high initial failure rates and eventual high failure rates due to aging and wear out, indicating a bathtub failure rate. The distributions that allow only monotone failure rates might not be adequate for modeling populations giving rise to such data. The major weakness of the Weibull distribution is its inability to accommodate non-monotone hazard rates, which has led to new generalizations of this distribution. One of the first extensions allowing for non-monotone hazard rates, including the bathtub shaped hazard rate function (hrf), is the *exponentiated Weibull* (ExpW) (Mudholkar and Srivastava 1993; Mudholkar et al. 1995; and Mudholkar et al. 1996) distribution. It has been well established in the literature that the ExpW distribution provides significantly better fits than the well-known exponential, gamma, Weibull and log-normal distributions. In the last paper, the authors presented a more flexible three-parameter *extended Weibull* (EW) model. Further, Shao et al. (2004) used this distribution to study flood frequency and Hao and Singh (2008) described some of its applications in hydrology. In this paper, we propose a further generalization by taking the EW distribution as the baseline model.

The three-parameter EW distribution is defined by the probability density function (pdf) and cumulative distribution function (cdf) (Mudholkar et al. 1996)

$$g_{\lambda,\alpha,\beta}(x) = \lambda\,\beta\,x^{\beta-1}\,\left(1+\alpha\lambda\,x^{\beta}\right)^{-\frac{1}{\alpha}-1},\ \alpha>0 \quad \text{and} \quad g_{\lambda,\beta}(x) = \lambda\,\beta\,x^{\beta-1}\,\mathrm{e}^{-\lambda x^{\beta}},\ \alpha=0, \tag{1}$$

and

$$G_{\lambda,\alpha,\beta}(x) = 1-\left(1+\alpha\lambda x^{\beta}\right)^{-\frac{1}{\alpha}},\ \alpha>0 \quad \text{and} \quad G_{\lambda,\beta}(x) = 1-\mathrm{e}^{-\lambda x^{\beta}},\ \alpha=0, \tag{2}$$

respectively, where $\lambda > 0$ is a scale parameter and $\alpha \geq 0$ and $\beta > 0$ are shape parameters. The support of the EW distribution is $\mathbb{R}^{+}$. The forms of the pdf and cdf when α goes to zero tend to those ones when $\alpha = 0$. Clearly, the cdf (2) extends the Weibull cdf and this fact justifies the terminology EW model. Due to the shape parameter α, more flexibility can be incorporated in model (1), which is very useful for lifetime data. The survival function associated to (1) is $S_{\lambda,\alpha,\beta}(x) = 1 - G_{\lambda,\alpha,\beta}(x)$ for $\alpha > 0$ and $S_{\lambda,\beta}(x) = 1 - G_{\lambda,\beta}(x)$ for $\alpha = 0$.

A family of univariate distributions generated by gamma random variables was proposed by Zografos and Balakrishnan (2009) and Ristic and Balakrishnan (2012). For any baseline cdf $G(x)$, $x \in \mathbb{R}$, they defined the *gamma-G* (denoted with the prefix "GG" for short) model with an extra parameter $a > 0$ by the pdf and cdf

$$f(x) = \frac{g(x)}{\Gamma(a)}\,\left\{-\log\left[1-G(x)\right]\right\}^{a-1} \tag{3}$$

and

$$F(x) = \frac{1}{\Gamma(a)}\int_{0}^{-\log[1-G(x)]} t^{a-1}\,\mathrm{e}^{-t}\,dt = \gamma_1(a, -\log\left[1-G(x)\right]), \tag{4}$$

respectively, where $g(x) = dG(x)/dx$, $\Gamma(a) = \int_0^{\infty} t^{a-1}\,\mathrm{e}^{-t}dt$ is the gamma function, $\gamma(a,z) = \int_0^{z} t^{a-1}\,\mathrm{e}^{-t}dt$ denotes the incomplete gamma function and $\gamma_1(a,z) = \gamma(a,z)/\,\Gamma(a)$ is the incomplete gamma function ratio.

The GG model has the same parameters of the parent G distribution plus one extra shape parameter $a > 0$. Each new GG distribution can be determined from a specified G model. For $a = 1$, the G distribution is a basic exemplar with a continuous crossover towards cases with different shapes (for example, a particular combination of skewness and kurtosis).

We introduce a new four-parameter model called the *"gamma extended Weibull"* ("GEW") distribution, which contains as special models some distributions such as the EW distribution. In fact, this model represents only a basic exemplar of the GEW distribution. We also study some of its mathematical properties. The paper is outlined as follows. In Section 2, we define the GEW distribution and provide some of its special cases. Further, two useful expansions for its density and cumulative distributions are derived in Section 3. In Section 4, we obtain its quantile function (qf). The ordinary moments and mean deviations are investigated in Section 5. Maximum likelihood estimation of the model parameters and some inferential tools are discussed in Section 6. In Section 7, we propose the *log-gamma extended Weibull* regression model, which can be applied for lifetime analysis. The importance of the new models is shown empirically by means of two real data sets in Section 8. Some conclusions are offered in Section 9.

2 The GEW distribution

By taking the pdf (1) and cdf (2) of the EW distribution with scale parameter $\lambda > 0$ and shape parameters $\alpha \geq 0$ and $\beta > 0$, the pdf and cdf of the GEW distribution can be obtained from Eqs. (3) and (4) (for $x > 0$) as

$$f(x) = f(x;\boldsymbol{\tau},a) = \begin{cases} \frac{\lambda\,\beta\,x^{\beta-1}\,(1+\alpha\lambda\,x^{\beta})^{-\frac{1}{\alpha}-1}}{\alpha^{a-1}\,\Gamma(a)}\left\{\log\left[\left(1+\alpha\lambda\,x^{\beta}\right)\right]\right\}^{a-1} & ,\ \alpha > 0, \\ \frac{\lambda^{a}\,\beta\,x^{\beta a-1}\,e^{-\lambda x^{\beta}}}{\Gamma(a)} & ,\ \alpha = 0, \end{cases} \tag{5}$$

$$F(x) = F(x;\boldsymbol{\tau},a) = \begin{cases} \gamma_1\left(a, \frac{1}{\alpha}\log\left[1+\alpha\lambda\,x^{\beta}\right]\right) & ,\ \alpha > 0, \\ \gamma_1\left(a, \lambda x^{\beta}\right), & ,\ \alpha = 0, \end{cases} \tag{6}$$

where $\boldsymbol{\tau} = (\alpha,\beta,\lambda)$. Clearly, if $\alpha \to 0$, the first expressions in (5) and (6) tend to the second ones in these equations. For $\alpha = 0$, the GEW model is identical to the *generalized gamma* distribution pioneered by Stacy (1962). Hereafter, a random variable X having pdf (5) is denoted by $X \sim GEW(\boldsymbol{\tau},a)$. Evidently, the density function (5) does not involve any complicated function and the EW distribution arises as the basic exemplar for $a = 1$. It is a positive point for the current generalization. The GEW model has several sub-models: the EW when $a = 1$, gamma Weibull (GW) when $\alpha = 0$, gamma extended exponential (GEE) when $\beta = 1$, extended exponential (EE) when $\beta = a = 1$, gamma exponential (GE) when $\alpha = 0$ and $\beta = 1$, Weibull (W) when $\alpha = 0$ and $a = 1$, and exponential when $\alpha = 0$ and $a = \beta = 1$. We show empirically the flexibility of the GEW distribution by comparing it with some of its sub-models by means of a real data set in Section 8. Plots of the density function and hrf of X for selected parameter values are displayed in Figs. 1 and 2, respectively.

The new distribution is easily simulated as follows: if V is a gamma random variable with parameter a, then

$$X = \left(\alpha^{-1}\,\lambda^{-1}\,\{\exp(\alpha V) - 1\}\right)^{1/\beta}$$

has the GEW$(\boldsymbol{\tau},a)$ distribution. This generating scheme is straightforward because of the existence of fast generators for gamma random variables.

3 Useful expansions

Expansions for (5) and (6) can be derived using the concept of exponentiated distributions. The *exponentiated extended Weibull* (EEW) distribution is defined by raising the cdf (2) to a power $a > 0$ and the associated random variable is denoted by $Y \sim$ EEW$(\boldsymbol{\tau},a)$. The cdf and pdf of Y are given by

$$H_a(x;\boldsymbol{\tau}) = \begin{cases} \left[1-\left(1+\alpha\lambda\,x^{\beta}\right)^{-\frac{1}{\alpha}}\right]^{a}, & \alpha > 0, \\ \left[1-\exp\left(-\lambda\,x^{\beta}\right)\right]^{a} & ,\ \alpha = 0, \end{cases}$$

and

$$h_a(x;\boldsymbol{\tau}) = \begin{cases} a\,\lambda\,\beta\,x^{\beta-1}\,\left(1+\alpha\lambda\,x^{\beta}\right)^{-\frac{1}{\alpha}-1}\left[1-\left(1+\alpha\lambda\,x^{\beta}\right)^{-\frac{1}{\alpha}}\right]^{a-1}, & \alpha > 0, \\ a\,\lambda\,\beta\,x^{\beta-1}\,e^{-\lambda x^{\beta}}\left[1-\exp\left(-\lambda\,x^{\beta}\right)\right]^{a-1} & ,\ \alpha = 0, \end{cases} \tag{7}$$

respectively. The properties of some exponentiated distributions have been investigated by several authors, see Mudholkar and Srivastava (1993) and Mudholkar et al. (1995) for exponentiated Weibull (ExpW), Gupta et al. (1998) for exponentiated Pareto, Gupta and

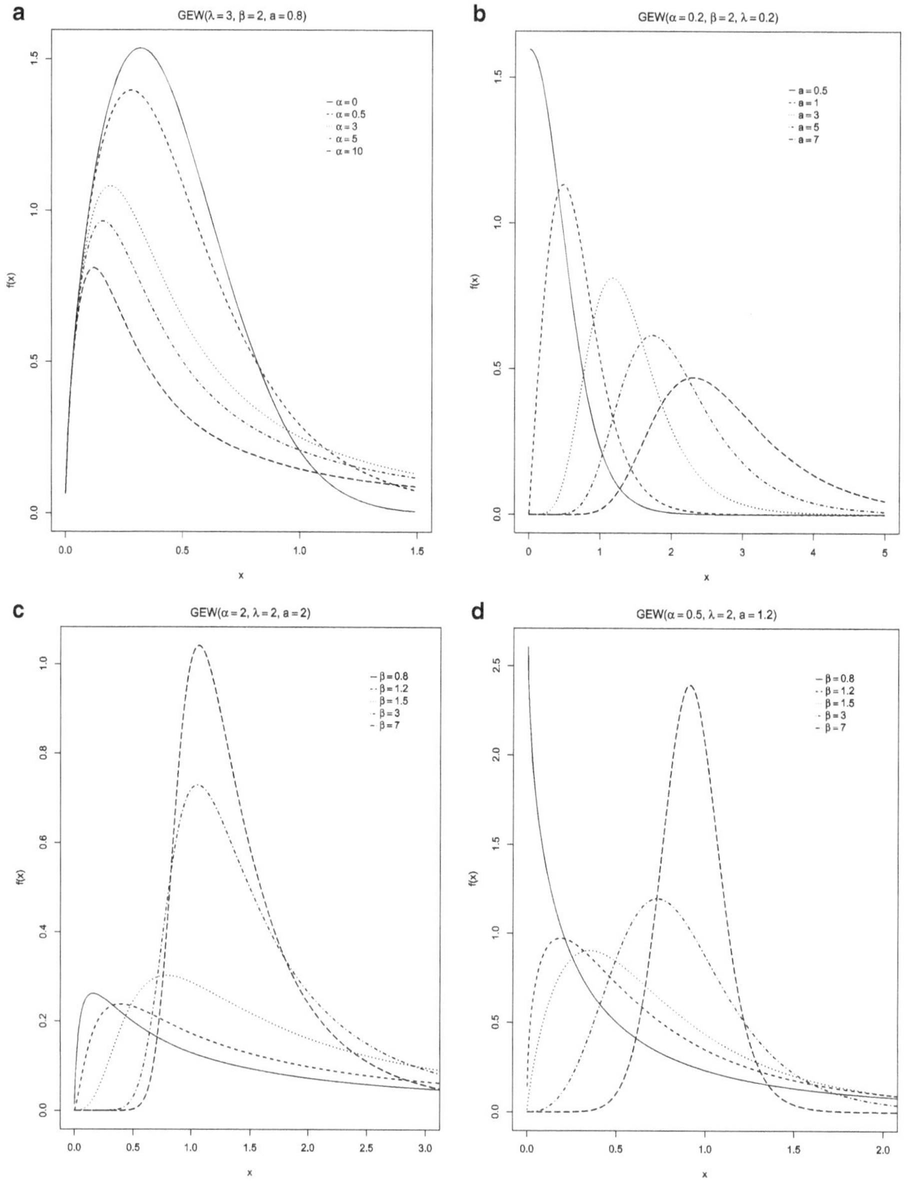

Fig. 1 Plots of the GEW pdf

Kundu (2001) for exponentiated exponential (ExpE) and Nadarajah and Gupta (2007) for exponentiated gamma (ExpG) distributions. More recently, Cordeiro et al. (2011) studied these properties for the exponentiated generalized gamma (ExpGG) distribution.

The following formula holds (http://functions.wolfram.com/ElementaryFunctions/Log/06/01/04/03/)

$$\left\{\frac{1}{\alpha}\log\left[1-\alpha\lambda\, x^{\beta}\right]\right\}^{a-1}$$
$$=(a-1)\sum_{k=0}^{\infty}\binom{k+1-a}{k}\sum_{j=0}^{k}\frac{(-1)^{j+k}\,p_{j,k}}{(a-1-j)}\binom{k}{j}\left[1-\left(1-\alpha\lambda\, x^{\beta}\right)^{-\frac{1}{\alpha}}\right]^{a+k-1},$$

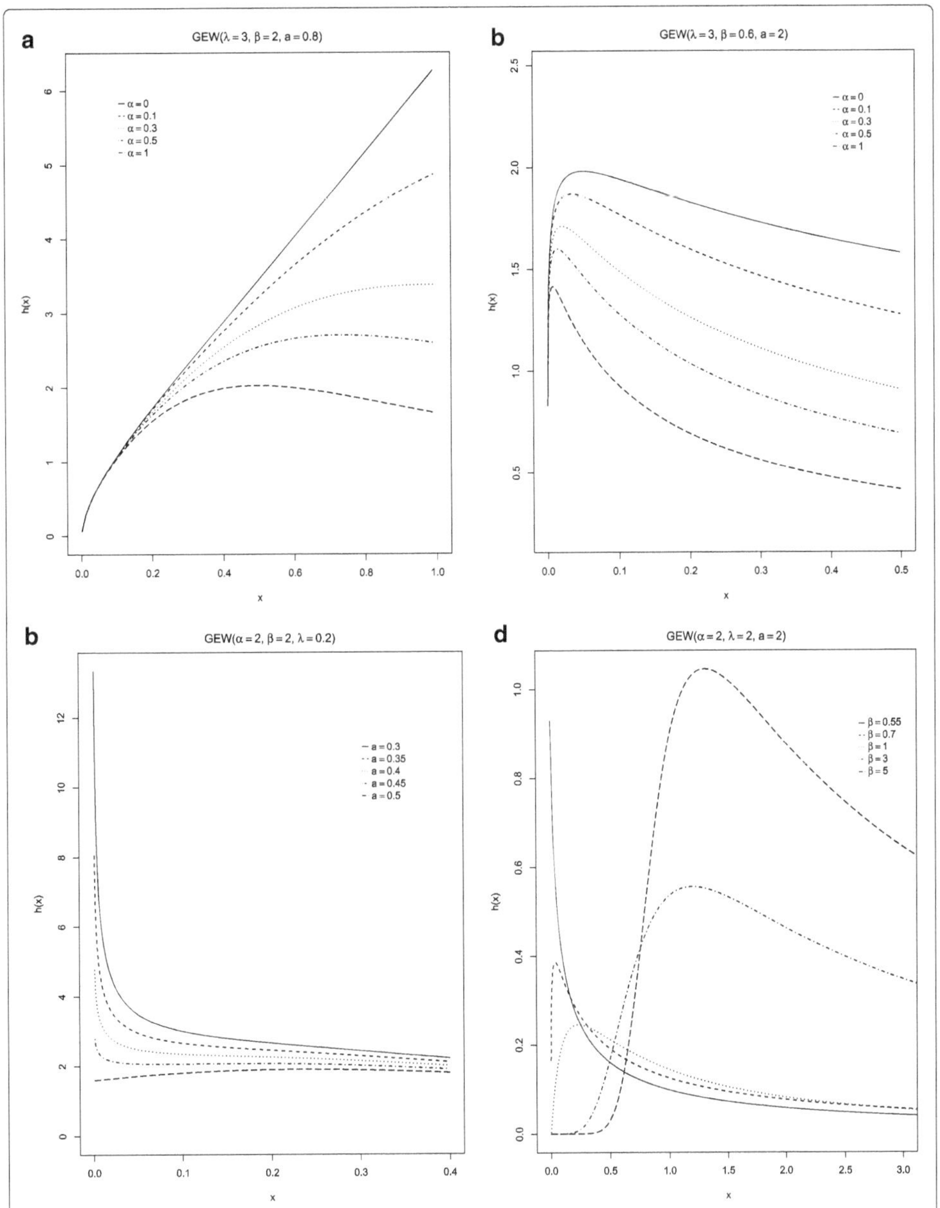

Fig. 2 Plots of the GEW hrf

where $a > 0$ is a real parameter and the constants $p_{j,k}$ can be determined recursively by

$$p_{j,k} = k^{-1} \sum_{m=1}^{k} \frac{(-1)^m \left[m(j+1) - k \right]}{(m+1)} \, p_{j,k-m}, \, j \geq 0,$$

for $k = 1, 2, \ldots$ and $p_{j,0} = 1$. Further, for any real parameter $a > 0$, we define

$$b_k = \frac{\binom{k+1-a}{k}}{(a+k)\Gamma(a-1)} \sum_{j=0}^{k} \frac{(-1)^{j+k} \, p_{j,k}}{(a-1-j)} \binom{k}{j}$$

and then the first equation in (5) can be expressed as

$$f(x) = \sum_{k=0}^{\infty} b_k \, h_{a+k}(x) \,, \tag{8}$$

where $h_{a+k}(x) = (a+k)\,\lambda\,\beta\,x^{\beta-1}\left(1+\alpha\lambda\,x^{\beta}\right)^{-1/\alpha-1}\left[1-\left(1+\alpha\lambda\,x^{\beta}\right)^{-\frac{1}{\alpha}}\right]^{a+k-1}$ denotes the EEW$(\boldsymbol{\tau}, a+k)$ density function. So, the GEW density function is a linear mixture of EEW densities.

Similarly, we can derive an expansion for the GW density when $\alpha = 0$. Using Eq. (5) for $\alpha = 0$, we obtain the same expression in (8), but the function $h_{a+k}(x)$ denotes, when $\alpha = 0$, the GEW$(\boldsymbol{\tau}, a+k)$ pdf given by the second equation of (7).

Inserting (7) in Eq. (8), we obtain

$$f_{\text{GEW}}(x;\boldsymbol{\tau},a) = \begin{cases} \sum\limits_{j=0}^{\infty} e_j\, g_{\lambda^\star,\alpha^\star,\beta}(x)\ , \text{ if } \ \alpha > 0\,, \\ \sum\limits_{j=0}^{\infty} e_j\, g_{\lambda^\star,\beta}(x) \quad , \text{ if } \ \alpha = 0\,, \end{cases} \tag{9}$$

where $e_j = \sum_{k=0}^{\infty}(-1)^j\,(a+k)\,b_k\binom{a+k-1}{j}$, $g_{\lambda^\star,\alpha^\star,\beta}(x)$ denotes the EW pdf with parameters $\lambda^\star = (j+1)\lambda$, $\alpha^* = \alpha/(j+1)$ and β and $g_{\lambda^\star,\beta}(x)$ denotes the Weibull pdf with parameters $\lambda^\star$ and β.

Equation (9) is the main result of this section. It reveals that the GEW density function is a linear mixture of EW (when $\alpha > 0$) and Weibull (when $\alpha = 0$) densities. So, several GEW structural properties can be obtained from those properties of the EW and Weibull distributions.

4 Quantile function

First, we consider the general case $\alpha > 0$. The GEW qf, say $Q(u) = F^{-1}(u)$, can be expressed in terms of the EW qf, say $Q_{EW}(\cdot)$. Inverting $F(x) = u$ given by (6), the qf of X (for $0 < u < 1$) follows as

$$F^{-1}(u) = Q_{GEW}(u) = Q_{EW}\left\{1 - \exp[-Q^{-1}(a, 1-u)]\right\}, \tag{10}$$

where $Q^{-1}(a, 1-u)$ is the inverse function of $Q(a,z) = 1 - \gamma_1(a, 1-u)$. A power series for it is given in the Wolfram website as

$$z = Q^{-1}(a, 1-u) = \sum_{i=0}^{\infty} a_i\, u^{i/a}\,,$$

where $a_0 = 0$, $a_1 = \Gamma(a+1)^{1/a}$, $a_2 = \Gamma(a+1)^{2/a}/(a+1)$, etc. Quantities of interest can be obtained from (10) by substituting appropriate values for u.

Further, the EW qf is given by

$$Q_{EW}(u) = \left\{\frac{1-(1-u)^{\alpha}}{\alpha\lambda(1-u)^{\alpha}}\right\}^{1/\beta}. \tag{11}$$

5 Moments and mean deviations

Some of the most important features and characteristics of a distribution can be studied through moments. We can write from Eq. (8)

$$\mu_r' = E(X^r) = \begin{cases} \alpha^{-\left(\frac{r}{\beta}+1\right)}\lambda^{-\frac{r}{\beta}}\sum\limits_{j=0}^{\infty} e_j^\star\, B\left(\frac{r}{\beta}+1, \frac{j+1}{\alpha}-\frac{r}{\beta}\right), \text{ if } \alpha > 0 \text{ and } r < \beta/\alpha\,, \\ \lambda^{-\frac{r}{\beta}}\,\Gamma\left(\frac{r}{\beta}+1\right)\sum\limits_{j=0}^{\infty}\frac{e_j}{(j+1)^{r/\beta}}, \qquad \text{if } \alpha = 0, \end{cases} \tag{12}$$

where $e_j^\star = (j+1)\,e_j$, $B(a,b) = \Gamma(a)\Gamma(b)/\Gamma(a+b)$ is the beta function and $\Gamma(a) = \int_0^\infty z^{a-1}\mathrm{e}^{-z}dz$ is the gamma function.

Established algebraic expansions to determine μ_r' can be more efficient than computing these moments directly by numerical integration of (5), which can be prone to rounding errors among others. The skewness γ_1 and kurtosis γ_2 of X are easily obtained from the ordinary moments using well-known formulae. Figure 3 displays some plots for the skewness and kurtosis of the GEW model.

For lifetime models, it is of interest to know the rth incomplete moment of X defined by $T_r(y) = \int_0^y x^r f(x)dx$. Based on (9), we obtain (for $\alpha > 0$)

$$T_r(y) = \sum_{j=0}^{\infty} e_j\,\lambda^\star\,\beta\,\rho\left(y; r, \alpha^\star\lambda^\star, {\alpha^\star}^{-1}\right), \tag{13}$$

where $\rho(y; r, p, q) = \int_0^y x^r\,(1+p\,x)^{-q-1}dx$ for $r, p, q > 0$. Using Maple, this integral can be determined as

$$\rho(y; r, p, q) = A(r,p,q)\left\{2\,y^{r-q}\;{}_2F_1\left(q^{-r}, q+1; q+1-r, -\frac{1}{py}\right)\left[B(r,p,q) + C(r,p,q)\right]\right\},$$

where ${}_2F_1$ is the hypergeometric function defined by

$${}_2F_1(a,b;c;x) = \frac{\Gamma(c)}{\Gamma(a)\Gamma(b)}\sum_{j=0}^{\infty}\frac{\Gamma(a+j)\Gamma(b+j)}{\Gamma(c+j)}\frac{x^j}{j!},$$

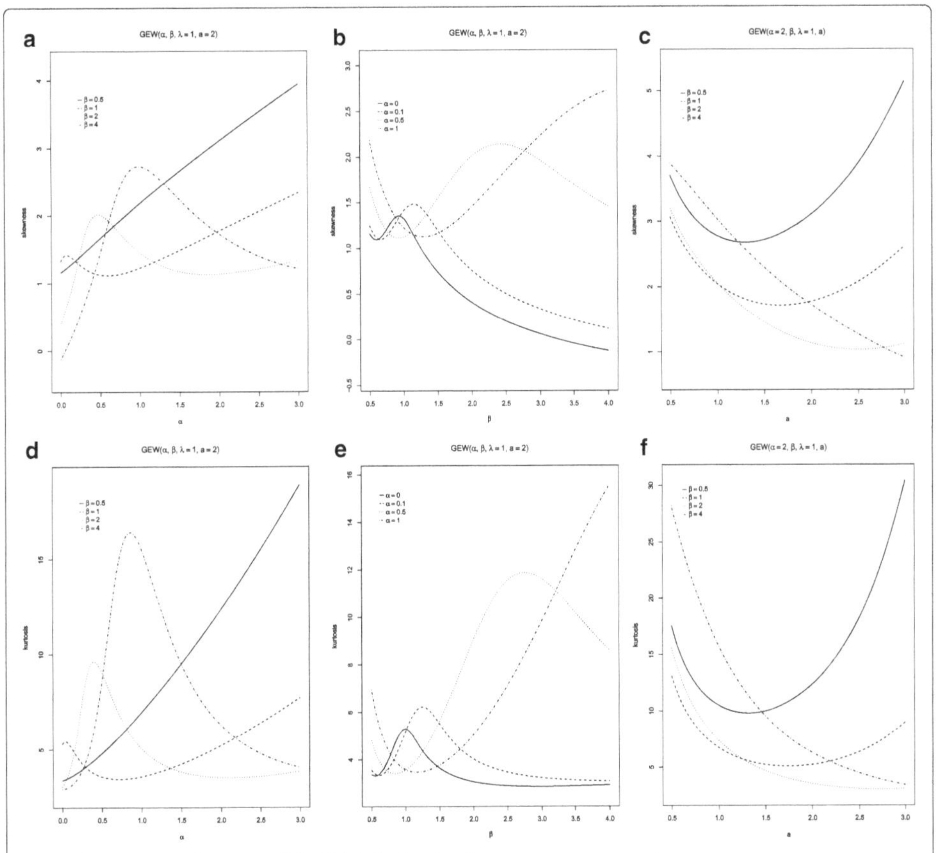

Fig. 3 Skewness and kurtosis of the GEW distribution for some parameter values

$$A(r,p,q) = \left\{p\, \sin[\pi(q-r)]\,(q-r)\,\Gamma(q+1)\,\Gamma(r+1-q)\right\}^{-1},$$

$$B(r,p,q) = p^{-q}\,\Gamma(q+1)\,\Gamma(r+1-q)\left[\cos(q\pi)\,\sin(\pi r) - \sin(q\pi)\,\cos(\pi r)\right],$$

and

$$C(r,p,q) = \pi\, p^{-r}\,\Gamma(r+1)(q-r)\,.$$

We obtain the incomplete moments by combining (13) and the expression for $\rho(y;r,p,q)$.

For $\alpha = 0$, the quantity $T_r(y)$ reduces to

$$T_r(y) = \lambda^{-r}\sum_{j=0}^{\infty}\frac{e_j}{(j+1)^r}\,\gamma(r+1,(j+1)\lambda y)\,.$$

An important application of $T_1(y)$ refers to the mean deviations about the mean and about the median, which can be used as measures of spread in a population. They are given by $\delta_1 = E\left(|X-\mu_1'|\right) = 2\mu_1'\,F\left(\mu_1'\right) - 2\,T_1\left(\mu_1'\right)$ and $\delta_2 = E(|X-m|) = \mu_1' - 2\,T_1(m)$, respectively, where the mean μ_1' is determined from (12) and $T_1(\cdot)$ comes from (13) with $r = 1$ as

$$T_1(y) = \lambda\sum_{j=0}^{\infty} e_j\,\rho\left(y;1,\alpha^\star\lambda^\star,{\alpha^\star}^{-1}\right).$$

Here, $\rho(y;1,p,q)$ becomes

$$\rho(y;1,p,q) = -\left[q(q-1)p^2\right]^{-1}\left[\frac{(pqy+1)}{(1+py)^q} - 1\right].$$

Further, for $\alpha = 0$, we have

$$T_1(q) = \sum_{j=0}^{\infty} e_j\left(\frac{1}{\lambda^\star}\right)^{1/\beta}\int_0^{\lambda^\star q^\beta} v^{1/\beta}\,\mathrm{e}^{-v}dv$$

and then $T_1(q)$ reduces to

$$T_1(q) = \sum_{j=0}^{\infty} e_j\left(\frac{1}{(j+1)\lambda}\right)^{1/\beta}\left[\Gamma\left(1+\frac{1}{\beta}\right) - \gamma\left(1+\frac{1}{\beta},(j+1)\lambda q^\beta\right)\right].$$

For a given probability π, both equations for $T_1(\cdot)$ can be used to determine the Lorenz and Bonferroni curves defined by $L(\pi) = T_1(q)/\mu_1'$ and $B(\pi) = T_1(q)/\left(\pi\,\mu_1'\right)$, respectively, where $q = Q_{GEW}(\pi)$ comes from Eq. (10) (see Fig. 4).

6 Estimation

The maximum likelihood method is used for estimating the parameters of the GEW model. We determine the maximum likelihood estimates (MLEs) from complete samples only. Let $x_1,\ldots,x_n$ be a sample of size n from the GEW(α,β,λ,a) distribution. The log-likelihood function for the vector of parameters $\boldsymbol{\theta} = (\alpha,\lambda,\beta,a)^T$ is given by $l(\boldsymbol{\theta}) = \sum_{i=1}^n \log[f(x_i)] = \sum_{i=1}^n l_i(\boldsymbol{\theta})$, where

$$l_i(\boldsymbol{\theta}) = \begin{cases} \log(\lambda) + \log(\beta) - \log[\Gamma(a)] - (a-1)\log(\alpha) + (\beta-1)\log(x_i) \\ -(1+1/\alpha)\log\left(1+\alpha\lambda x_i^\beta\right) + (a-1)\log\left[\log\left(1+\alpha\lambda x_i^\beta\right)\right] & ,\ \text{if}\quad \alpha > 0, \\ a\log(\lambda) + \log(\beta) - \log[\Gamma(a)] + (\beta a-1)\log(x_i) - \lambda x_i^\beta & ,\ \text{if}\quad \alpha = 0. \end{cases}$$

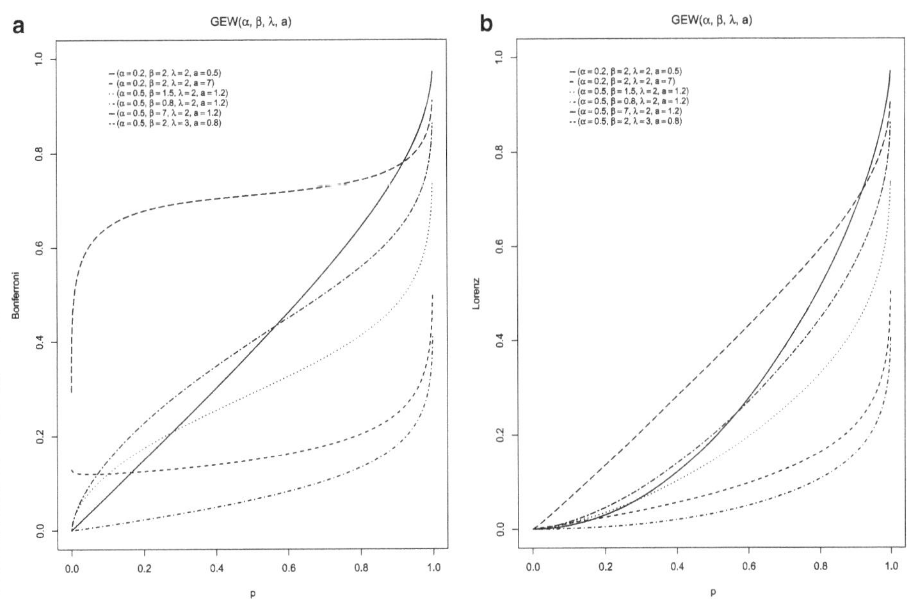

Fig. 4 Plots of the (**a**) Bonferroni and (**b**) Lorenz curves

The expression for $l(\boldsymbol{\theta})$ when $\alpha = 0$ gives the log-likelihood function for the GG distribution, a sub-model of the GEW distribution. Next, we maximize $l(\boldsymbol{\theta}) = \sum_{i=1}^{n} l_i(\boldsymbol{\theta})$ for $\alpha > 0$.

The components of the score vector $U(\boldsymbol{\theta})$ are given by $\sum_{i=1}^{n} \partial l_i(\boldsymbol{\theta})/\partial\theta_j$, $j = 1, \ldots, 4$, where the θ_j's denote the parameters α, β, λ or a. The elements $\partial l_i(\boldsymbol{\theta})/\partial\theta_j$, $j = 1, \ldots, 4$, are given below

$$
\begin{aligned}
\partial l_i(\boldsymbol{\theta})/\partial\alpha &= -\frac{a-1}{\alpha} + \frac{\log\left(1+\alpha\,\lambda\,x_i^{\beta}\right)}{\alpha^2} - \frac{\left(\frac{1}{\alpha}+1\right)\lambda\,x_i^{\beta}}{1+\alpha\,\lambda\,x_i^{\beta}} + \frac{(a-1)\,\lambda\,x_i^{\beta}}{\left(1+\alpha\,\lambda\,x_i^{\beta}\right)\log\left(1+\alpha\,\lambda\,x_i^{\beta}\right)},\\
\partial l_i(\boldsymbol{\theta})/\partial\beta &= \frac{1}{\beta} + \log(x_i) - \frac{\left(\frac{1}{\alpha}+1\right)\alpha\,\lambda\,x_i^{\beta}\,\log(x_i)}{1+\alpha\,\lambda\,x_i^{\beta}} + \frac{(a-1)\,\alpha\,\lambda\,x_i^{\beta}\,\log(x_i)}{\left(1+\alpha\,\lambda\,x_i^{\beta}\right)\log\left(1+\alpha\,\lambda\,x_i^{\beta}\right)},\\
\partial l_i(\boldsymbol{\theta})/\partial\lambda &= \frac{1}{\lambda} - \frac{\left(\frac{1}{\alpha}+1\right)\alpha\,x_i^{\beta}}{1+\alpha\,\lambda\,x_i^{\beta}} + \frac{(a-1)\,\alpha\,x_i^{\beta}}{\left(1+\alpha\,\lambda\,x_i^{\beta}\right)\log\left(1+\alpha\,\lambda\,x_i^{\beta}\right)},\\
\partial l_i(\boldsymbol{\theta})/\partial a &= -\psi(a) - \log(\alpha) + \log\left[\log\left(1+\alpha\,\lambda\,x_i^{\beta}\right)\right],
\end{aligned}
$$

where $\psi(\cdot)$ is the digamma function.

Setting these equations to zero and solving them simultaneously yield the MLEs of the four parameters. They can be solved numerically by using the R-language or any iterative methods such as the NR (Newton-Raphson), BFGS (Broyden-Fletcher-Goldfarb-Shanno), BHHH (Berndt-Hall-Hall-Hausman), NM (Nelder-Mead), SANN (Simulated-Annealing) and Limited-Memory Quasi-Newton code for Bound-Constrained Optimization (L-BFGS-B). Regarding the initial values for the parameters, several authors (Ugray et al. 2007; Glover 1998; Dixon and Szegö 1978; Varadhan and Gilbert 2009) suggest to find local maxima starting from widely varying starting points and then pick the maximum out of these, i.e., they suggest the use of multiple starting points (commonly referred to as multistart optimization). We followed their advice in this paper. As a useful tool we suggest the use of the R rotine `multiStart` of R package `BB` (for solving and optimizing large-scale nonlinear systems) for dealing with multiple starting points to obtain multiple solutions and to test sensitivity to starting values.

For interval estimation of the model parameters, we require the 4×4 total observed information matrix $J(\theta)$. The elements of $J(\theta) = \{J_{rs}\}$, where $r, s \in \{\alpha, \beta, \lambda, a\}$, can be obtained from the authors upon request. The multivariate normal $N_4\left(0, J(\widehat{\boldsymbol{\theta}})^{-1}\right)$ distribution, where $J(\widehat{\boldsymbol{\theta}})^{-1}$ is the inverse observed matrix evaluated at $\boldsymbol{\theta} = \widehat{\boldsymbol{\theta}}$, can be used to construct approximate confidence intervals for the parameters.

The likelihood ratio (LR) statistic is useful for comparing the GEW distribution with some of its special models. We can evaluate the maximum values of the unrestricted and restricted log-likelihoods to obtain LR statistics for testing some of its sub-models. In any case, hypothesis tests of the type $H_0 : \boldsymbol{\psi} = \boldsymbol{\psi}_0$ versus $H : \boldsymbol{\psi} \neq \boldsymbol{\psi}_0$, where $\boldsymbol{\psi}$ is a vector formed with some components of $\boldsymbol{\theta}$ and $\boldsymbol{\psi}_0$ is a specified vector, can be performed using LR statistics.

7 The LGEW regression model

If X is a random variable having the GEW density function (5), then $Y = \log(X)$ has the *log-gamma extended Weibull* (LGEW) distribution. The density function of Y, parameterized in terms of $\beta = \sigma^{-1}$ and $\lambda = \mathrm{e}^{-\mu/\sigma}$, is given by

$$f(y; a, \alpha, \mu, \sigma) = \begin{cases} \dfrac{\exp\left(\frac{y-\mu}{\sigma}\right)\left[1+\alpha\exp\left(\frac{y-\mu}{\sigma}\right)\right]^{\left(-\frac{1}{\alpha}-1\right)}}{\sigma\,\alpha^{a-1}\Gamma(a)} \left\{\log\left[1+\alpha\exp\left(\frac{y-\mu}{\sigma}\right)\right]\right\}^{a-1}, & \text{if } \alpha > 0, \\ \dfrac{\exp\left[a\left(\frac{y-\mu}{\sigma}\right)-\exp\left(\frac{y-\mu}{\sigma}\right)\right]}{\sigma\,\Gamma(a)}, & \text{if } \alpha = 0, \end{cases} \tag{14}$$

where $y \in \mathbb{R}$, $a > 0$ and $\alpha > 0$ are shape parameters, $\sigma > 0$ and $\mu \in \mathbb{R}$. If Y is a random variable having density function (14), we write $Y \sim \mathrm{LGEW}(a, \alpha, \mu, \sigma)$. Plots of the density function of Y for selected parameter values are displayed in Fig. 5. They reveal great flexibility for different values of the shape parameters a and α.

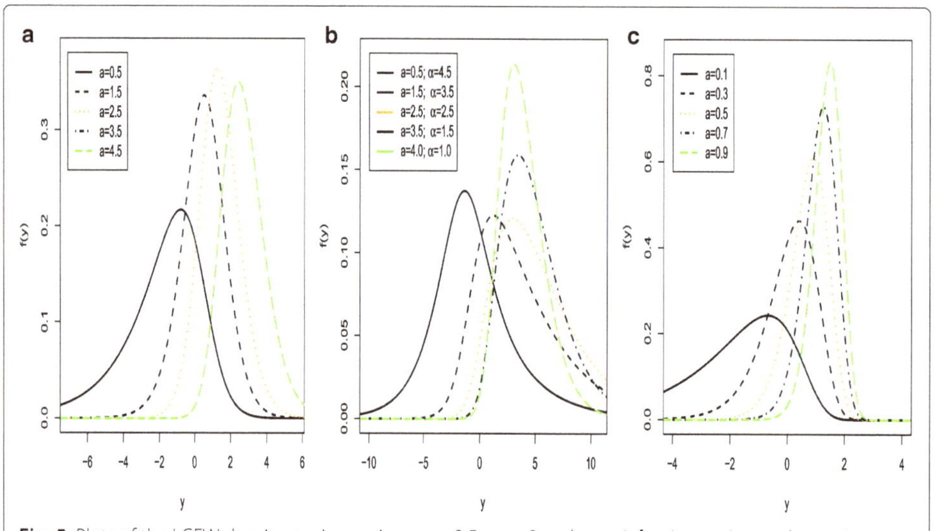

Fig. 5 Plots of the LGEW density: **a** a increasing, $\alpha = 0.5$, $\mu = 0$ and $\sigma = 1$; **b** a increasing, α decreasing, $\mu = 0$ and $\sigma = 1$; and **c** $\alpha = 0$ and a increasing, $\mu = 0$ and $\sigma = 1$

Thus,

$$\text{if} \quad X \sim \text{GEW}(a,\alpha,\beta,\lambda) \quad \text{then} \quad Y = \log(X) \sim \text{LGEW}(a,\alpha,\mu,\sigma).$$

The corresponding survival function of Y is given by

$$S(y;a,\alpha,\mu,\sigma) = \begin{cases} 1-\gamma_1\left(a,\frac{1}{\alpha}\log\left[1+\alpha\exp\left(\frac{y-\mu}{\sigma}\right)\right]\right) &, \text{if} \quad \alpha>0, \\ 1-\gamma_1\left(a,\exp\left(\frac{y-\mu}{\sigma}\right)\right) &, \text{if} \quad \alpha=0. \end{cases}$$

The pdf of the standardized random variable $Z=(Y-\mu)/\sigma$ is given by

$$\pi(z;a,\alpha) = \begin{cases} \frac{\exp(z)[1+\alpha\exp(z)]^{-\frac{1}{\alpha}-1}}{\alpha^{a-1}\Gamma(a)}\left\{\log[\,1+\alpha\exp(z)]\,\right\}^{a-1} &, \text{if} \quad \alpha>0, \\ \frac{\exp[a\,z-\exp(z)]}{\Gamma(a)} &, \text{if} \quad \alpha=0. \end{cases} \tag{15}$$

The special case $\alpha = 0$ and $a = 1$ refers to the log-Weibull (LW) (or extreme-value) distribution and, for $\alpha = 0$, we obtain the log-gamma-Weibull (LGW) model.

In many practical applications, the lifetimes x_i are affected by explanatory variables such as the cholesterol level, blood pressure and many others. Let $\mathbf{v}_i = (v_{i1},\ldots,v_{ip})^T$ be the explanatory variable vector associated with the ith response variable y_i for $i = 1,\ldots,n$. Consider a sample $(y_1,\mathbf{v}_1),\ldots,(y_n,\mathbf{v}_n)$ of n independent observations, where each random response is defined by $y_i = \min\{\log(x_i),\log(c_i)\}$, and $\log(x_i)$ and $\log(c_i)$ are the log-lifetime and log-censoring, respectively. We consider non-informative censoring such that the observed lifetimes and censoring times are independent.

We construct a linear regression model for the response variable y_i based on the LGEW distribution given by

$$y_i = \mathbf{v}_i^T\boldsymbol{\beta} + \sigma z_i,\ i=1,\ldots,n, \tag{16}$$

where the random error z_i has the density function (15), $\boldsymbol{\beta} = \left(\beta_1,\ldots,\beta_p\right)^T$, $\sigma > 0$, $a > 0$ and $\alpha > 0$ are unknown scalar parameters and $\mathbf{v}_i$ is the vector of explanatory variables modeling the location parameter $\mu_i = \mathbf{v}_i^T\boldsymbol{\beta}$. Hence, the location parameter vector $\boldsymbol{\mu} = (\mu_1,\ldots,\mu_n)^T$ of the LGEW model has a linear structure $\boldsymbol{\mu} = \mathbf{v}^T\boldsymbol{\beta}$, where $\mathbf{v} = (\mathbf{v}_1,\ldots,\mathbf{v}_n)^T$ is a known model matrix. The log-Weibull (or the extreme value) regression model is defined by (16) with $\alpha = 0$ and $a = 1$.

Let F and C be the sets of individuals for which y_i is the log-lifetime or log-censoring, respectively. The total log-likelihood function for the model parameters $\boldsymbol{\theta} = \left(a,\alpha,\sigma,\boldsymbol{\beta}^T\right)^T$ can be obtained from Eqs. (15) and (16) as

$$l(\boldsymbol{\theta}) = \begin{cases} -q\{\log(\sigma)+(a-1)\log(\alpha)-\log[\,\Gamma(a)]\,\}+\sum\limits_{i\in F} z_i \\ -\left(\frac{1}{\alpha}+1\right)\sum\limits_{i\in F}\log\left[1+\alpha\exp(z_i)\right] \\ +(a-1)\sum\limits_{i\in F}\log\{\log[\,1+\alpha\exp(z_i)]\,\} \\ +\sum\limits_{i\in C}\log\left\{1-\gamma_1\left(a,\frac{1}{\alpha}\log[\,1+\alpha\exp(z_i)]\,\right)\right\} &, \text{if} \quad \alpha>0, \\ -q\{\log(\sigma)+\log[\,\Gamma(a)]\,\}+a\sum\limits_{i\in F} z_i-\sum\limits_{i\in F}\exp(z_i) \\ +\sum\limits_{i\in C}\log\{1-\gamma_1(a,\exp[\,z_i])\} &, \text{if} \quad \alpha=0, \end{cases} \tag{17}$$

where q is the observed number of failures and $z_i = \left(y_i - \mathbf{v}_i^T \boldsymbol{\beta}\right)/\sigma$. The MLE $\widehat{\boldsymbol{\theta}}$ of $\boldsymbol{\theta}$ can be obtained by maximizing the log-likelihood function (17). We use the procedure NLMixed in SAS to evaluate the estimate $\widehat{\boldsymbol{\theta}}$. Initial values for $\boldsymbol{\beta}$ and σ are taken from the fit of the LW regression model with $\alpha = 0$ and $a = 1$.

From the fitted model (16), the survival function for y_i can be estimated by

$$\hat{S}\left(y_i; \hat{a}, \hat{\alpha}, \hat{\sigma}, \widehat{\boldsymbol{\beta}}^T\right) = \begin{cases} 1 - \gamma_1\left(\hat{a}, \frac{1}{\hat{\alpha}} \log\left[1 + \hat{\alpha}\exp\left(\frac{y_i - \mathbf{v}_i^T \widehat{\boldsymbol{\beta}}}{\hat{\sigma}}\right)\right]\right) & , \text{ if } \quad \alpha > 0, \\ 1 - \gamma_1\left(\hat{a}, \exp\left[\frac{y_i - \mathbf{v}_i^T \hat{\boldsymbol{\beta}}}{\hat{\sigma}}\right]\right) & , \text{ if } \quad \alpha = 0. \end{cases} \tag{18}$$

Under general regularity conditions, the asymptotic distribution of $(\widehat{\boldsymbol{\theta}} - \boldsymbol{\theta})$ is multivariate normal $N_{p+3}\,(0, K(\boldsymbol{\theta})^{-1})$, where $K(\boldsymbol{\theta})$ is the $(p+3) \times (p+3)$ expected information matrix. The asymptotic covariance matrix $K(\boldsymbol{\theta})^{-1}$ of $\widehat{\boldsymbol{\theta}}$ can be estimated by $\widehat{\boldsymbol{\theta}}^{-1}$ and then the inference on the parameter vector $\boldsymbol{\theta}$ can be based on the normal approximation $N_{p+3}\left(0, J(\widehat{\boldsymbol{\theta}})^{-1}\right)$ for $\widehat{\boldsymbol{\theta}}$. In fact, an $100(1-\eta)\%$ asymptotic confidence interval for each parameter θ_r is given by

$$ACI_r = \left(\hat{\theta}_r - z_{\eta/2}\sqrt{-\widehat{J}^{r,r}}, \hat{\theta}_r + z_{\eta/2}\sqrt{-\widehat{J}^{r,r}}\right),$$

where $-\widehat{J}^{r,r}$ represents the rth diagonal element of $J(\widehat{\boldsymbol{\theta}})^{-1}$ and $z_{\eta/2}$ is the quantile $1 - \eta/2$ of the standard normal distribution. The LR statistic can be used to discriminate between the LGEW and LW regression models since they are nested models. In this case, the hypotheses to be tested are $H_0 : (a, \alpha)^T = (1, 0)^T$ versus $H_1 : H_0$ *is not true*, and the LR statistic reduces to $w = 2\left\{l(\widehat{\boldsymbol{\theta}}) - l(\widetilde{\boldsymbol{\theta}})\right\}$, where $\widetilde{\boldsymbol{\theta}}$ is the MLE of $\boldsymbol{\theta}$ under H_0. The null hypothesis is rejected if $w > \chi^2_{2,1-\eta}$, where $\chi^2_{2,1-\eta}$ is the quantile of the chi-square distribution with two degrees of freedom.

8 Applications

8.1 The GEW model

In the first application, we use the warp break data for six types of weaving warps discussed by Tippett (1950, p. 105). We describe his experiment: "The results of a weaving experiment was conducted in a factory. There were 6 lots of warp yarn labelled respectively *AL*, *AM*, etc. They were spun from two growths of cotton, *A* and *B*, and each cotton was spun to three twists (i.e., the number of turns in the yarn per inch): low (*L*), medium (*M*), and high (*H*). The combination of these three factors give 6 kinds of yarn, which are the experimental treatments. From each yarn were prepared 9 warps (a warp is a quantity of warp yarn that goes into one loom as a unit), and, as a loom came available in the course of events, a warp chosen at random from the 54 was assigned to it, until ultimately all 54 were disposed of. More than one warp was woven in some looms, but that did not affect the randomness of the distribution. The number of warp threads that broke during the weaving of each warp was counted and expressed as a rate of so many breaks per unit of warp."

We analyze the warp breakage rates for individual warps disregarding the factors. First, we fit the GEW model and some of its sub-models to these data by the method of maximum likelihood. Afterwards, we compare the GEW model with some four-parameter competitive models.

The GEW model, its sub-models and some competitors

We compare the GEW model with some of its sub-models and we also compare it with some competitor models: the Burr XII geometric (BXIIG) (Silva and Cordeiro 2015), the beta log-normal (BLN) (Castellares et al. 2013) and the exponentiated Weibull-Poisson (EWP) (Mahmoudi and Sepahdar 2013) distributions. The MLEs of the models parameters and the Akaike Information Criterion (AIC) statistic for some models fitted to the data are listed in Table 1. In such table the parameters used for the competitor models are the same in the cited references. However, for the BXIIG we included the scale parameter s that was not considered in Silva and Cordeiro (2015). Even though it is expected that the GEW model will provide a better fit than its sub-models, we can verify whether or not there are significant differences. Some model comparison is given in Table 2.

The histogram of the data and the plots of the fitted densities for some models are displayed in Fig. 6. They indicate that the new distribution provides a better fit than other sub-models. A comparison of the new distribution with four sub-models using LR statistics is performed in Table 2. These statistics indicate that the new distribution is the most adequate model to describe the data.

In the case of the competitor models, the AIC values suggest that the GEW model is quite comparable to the BLN and EWP models (being slightly superior) and it is a bit better than the BXIIG model. Figure 7 displays the histogram of the data and the fitted GEW, BXIIG, BLN and EWP densities. The last comments about the AIC are in agreement with the general notion that the GEW density provides the best fit for the current data, although all of them are very close.

The required numerical evaluations are implemented by using an R script (sub-routine `nlminb` that can be found at `https://cran.r-project.org`). The `nlminb()` is a derivative-free method for function minimization as it requires only the function to be

Table 1 MLEs of the model parameters for the warp breakage rate data, the corresponding SEs (given in parentheses) and the AIC measure

Model	α	λ	β	a	AIC
GEW	0.895137	0.000026	9.792369	23.609396	420.257020
	(0.042691)	(0.000011)	(0.714861)	(2.615331)	
GW	0	0.008219	1.622223	2.061746	425.080529
	(-)	(0.000971)	(0.025583)	(0.071981)	
EW	0.104458	0.000192	2.499730	1	427.270840
	(0.025452)	(0.000032)	(0.052776)	(-)	
GEE	0.000961	0.190944	1	5.364537	421.924911
	(0.013883)	(0.026556)	(-)	(0.502296)	
GE	0	0.250448	1	6.896945	422.114535
	(-)	(0.006646)	(-)	(0.176429)	
Model	k	c	θ	s	AIC
BXIIG	1.430692	3.850236	0.885467	49.00841	422.653656
	(0.530019)	(0.094952)	(0.055033)	(11.093854)	
Model	μ	σ	a	b	AIC
BLN	1.290503	1.322159	30.985635	2.704164	420.316246
	(0.668315)	(0.373892)	(21.297427)	(1.327947)	
Model	α	θ	β	γ	AIC
EWP	14.50818	0.90275	0.13294	0.84268	420.51673
	(2.69087)	(0.27133)	(0.02132)	(0.04587)	

Table 2 LR tests for the warp breakage rate data

Model	Hypotheses	Statistic LR	*p*-value
GEW vs *GW*	$H_0 : \alpha = 0$ vs $H_1 : H_0$ is false	6.823509	0.008996563
GEW vs *EW*	$H_0 : a = 1$ vs $H_1 : H_0$ is false	9.013820	0.002679457
GEW vs *GEE*	$H_0 : \beta = 1$ vs $H_1 : H_0$ is false	3.667891	0.055470343
GEW vs *GE*	$H_0 : \alpha = 0, \beta = 1$ vs $H_1 : H_0$ is false	5.857515	0.053463433

evaluated. Other R routine choices such as the `optimx` package (Nash and Varadhan, 2011) provided similar results. The R rotine `multiStart` of R package `BB` is also used in this work. The data set `warpbreaks` is available in the R data frame.

Some estimation issues

We now discuss some estimation issues related to the GEW distribution. The GEW density is given by (5) and, when $\alpha \to 0$, the first expression in (5) tends to the second one. For $\alpha = 0$, the GEW distribution is identical to the generalized gamma distribution. Therefore, according to Theorem 2.1 in Mäkeläinen et al. (1981), the GEW density is not constant on the boundary since

$$\lim_{\alpha \to 0} f(x; \tau, a) = \frac{\lambda^a \, \beta \, x^{\beta a - 1} \, e^{-\lambda x^\beta}}{\Gamma(a)}.$$

This fact implies that the MLE $\hat{\boldsymbol{\theta}}$ of $\boldsymbol{\theta} = (\alpha, \lambda, \beta, a)$ of the GEW model may not be unique.

As mentioned before, in order to handle the case of multiple solutions we adopt different initial points, say k, in the maximization procedure and use the estimate which maximizes the log-likelihood among these k values. For the current data sets, we have always been able to find a solution.

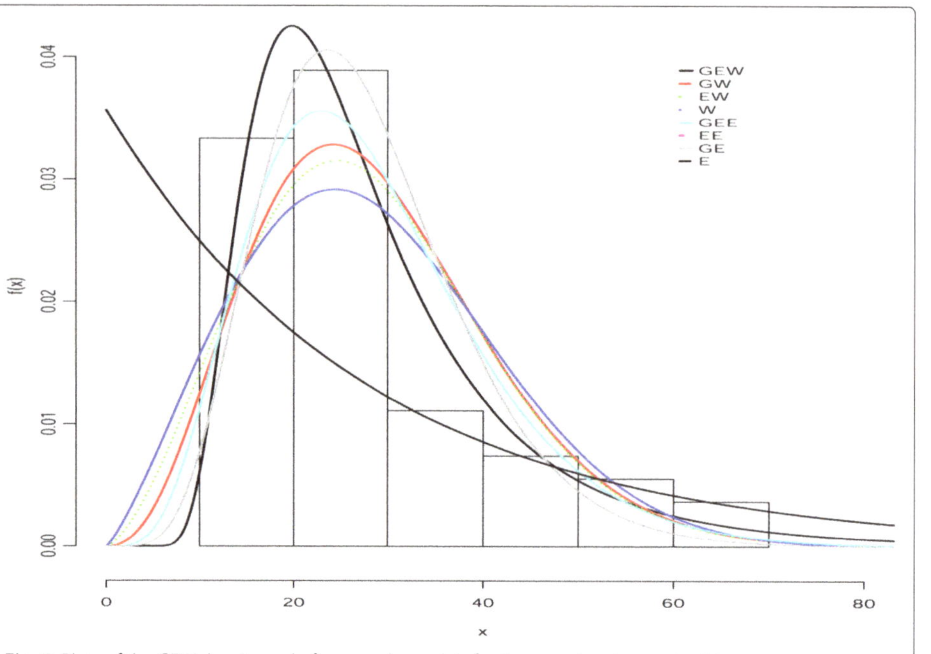

Fig. 6 Plots of the GEW density and of some sub-models for the warp breakage rate data

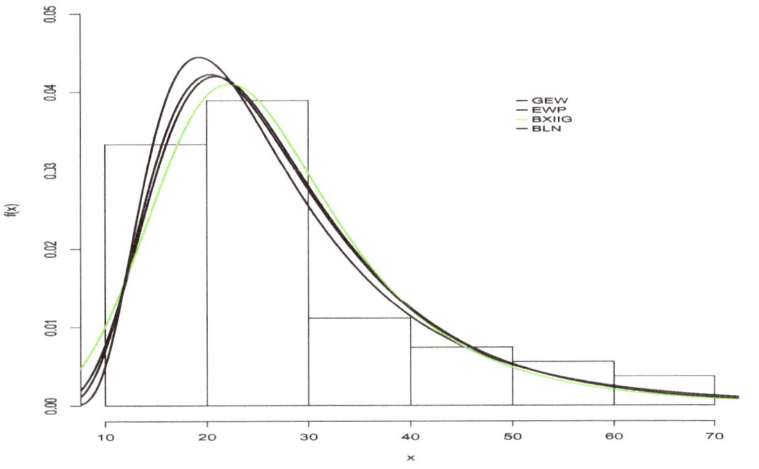

Fig. 7 Plots of some four-parameter densities for the warp breakage rate data

As mentioned above, the GEW distribution has different expressions depending on whether $\alpha > 0$ or $\alpha = 0$. In order to study possible estimation difficulties for this model when the true value of α is either zero or around it, we simulate data from the gamma generalized distribution, i.e., by taking $\alpha = 0$ for some choices of the parameters a, λ and β and then use the GEW model in the estimation procedure. We note that the GEW model is able to estimate all the parameter values correctly. As expected, for larger samples, the estimated values tend to be closer to the true values. Of course, for any similar problem, the program may fail to converge for some choices of initial values.

We also perform some additional analyses in order to evaluate the GEW robustness. We simulate two different data sets from the beta-Weibull distribution (which does not belong to the GEW family) and use them to fit the GEW distribution. The resulting fits (not shown here) indicate that the GEW distribution is very robust. In fact, the estimated GEW densities capture the main aspects of the beta-Weibull generated data in the sense that they are able to correctly locate the mode of the histograms obtained from the generated data, inflection points and other data characteristics.

Regarding the impact of the sample size on the estimated values of the parameters, a simulation study is conducted in which we generate GEW data considering $(\alpha, \lambda, \beta, a) = (0.895137, 0.000026, 9.792369, 23.609396)$ as the true parameter values. We simulate data by taking $k = 500$ replicates for each of the sample sizes $n = 50, 100, 500, 1000, 5000$ and $n = 10,000$. For a given sample size and $k = 500$ estimated values we evaluate the average of those estimated vectors and the mean squared errors (MSEs). We can conclude that the estimated expected vector does approach the true vector, but the MSE decreases slowly.

8.2 The LGEW model

We now illustrate some of the ideas and methodology of regression models for the LGEW model using the data set from a two-arm clinical trial considered earlier by Efron (1988) and Mudholkar et al. (1996). Efron observed that the empirical hazard functions for both samples start near zero, suggesting an initial high-risk period in the beginning, a decline for a while, and then stabilization after about one year. Specifically, Efron's data from a head and neck cancer clinical trial consist of survival times of 51 patients in arm A who were given radiation therapy and 45 patients in arm B who were given radiation plus chemotherapy. Nine patients in arm A and 14 patients in arm B were lost to follow-up and were regarded as censored. In this paper, we consider only one predictor: ($v1$): two-Arm ($Arm\,A = 0, Arm\,B = 1$).

We fit the LGEW regression model

$$y_i = \beta_0 + \beta_1 v_{i1} + \sigma z_i, \tag{19}$$

where the errors $z_1, \ldots, z_{96}$ are independent random variables with density function (15).

An alternative approach for modeling these data can be provided by the LW distribution. There are various extensions of this lifetime distribution; see, for example, the log-beta Weibull (LBW) (Cordeiro et al. 2013) and Kumaraswamy-logistic (KwL) (Santana et al. 2012) distributions, among others. The LBW and KwL density functions are given below:

- **LBW distribution**

$$f(y) = \frac{1}{\sigma B(a,b)} \exp\left\{\left(\frac{y-\mu}{\sigma}\right) - b\exp\left(\frac{y-\mu}{\sigma}\right)\right\}\left\{1 - \exp\left[-\exp\left(\frac{y-\mu}{\sigma}\right)\right]\right\}^{a-1},$$

where $-\infty < y < \infty, \sigma > 0$ and $-\infty < \mu < \infty$. See, for example, more details and properties in Cordeiro et al. (2013) and Ortega et al. (2015).

- **KwL distribution**

$$f(y) = \frac{a\,b}{\sigma} \exp\left[a\left(\frac{y-\mu}{\sigma}\right)\right]\left[1 + \exp\left(\frac{y-\mu}{\sigma}\right)\right]^{-(a+1)} \times \left\{1 - \left[1 - \frac{1}{1+\exp\left(\frac{y-\mu}{\sigma}\right)}\right]^{a}\right\}^{b-1},$$

where $-\infty < y < \infty, \sigma > 0$ and $-\infty < \mu < \infty$. Some applications of the KwL distribution are discussed in Santana et al. (2012) and Nadarajah et al. (2012).

Taking $\mu = \beta_0 + \beta_1 v_{i1}$ for the location parameter, σ is the dispersion parameter and a and b are shape parameters.

Table 3 MLEs of the parameters from some fitted regression models to the Efron's data, the corresponding SEs (given in parentheses), p-values in [.] and the AIC and BIC measures

Model	a	α	σ	β_0	β_1	AIC	BIC
LGEW	0.3101	32.5577	0.1420	5.6000	−0.0868	281.4	294.2
	(0.0904)	(21.3182)	(0.0405)	(0.1706)	(0.1654)		
				[<0.0001]	[0.6010]		
LGW	0.0264	0	0.0510	8.0113	−0.6274	339.7	349.9
	(0.0079)		(0.0152)	(0.1458)	(0.1608)		
				[<0.0001]	[0.0002]		
LW	1	0	1.1800	6.7873	−0.7586	312.6	320.3
			(0.1082)	(0.2088)	(0.2803)		
				[<0.0001]	[0.0089]		
	a	b	σ	β_0	β_1	AIC	BIC
LBW	167.06	14.0059	12.8338	−6.0419	−0.5836	299.4	312.2
	(0.3987)	(0.0682)	(1.1065)	(1.0492)	(0.2726)		
				[<0.0001]	[0.0348]		
KwL	16.2819	312.91	5.8106	1.6912	−0.6834	308.0	320.8
	(2.6189)	(24.86)	(0.3258)	(6.0235)	(0.2792)		
				[0.7795]	[0.0162]		

Table 4 LR tests for the Efron's data

Model	Hypotheses	Statistic LR	p-value
LGEW vs LGW	$H_0 : \alpha = 0$ vs $H_1 : H_0$ is false	60.3	<0.0001
LGEW vs LW	$H_0 : (a, \alpha)^T = (1, 0)^T$ vs $H_1 : H_0$ is false	35.2	<0.0001

Table 3 lists the MLEs of the parameters for the fitted LGEW, LGW, LW, LBW and KwL regression models to the current data using the NLMixed procedure in SAS. Initial values for $\boldsymbol{\beta}$ and σ are taken from the fitted LW regression model with $\alpha = 0$ and $a = 1$.

A comparison of the new distribution with two of its sub-models using LR statistics is performed in Table 4. These statistics indicate that the LGEW regression model is the most adequate model to explain the data.

The LGEW model involves an extra parameter, which gives it more flexibility to fit the data. We note from the fitted LGEW regression model that the dummy variable v_1 is not significant at 5 %. We note that there is no significant difference between "Arm A" and "Arm B" clinical trial for the survival times. The LGEW regression model outperforms the other models irrespective of the criteria and it can be used effectively in the analysis of these data.

A graphical comparison of the fitted LGEW, LGW and LW models (see Fig. 8) indicates that the LGEW model provides a superior fit. The curves displayed in Fig. 8 represent the empirical survival function and the estimated survival functions obtained from Eq. (18). Finally, we conclude that the LGEW regression model provides a good fit to these data.

9 Conclusions

We introduce a new model named the gamma extended Weibull (GEW) distribution and study some of its structural properties. It generalizes some important distributions in the literature and provides means of its continuous extension to still more complex situations. The new model contains several distributions as special models including the extended Weibull (Mudholkar et al. 1996), gamma Weibull (Zografos and Balakrishnan 2009) and

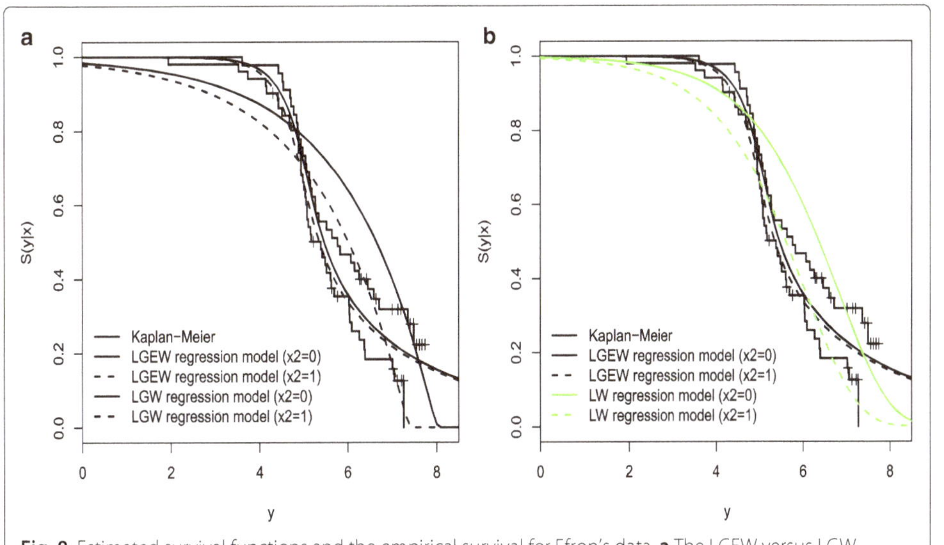

Fig. 8 Estimated survival functions and the empirical survival for Efron's data. **a** The LGEW versus LGW regression models. **b** The LGEW versus LW regression models

generalized gamma (Stacy 1962). We derive explicit expressions for the density function, ordinary and incomplete moments, quantile function and mean deviations. The model parameters are estimated by maximum likelihood. The usefulness of the GEW distribution is illustrated by means of an application to real data, where we show empirically that it gives a better fit than some of its sub-models. We also propose the log-gamma extended Weibull (LGEW) regression model, which has greater flexibility as shown by means of an application to real data.

Competing interests

The authors declare that they have no competing interests.

Authors' contributions

GMC proposed the model, wrote sections 1, 2, 4, 5, 9 (1, 2 and 4 jointly with MCSL, and 5 jointly with AEG) and drafted the manuscript. MCSL wrote and obtained the results in sections 1 to 4 (1, 2 and 4 jointly with GMC). AEG wrote and obtained the results in sections 5 (jointly with GMC), 6 and 8.1 (jointly with CQS). CQS wrote and obtained the results in Sections 6 and 8.1 (jointly with AEG), and the plots in Fig. 1 to 4, 6 and 7. EMMO proposed the log gamma extended Weibull regression model described in section 7, wrote and obtained the results in that section and in section 8.2, and also the plots in Fig. 5 and 8.

Author details

[1]Departamento de Estatística, Universidade Federal de Pernambuco, 52171-900 Recife, PE, Brazil. [2]Departamento de Estatística, Universidade de Brasília, 70910-900 Brasília, DF, Brazil. [3]Departamento de Ciências Exatas, Universidade de São Paulo, 13418-900 Piracicaba, SP, Brazil.

References

Castellares, F, Montenegro, LC, Cordeiro, GM: The beta log-normal distribution. J. Stat. Comput. Simul. **83**(2), 203–228 (2013)

Cordeiro, GM, Nadarajah, S, Ortega, EMM: General results for the beta Weibull distribution. J. Stat. Comput. Sim. **83**, 1082–1114 (2013)

Cordeiro, GM, Ortega, EMM, Silva, GO: The exponentiated generalized gamma distribution with application to lifetime data. J. Stat. Comput. Simul. **81**, 827–842 (2011)

Dixon, LCW, Szegö, GP: The Global Optimization Problem: an Introduction. In: Dixon, LCW, Szegö, GP (eds.) Towards Global Optimisation, Vol. 2, pp. 1–15. North Holland, Amsterdam, (1978)

Efron, B: Logistic regression, survival analysis, and the Kaplan-Meier curve. J. Am. Stat. Assoc. **83**, 414–425 (1988)

Glover, F: A template for scatter search and path relinking. In: Hao, J-K, Lutton, E, Ronald, E, Schoenauer, M, Snyers, D (eds.) Artificial Evolution. Lecture Notes in Computer Science, 1363, pp. 13–54. Springer, Berlin/Heidelberg, (1998)

Gupta, RC, Gupta, RD, Gupta, PL: Modeling failure time data by Lehman alternatives. Commun. Stat. Theor. M. **27**, 887–904 (1998)

Gupta, RD, Kundu, D: Exponentiated exponential family: an alternative to gamma and Weibull distributions. Biometrical J. **43**, 117–130 (2001)

Hao, Z, Singh, VP: Entropy-based parameter estimation for extended Burr XII distribution. Stoch. Env. Res. Risk A. **23**, 1113–1122 (2008)

Mahmoudi, E, Sepahdar, A: Exponentiated Weibull-Poisson distribution: Model, Properties and Applications. Math. Comput. Simul. **92**, 76–97 (2013)

Mäkeläinen, T, Schmidt, K, Styan, GPH: On the existence and uniqueness of the maximum likelihood estimate of a vVector-valued parameter in fixed-size. Ann. Stat. **9**, 758–767 (1981)

Mudholkar, GS, Srivastava, DK: Exponentiated Weibull family for analyzing bathtub failure-rate data. IEEE T. Reliab. **42**, 299–302 (1993)

Mudholkar, GS, Srivastava, DK, Freimer, M: The exponentiated Weibull family: a reanalysis of the bus-motor-failure data. Technometrics. **37**, 436–445 (1995)

Mudholkar, GS, Srivastava, DK, Kollia, GD: A generalization of the Weibull distribution with application to the analysis of survival data. J. Am. Stat. Assoc. **91**, 1575–1583 (1996)

Nadarajah, S, Cordeiro, GM, Ortega, EMM: General results for the Kumaraswamy-G distribution. J. Stat. Comput. Sim. **82**, 951–979 (2012)

Nadarajah, S, Gupta, AK: The exponentiated gamma distribution with application to drought data. Calcutta Stat. Assoc. **59**, 29–54 (2007)

Nash, JC, Varadhan, R: Unifying optimization algorithms to aid software system users: optimx for R. J. Stat. Software. **43**, 1–14 (2011)

Ortega, EMM, Cordeiro, GM, Campelo, AK, Kattam, MW, Cancho, VG: A power series beta Weibull regression model for predicting breast carcinoma. Stat. Med. **34**, 1366–1388 (2015)

Ristic, MM, Balakrishnan, N: The gamma exponentiated exponential distribution. J. Stat. Comput. Simul. **82**, 1191–1206 (2012)

Santana, TVF, Ortega, EMM, Cordeiro, GM, Silva, GO: The Kumaraswamy-Log-Logistic Distribution. J. Stat. Theory Appl. **11**, 265–291 (2012)

Shao, Q, Wong, H, Xia, J, Wai-Cheung, I: Models for extremes using the extended three-parameter Burr XII system with application to flood frequency analysis. Hydrolog. Sci. J. **49**, 685–702 (2004)

Silva, RB, Cordeiro, GM: The Burr XII power series distributions: A new compounding family. Stat. Braz. J. Probab. Stat. **29**(3), 565–589 (2015)
Stacy, EW: A generalization of the gamma distribution. Ann. Math. Stat. **33**, 1187–1192 (1962)
Tippett, LHC: Technological Applications of Statistics. Wiley, New York (1950)
Ugray, Z, Lasdon, L, Plummer, J, Glover, F, Kelly, J, Marti, R: Scatter search and local NLP solvers: a multistart framework for global optimization. INFORMS J. Comput. **19**(3), 328–340 (2007)
Varadhan, R, Gilbert, PD: BB: An R package for solving a large system of nonlinear equations and for optimizing a high-dimensional nonlinear objective function. J. Stat. Softw. **32**, 4 (2009)
Zografos, K, Balakrishnan, N: On families of beta- and generalized gamma-generated distributions and associated inference. Statist. Method. **6**, 344–362 (2009)

6

Exponentiated Marshall-Olkin family of distributions

Cícero R. B. Dias[1], Gauss M. Cordeiro[1], Morad Alizadeh[3], Pedro Rafael Diniz Marinho[1*]
and Hemílio Fernandes Campos Coêlho[2]

*Correspondence:
pedro.rafael.marinho@gmail.com
[1]Department of Statistics, Federal University of Pernambuco, Recife, Brazil
Full list of author information is available at the end of the article

Abstract
We study general mathematical properties of a new class of continuous distributions with three extra shape parameters called the exponentiated Marshal-Olkin family of distributions. Further, we present some special models of the new class and investigate the shapes and derive explicit expressions for the ordinary and incomplete moments, quantile and generating functions and probability weighted moments. We discuss the estimation of the model parameters by maximum likelihood and show empirically the potentiality of the family by means of two applications to real data.

Keywords: Generalized exponential geometric distribution, Generated family, Maximum likelihood, Moment

1 Introduction

In the past few years, several ways of generating new distributions from classic ones were developed and discussed. Eugene et al. (2002) defined a class of beta-generated distribution. Jones (2004) studied a family of distributions that arises naturally from the distribution of the order statistics and introduced general properties of the proposed class of distributions. Zografos and Balakrishnan (2009) proposed the gamma-generated family of distributions. Later, Cordeiro and de Castro (2011) defined the Kumaraswamy family. Recently, Alzaatreh et al. (2013) proposed a new technique to derive wider families by using any probability density function (pdf) as a generator. This generator called the T-X family of distributions has cumulative distribution function (cdf) defined by

$$F(x) = \int_{a}^{W[G(x)]} r(t)dt, \tag{1}$$

where $G(x)$ is the cdf of a random variable X, $r(t)$ is the pdf of a random variable T defined on $[a, b]$ and $W[G(x)]$ is a function of $G(x)$, which satisfies the following conditions:

- $W[G(x)] \in [a, b]$;
- $W[G(x)]$ is differentiable and monotonically non-decreasing;
- $W[G(x)] \to a$ as $x \to -\infty$ and $W[G(x)] \to b$ as $x \to \infty$.

Following Alzaatreh et al. (2013) and replacing $r(t)$ by the generalized exponential-geometric (GEG) density function (Silva et al. 2010), where $T \in [0, \infty)$, and using $W[G(x)] = -\log[1 - G(x)]$, we define the cdf of a new wider family by

$$F(x)=\int_0^{-\log[1-G(x;\boldsymbol{\xi})]}\frac{\alpha\lambda(1-p)e^{-\lambda t}[1-e^{-\lambda t}]^{\alpha-1}}{[1-pe^{-\lambda t}]^{\alpha+1}}dt=\left\{\frac{1-[1-G(x;\boldsymbol{\xi})]^{\lambda}}{1-p[1-G(x;\boldsymbol{\xi})]^{\lambda}}\right\}^{\alpha}, \tag{2}$$

where $G(x;\boldsymbol{\xi})$ is the baseline cdf depending on a parameter vector $\boldsymbol{\xi}$ and $\alpha>0, \lambda>0$ and $p<1$ are three additional shape parameters. For each baseline G, the *exponentiated Marshall-Olkin-G* ("EMO-G" for short) distribution is defined by the cdf (2). The EMO family includes as special cases the *exponentiated generalized class of distributions* (Cordeiro et al. 2013), the proportional and reversed hazard rate models, the Marshall-Olkin family and other sub-families. Some special models are listed in Table 1, where $G(x)=G(x;\boldsymbol{\xi})$.

Furthemore, the basic motivations for using the EMO-G family in practice are the following:

i. to make the kurtosis more flexible compared to the baseline model;
ii. to produce a skewness for symmetrical distributions;
iii. to construct heavy-tailed distributions for modeling real data;
iv. to generate distributions with symmetric, left-skewed, right-skewed or reversed-J shape;
v. to define special models with all types of the hrf;
vi. to provide consistently better fits than other generated models under the same baseline distribution.

This paper is organized as follows. In Section 2, we define the new family of distributions and provide a physical interpretation. Five of its special distributions are discussed in this section. In Section 3, some properties of the EMOG family are presented. The shape of the density and hazard rate functions are described analytically, two useful linear mixtures are provided. We derive a power series for the quantile function (qf) and we provide two general formulae for the moments. The incomplete moments are investigated and we derive the moment generating function (mgf) and determine the mean deviations. Estimation of the model parameters by maximum likelihood is performed in Section 4. Applications to two real data sets illustrate the performance of the EMO family in Section 5. The paper is concluded in Section 6.

Table 1 Some special models

α	λ	p	$G(x)$	Reduced distribution
-	-	0	$G(x)$	Exponentiated Generalized Class of Distributions (Cordeiro et al. 2013)
1	-	-	$G(x)$	Marshal-Olkin family of distributions (Marshall and Olkin 1997)
1	-	0	$G(x)$	Proportional hazard rate model (Gupta and Gupta 2007)
-	1	0	$G(x)$	Proportional reversed hazard rate model (Gupta and Gupta 2007)
1	1	0	$G(x)$	$G(x)$
1	-	-	$1-e^{-x}$	Exponential - Geometric distribution (Adamidis and Loukas 1998)
-	-	-	$1-e^{-x}$	Generalized Exponential - Geometric distribution (Silva et al. 2010)
1	-	-	$1-e^{-\beta x^{\gamma}}$	Weibull-Geometric distribution (Barreto-Souza et al. 2011)
-	-	-	$1-e^{-\beta x^{\gamma}}$	Exponentiated Weibull-Geometric distribution (Mahmoudi and Shiran 2012)

2 The new family

The density function corresponding to (2) is given by

$$f(x) = \alpha\,\lambda\,(1-p)\,g(x;\boldsymbol{\xi})\,[1-G(x;\boldsymbol{\xi})]^{\lambda-1}\,\frac{\{1-[1-G(x;\boldsymbol{\xi})]^{\lambda}\}^{\alpha-1}}{\{1-p[1-G(x;\boldsymbol{\xi})]^{\lambda}\}^{\alpha+1}}, \tag{3}$$

where $g(x;\boldsymbol{\xi})$ is the baseline pdf. This density function will be most tractable when the functions $G(x)$ and $g(x)$ have simple analytic expressions. Hereafter, a random variable X with density function (3) is denoted by $X \sim$ EMO-G$(p,\alpha,\lambda,\boldsymbol{\xi})$. Henceforth, we can omit sometimes the dependence on the baseline vector $\boldsymbol{\xi}$ of parameters and write simply $G(x) = G(x;\boldsymbol{\xi}), f(x) = f(x;p,\alpha,\lambda,\boldsymbol{\xi})$, etc.

A physical interpretation of the EMO-G distribution can be given as follows. Consider a system formed by α independent components having the Marshall-Olkin cdf given by

$$H(x) = \frac{1-[1-G(x)]^{\lambda}}{1-p[1-G(x)]^{\lambda}}.$$

Suppose the system fails if all of the α components fail and let X denote the lifetime of the entire system. Then, the cdf of X is

$$F(x) = H(x)^{\alpha} = \left\{\frac{1-[1-G(x;\boldsymbol{\xi})]^{\lambda}}{1-p[1-G(x;\boldsymbol{\xi})]^{\lambda}}\right\}^{\alpha}.$$

The hazard rate function (hrf) of X becomes

$$h(x) = \frac{\alpha\,\lambda\,(1-p)\,g(x;\boldsymbol{\xi})\,[1-G(x;\boldsymbol{\xi})]^{\lambda-1}\,\{1-[1-G(x;\boldsymbol{\xi})]^{\lambda}\}^{\alpha-1}}{\{\{1-p[1-G(x;\boldsymbol{\xi})]^{\lambda}\}^{\alpha}-\{1-[1-G(x;\boldsymbol{\xi})]^{\lambda}\}^{\alpha}\}\,\{1-p[1-G(x;\boldsymbol{\xi})]^{\lambda}\}}. \tag{4}$$

The EMO family of distributions is easily simulated by inverting (2): if u has a uniform $U(0,1)$ distribution, the solution of the nonlinear equation

$$X = G^{-1}\left[1-\left(\frac{1-u^{1/\alpha}}{1-p\,u^{1/\alpha}}\right)^{1/\lambda}\right] \tag{5}$$

follows the density function (3).

2.1 Special EMO distributions

For $p = 0$, we obtain, as a special case of (3), the exponentiated generalized class (Cordeiro et al. 2013) of distributions, which provides greater flexibility of its tails and can be applied in many areas of engineering and biology. Here, we present some special cases of the EMO family since it extends several useful distributions in the literature. For all cases listed below, $p \in (0,1)$, $\alpha > 0$ and $\lambda > 0$. These cases are defined by taking $G(x)$ and $g(x)$ to be the cdf and pdf of a specified distribution. The general form of the pdf of the special EMO distributions can be expressed as:

$$f(x) = q(\theta_1,\theta_2,\ldots,\theta_m)g(x)\,[1-G(x)]^{\lambda-1}\,\frac{\{1-[1-G(x)]^{\lambda}\}^{\alpha-1}}{\{1-p\,[1-G(x)]^{\lambda}\}^{\alpha+1}},$$

where $q(\theta_1,\theta_2,\ldots,\theta_m)$ is defined as a function of m parameters of the special EMO distribution. We list some special EMO distributions in Table 2, where the letters N, Fr, Ga, B, and Gu stand for the normal, Fréchet, gamma, beta and Gumbel baselines, respectively.

Table 2 Special EMO distributions

Distribution	$q(\cdot)$	$G(x)$	$g(x)$
$EMON(p,\alpha,\lambda,\mu,\sigma^2)$	$\frac{\alpha\lambda(1-p)}{\sigma}$	$\phi\left(\frac{x-\mu}{\sigma}\right)$	$\Phi\left(\frac{x-\mu}{\sigma}\right)$
$EMOFr(p,\alpha,\lambda,\beta,\sigma)$	$\alpha\lambda(1-p)\beta$	$\exp\left\{-\left(\frac{\sigma}{x}\right)^{\beta}\right\}$	$\sigma^{\beta}x^{-\beta-1}\exp\left\{-\left(\frac{\sigma}{x}\right)^{\beta}\right\}$
$EMOGa(p,\alpha,\lambda,a,b)$	$\frac{\alpha\lambda(1-p)b^{a}}{\Gamma(a)}$	$\frac{\gamma(a,bx)}{\Gamma(a)}$	$\frac{b^{a}}{\Gamma(a)}x^{a-1}e^{-bx}$
$EMOB(p,\alpha,\lambda,a,b)$	$\frac{\alpha\lambda(1-p)}{B(a,b)}$	$\frac{\int_0^x w^{a-1}(1-w)^{b-1}dw}{B(a,b)}$	$\frac{1}{B(a,b)}x^{a-1}(1-x)^{b-1}$
$EMOGu(p,\alpha,\lambda,\mu,\sigma)$	$\frac{\alpha\lambda(1-p)}{\sigma}$	$\exp\left\{-\exp\left[-\left(\frac{x-\mu}{\sigma}\right)\right]\right\}$	$\exp\left\{-\exp\left[-\frac{(x-\mu)}{\sigma}\right]-\frac{(x-\mu)}{\sigma}\right\}$

For the EMON distribution, the parameter σ has the same dispersion property as in the normal density. For the EMOB distribution, the beta distribution corresponds to the limiting case: $p \to 0$ and $\alpha = \lambda = 1$. For the EMOFr distribution, we have the classical Fréchet distribution when $p = 0$ and $\alpha = \lambda = 1$. The Kumaraswamy beta (KwB) and Kumaraswamy-gamma (KwGa) distributions can be obtained from the EMOB and EMOGa models when $p \to 0$. Plots of these EMOG density functions are displayed in Figs. 1, 2, 3, 4 and 5.

3 Some properties of the EMOG family

We investigate some properties of the EMOG in this section.

3.1 Asymptotic and shapes

Proposition 1 *Let $a = \inf\{x|\, G(x) > 0\}$. The asymptotics of Eqs.* (2), (3) *and* (4) *when $x \to a$ are given by*

$$F(x) \sim [\lambda\, G(x)]^{\alpha},$$
$$f(x) \sim \alpha\,\lambda^{\alpha}\, g(x)\, G(x)^{\alpha-1},$$
$$h(x) \sim \alpha\,\lambda^{\alpha}\, g(x)\, G(x)^{\alpha-1}.$$

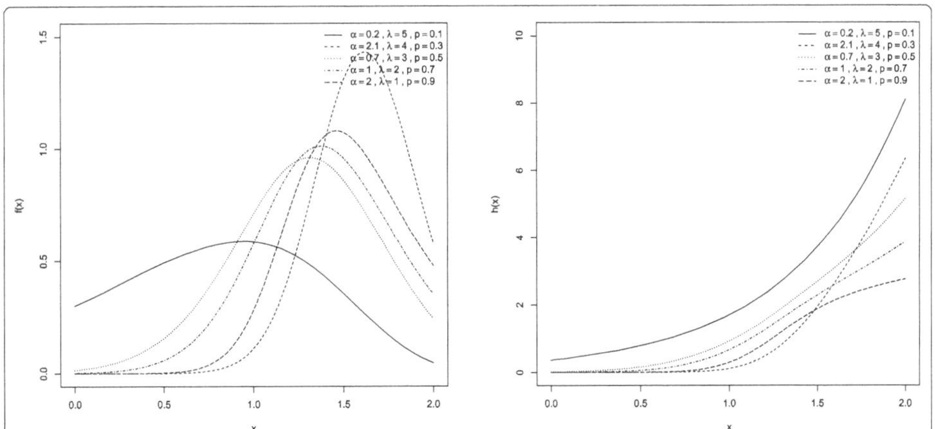

(a) $\mu = 0.5$ and $\sigma = 0.5$ (b) $\mu = 0.5$ and $\sigma = 0.5$

Fig. 1 Plots of the EMON pdf (**a**) and EMON hrf (**b**) for some parameter values

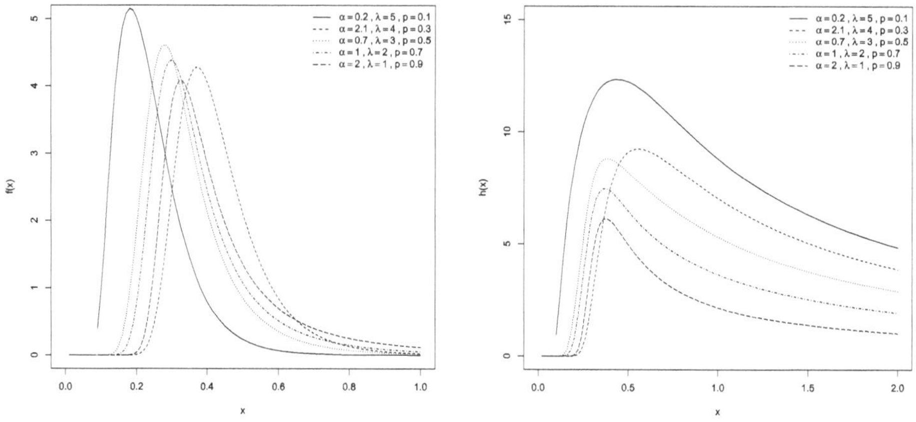

(a) $\beta = 0.5$ and $\sigma = 1.5$ (b) $\beta = 0.5$ and $\sigma = 1.5$

Fig. 2 Plots of the EMOFr pdf (**a**) and EMOFr hrf (**b**) for some parameter values

Proposition 2 *The asymptotics of Eqs.* (2), (3) *and* (4) *when* $x \to \infty$ *are given by*

$$\begin{aligned} 1 - F(x) &\sim \alpha\, \bar{G}(x)^{\lambda}, \\ f(x) &\sim \alpha\, \lambda\, g(x)\, \bar{G}(x)^{\lambda-1}, \\ h(x) &\sim \frac{\lambda\, g(x)}{\bar{G}(x)}. \end{aligned}$$

The shapes of the density and hazard rate functions can be described analytically. The critical points of the EMO-G density function are the roots of the equation

$$\begin{aligned} \frac{d\,\log[f(x)]}{dx} &= \frac{g'(x)}{g(x)} + (1-\lambda)\frac{g(x)}{1-G(x)} \\ &\quad - \lambda g(x)[1-G(x)]^{\lambda-1}\left\{\frac{1-\alpha}{1-[1-G(x)]^{\lambda}} + \frac{p(\alpha+1)}{1-p[1-G(x)]^{\lambda}}\right\} = 0 \end{aligned} \tag{6}$$

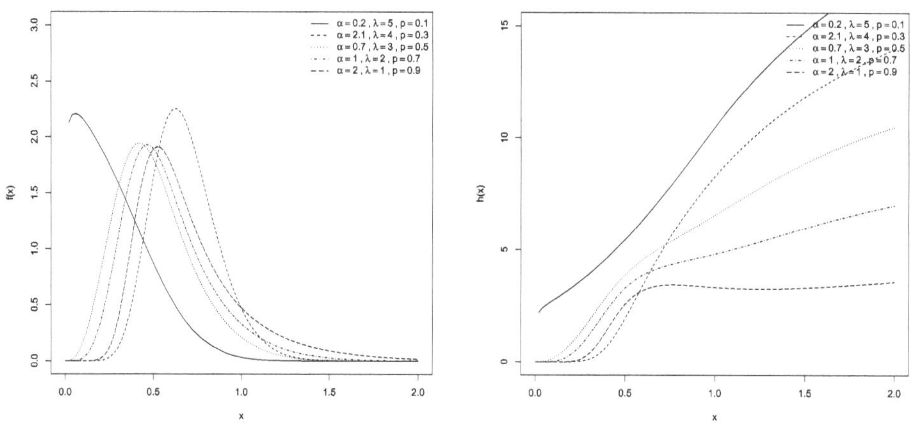

(a) $a = 5.5$ and $b = 5.5$ (b) $a = 5.5$ and $b = 5.5$

Fig. 3 Plots of the EMOGa pdf (**a**) and EMOGa hrf (**b**) for some parameter values

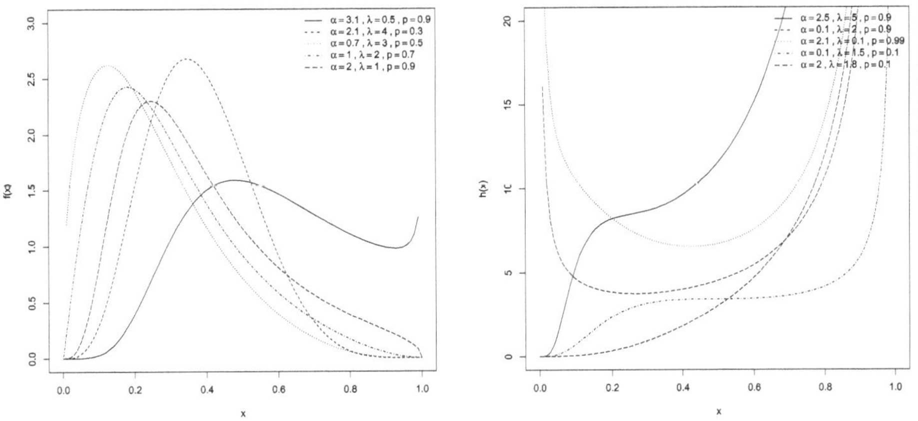

(a) $a = 2.0$ and $b = 1.5$ (b) $a = 2.0$ and $b = 1.5$

Fig. 4 Plots of the EMOB pdf (**a**) and EMOB hrf (**b**) for some parameter values

that corresponds to points where $f'(x) = 0$. There may be more than one root to (6). Let $\lambda(x) = d^2 \log[f(x)]\,/dx^2$. We have

$$\begin{aligned}\lambda(x) &= \frac{g''(x)g(x) - g'(x)^2}{g^2(x)} + (1-\lambda)\frac{g'(x)[1-G(x)] + g^2(x)}{[1-G(x)]^2} + \lambda(\alpha - 1) \\ &\times \left\{ g'(x)\frac{[1-G(x)]^{\lambda-1}}{1-[1-G(x)]^{\lambda}} - (\lambda-1)g^2(x)\frac{[1-G(x)]^{\lambda-2}}{1-[1-G(x)]^{\lambda}} - \lambda g^2(x)\frac{[1-G(x)]^{2\lambda-2}}{\{1-[1-G(x)]^{\lambda}\}^2} \right\} \\ &- p\lambda(\alpha+1)\left\{ g'(x)\frac{[1-G(x)]^{\lambda-1}}{1-p[1-G(x)]^{\lambda}} - (\lambda-1)g^2(x)\frac{[1-G(x)]^{\lambda-2}}{1-p[1-G(x)]^{\lambda}} \right. \\ &\left. - p\lambda g^2(x)\frac{[1-G(x)]^{2\lambda-2}}{\{1-p[1-G(x)]^{\lambda}\}^2} \right\}.\end{aligned} \tag{7}$$

If $x = x_0$ is a root of (6) then it corresponds to a local maximum if $\lambda(x) > 0$ for all $x < x_0$ and $\lambda(x) < 0$ for all $x > x_0$. It corresponds to a local minimum if $\lambda(x) < 0$ for all

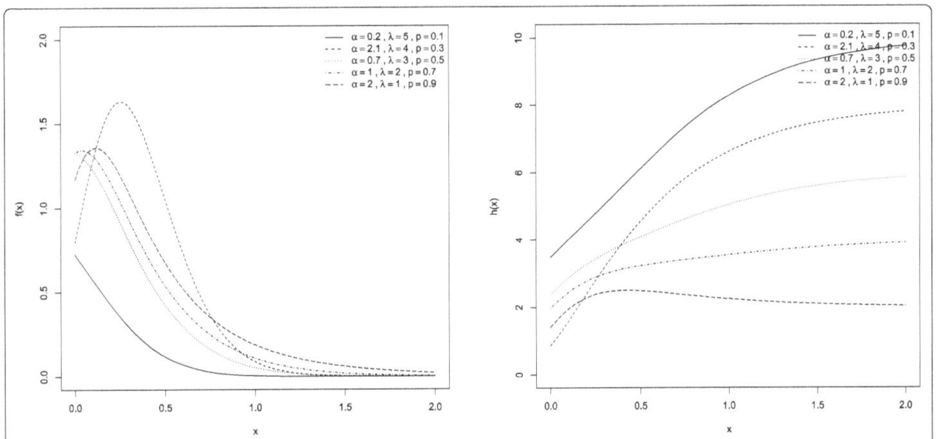

(a) $\mu = 0.5$ and $\sigma = 1.5$ (b) $\mu = 0.5$ and $\sigma = 1.5$

Fig. 5 Plots of the EMOGu pdf (**a**) and EMOGu hrf (**b**) for some parameter values

$x < x_0$ and $\lambda(x) > 0$ for all $x > x_0$. It gives a point of inflexion if either $\lambda(x) > 0$ for all $x \neq x_0$ or $\lambda(x) < 0$ for all $x \neq x_0$.

The critical points of the (hrf) of X are obtained from the equation

$$\frac{d\log[h(x)]}{dx} = \frac{g'(x)}{g(x)} + (1-\lambda)\frac{g(x)}{1-G(x)} + \lambda(\alpha-1)\frac{g(x)[1-G(x)]^{\lambda-1}}{1-[1-G(x)]^{\lambda}} - \lambda p\frac{g(x)[1-G(x)]^{\lambda-1}}{1-p[1-G(x)]^{\lambda}}$$
$$-\frac{p\alpha\lambda g(x)[1-G(x)]^{\lambda-1}\{1-[1-G(x)]^{\lambda}\}^{\alpha-1} - \lambda\alpha g(x)[1-G(x)]^{\lambda-1}\{1-p[1-G(x)]^{\lambda}\}^{\alpha-1}}{\{1-p[1-G(x)]^{\lambda}\}^{\alpha} - \{1-[1-G(x)]^{\lambda}\}^{\alpha}} = 0. \tag{8}$$

There may be more than one root to (8). Let $\tau(x) = d^2\log[h(x)]/dx^2$. If $x = x_0$ is a root of (8) then it refers to a local maximum if $\tau(x) > 0$ for all $x < x_0$ and $\tau(x) < 0$ for all $x > x_0$. It corresponds to a local minimum if $\tau(x) < 0$ for all $x < x_0$ and $\tau(x) > 0$ for all $x > x_0$. It gives an inflexion point if either $\tau(x) > 0$ for all $x \neq x_0$ or $\tau(x) < 0$ for all $x \neq x_0$.

3.2 Linear mixtures

We can demonstrate that the cdf (2) of X admits the expansion

$$F(x) = \sum_{k=0}^{\infty} b_k\, H_k(x;\boldsymbol{\xi}), \tag{9}$$

where $b_k = \sum_{i,j=0}^{\infty} w_{i,j,k}$,

$$w_{i,j,k} = w_{i,j,k}(\alpha,\lambda,p) = (-1)^{i+j+k}\binom{-\alpha}{i}\binom{\alpha}{j}\binom{(i+j)\lambda}{k}p^i,$$

and $H_k(x;\boldsymbol{\xi}) = G(x;\boldsymbol{\xi})^k$ denotes the exponentiated-G ("exp-G") cdf with power parameter k.

The density function of X can be expressed as an infinite linear mixture of exp-G density functions

$$f(x) = \sum_{k=0}^{\infty} b_{k+1}\, h_{k+1}(x;\boldsymbol{\xi}), \tag{10}$$

where (for $k \geq 0$) $h_{k+1}(x;\boldsymbol{\xi}) = (k+1)\,g(x;\boldsymbol{\xi})\,G(x;\boldsymbol{\xi})^k$ denotes the density function of the random variable $Y_{k+1} \sim$ exp-G$(k+1)$. Equation (10) reveals that the EMO-G density function is a linear mixture of exp-G density functions. Thus, some of its mathematical properties can be derived directly from those properties of the exp-G distribution. For example, the ordinary and incomplete moments and (mgf) of X can be obtained from those quantities of the exp-G distribution. Some structural properties of the exp-G distributions are well-defined by Mudholkar and Hutson (1996), Gupta and Kundu (2001) and Nadarajah and Kotz (2006), among others.

The formulae derived throughout the paper can be easily handled in most symbolic computation software platforms such as Maple, Mathematica and Matlab. These platforms have currently the ability to deal with analytic expressions of formidable size and complexity. Established explicit expressions to calculate statistical measures can be more efficient than computing them directly by numerical integration. The infinity limit in these sums can be substituted by a large positive integer such as 20 or 30 for most practical purposes.

3.3 Quantile power series

We obtain explicit expressions for the moments and generating function of the EMO family using a power series for the qf $x = Q(u) = F^{-1}(u)$ of X by expanding (5). If the G qf, say $Q_G(u)$, does not have a closed-form expression, this function can usually be expressed as a power series

$$Q_G(u) = \sum_{i=0}^{\infty} a_i\, u^i, \tag{11}$$

where the coefficients a_i's are suitably chosen real numbers depending on the parameters of the parent distribution. For several important distributions such as the normal, Student t, gamma and beta distributions, $Q_G(u)$ does not have explicit expressions but it can be expanded as in Eq. (11).

We use throughout the paper a result of Gradshteyn and Ryzhik (2000) for a power series raised to a positive integer n (for $n \geq 1$)

$$Q_G(u)^n = \left(\sum_{i=0}^{\infty} a_i\, u^i\right)^n = \sum_{i=0}^{\infty} c_{n,i}\, u^i, \tag{12}$$

where the coefficients $c_{n,i}$ (for $i = 1, 2, \ldots$) are easily determined from the recurrence equation, with $c_{n,0} = a_0^n$,

$$c_{n,i} = (i\, a_0)^{-1} \sum_{m=1}^{i} [\, m(n+1) - i]\; a_m\, c_{n,i-m}. \tag{13}$$

Clearly, $c_{n,i}$ can be easily evaluated numerically from $c_{n,0}, \ldots, c_{n,i-1}$ and then from the quantities $a_0, \ldots, a_i$.

Next, we derive an expansion for the argument of $Q_G(\cdot)$ in Eq. (5)

$$A = 1 - \frac{(1 - u^{1/\alpha})^{1/\lambda}}{(1 - p\, u^{1/\alpha})^{1/\lambda}}.$$

Using the the generalized binomial expansion four times since $u \in (0, 1)$, we can write

$$A = \sum_{r,s,t=0}^{\infty} \sum_{m=0}^{t} (-1)^{r+s+t+m}\, p^r \binom{-\lambda^{-1}}{r} \binom{\lambda^{-1}}{s} \binom{(r+s)\alpha^{-1}}{t} \binom{t}{m} u^m$$

Then, the qf of X can be expressed from (5) as

$$Q(u) = Q_G\left(\sum_{m=0}^{\infty} \delta_m\, u^m\right), \tag{14}$$

where

$$\delta_m = \begin{cases} 1 - \sum_{r,s,t=0}^{\infty} (-1)^{r+s+t}\, p^r \binom{-\lambda^{-1}}{r} \binom{-\lambda^{-1}}{s} \binom{(r+s)\alpha^{-1}}{t}, & m = 0, \\ \sum_{r,s,t=0}^{\infty} (-1)^{r+s+t+m}\, p^r \binom{-\lambda^{-1}}{r} \binom{\lambda^{-1}}{s} \binom{(r+s)\alpha^{-1}}{t} \binom{t}{m}, & m > 0. \end{cases}$$

By combining (11) and (14), we have

$$Q(u) = \sum_{i=0}^{\infty} a_i \left(\sum_{m=0}^{\infty} \delta_m\, u^m\right)^i,$$

and then using (12) and (13),

$$Q(u) = \sum_{m=0}^{\infty} e_m\, u^m, \tag{15}$$

where $e_m = \sum_{i=0}^{\infty} a_i \, d_{i,m}$, $d_{i,0} = \delta_0^i$ and, for $m > 1$,

$$d_{i,m} = (m\,\delta_0)^{-1} \sum_{n=1}^{m} [\, n(i+1) - m]\ \delta_n \, d_{i,m-n}.$$

Equation (15) is the main result of this section. It allows to obtain various mathematical quantities for the EMO-G family as can be seen in the next sections. Note that

$$Q(u)^r = \left(\sum_{m=0}^{\infty} e_m u^m \right)^r = \sum_{m=0}^{\infty} f_{r,m} \, u^m, \tag{16}$$

where $f_{r,m}$ is obtained from the e_m's using (13).

The effects of the shape parameters on the skewness and kurtosis can be determined from quantile measures. The shortcomings of the classical kurtosis measure are well-known. The Bowley skewness (Kenney and Keeping 1962) is one of the earliest skewness measures defined by the average of the quartiles minus the median divided by half the interquartile range, namely

$$B = \frac{Q\left(\frac{3}{4}\right) + Q\left(\frac{1}{4}\right) - 2Q\left(\frac{1}{2}\right)}{Q\left(\frac{3}{4}\right) - Q\left(\frac{1}{4}\right)}.$$

Since only the middle two quartiles are considered and the outer two quartiles are ignored, this adds robustness to the measure. The Moors kurtosis (Moors 1998) is based on octiles

$$M = \frac{Q\left(\frac{3}{8}\right) - Q\left(\frac{1}{8}\right) + Q\left(\frac{7}{8}\right) - Q\left(\frac{5}{8}\right)}{Q\left(\frac{6}{8}\right) - Q\left(\frac{2}{8}\right)}.$$

These measures are less sensitive to outliers and they exist even for distributions without moments. In Figs. 6 and 7, we plot the measures B and M for the EMOFr and EMON distributions (discussed in Section 2), respectively. These plots reveal how both measures B and M vary on the shape parameters.

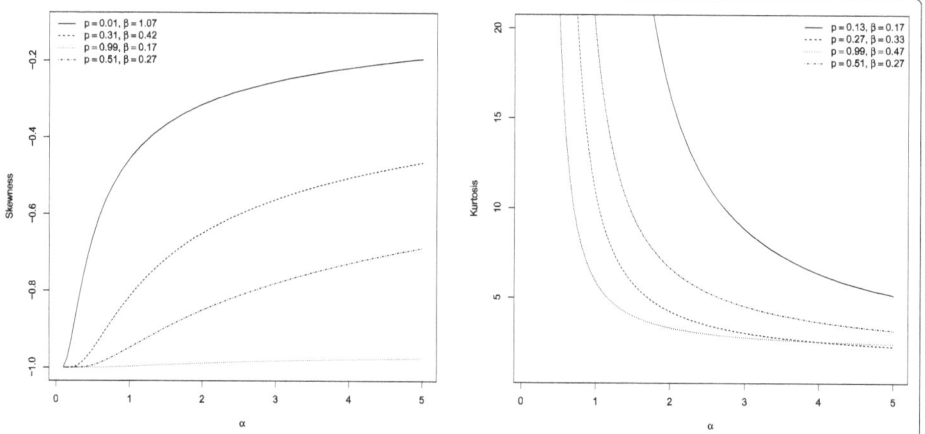

(a) $\sigma = 1.0$ and $\lambda = 1.0$ (b) $\sigma = 1.0$ and $\lambda = 1.0$

Fig. 6 Skewness (**a**) and Kurtosis (**b**) of the EMOFr distribution

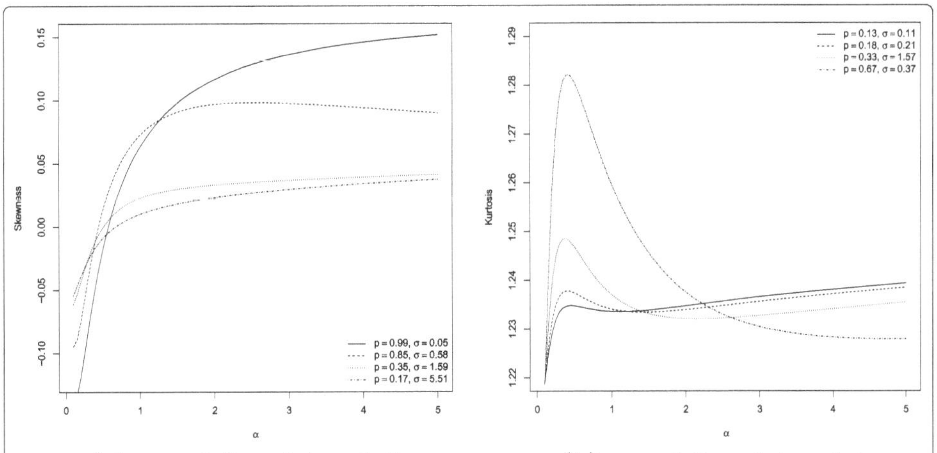

(a) $\mu = 3.3$ and $\lambda = 1.0$ (b) $\mu = 3.3$ and $\lambda = 1.0$

Fig. 7 Skewness (**a**) and Kurtosis (**b**) of the EMON distribution

3.4 Moments

Hereafter, we shall assume that G is the cdf of a random variable Z and that F is the cdf of a random variable X having density function (3). The rth ordinary moment of X can be obtained from the (r, k)th Probability Weighted Moment (PWM) of Z defined by

$$\tau_{r,k} = \mathrm{E}[Z^r\, G(Z)^k] = \int_{-\infty}^{\infty} z^r\, G(z)^k\, g(z) dz. \tag{17}$$

In fact, we have

$$\mathrm{E}(X^r) = \sum_{k=0}^{\infty} (k+1)\, b_{k+1}\, \tau_{r,k}. \tag{18}$$

Thus, the moments of any EMO-G distribution can be expressed as an infinite linear combination of the PWMs of G. A second formula for $\tau_{r,k}$ can be based on the parent qf $Q_G(u) = G^{-1}(u)$. Setting $G(x) = u$, we obtain

$$\tau_{r,k} = \int_0^1 Q_G(u)^r\, u^k du, \tag{19}$$

where the integral follows from (16) as

$$\mathrm{E}(X^r) = \int_0^1 Q(u)^r du = \sum_{m=0}^{\infty} \frac{f_{r,m}}{m+1}. \tag{20}$$

The PWMs for some well-known distributions will be determined in the following sections using alternatively Eqs. (17) and (19).

The central moments (μ_s) and cumulants (κ_s) of X can be obtained from Eqs. (18) and (20) as

$$\mu_s = \sum_{j=0}^{s} (-1)^j \binom{s}{j} \mu_1^{\prime s} \mu'_{s-j}, \qquad \kappa_s = \mu'_s - \sum_{j=1}^{s-1} \binom{s-1}{j-1} \kappa_j \mu'_{s-j},$$

where $\kappa_1 = \mu'_1$. The skewness and kurtosis measures can be calculated from the ordinary moments using well-known relationships. The pth descending factorial moment of X is

$$\mu'_{(p)} = \mathrm{E}[X^{(p)}] = \mathrm{E}[X(X-1)\times\cdots\times(X-p+1)] = \sum_{k=0}^{p} s(p,k)\mu'_k,$$

where $s(r,k) = (k!)^{-1}[d^k x^{(r)}/dx^k]_{x=0}$ is the Stirling number of the first kind. So, we can obtain the factorial moments from the ordinary moments given before.

3.5 Incomplete moments

The nth incomplete moment of X is defined as $m_r(y) = \int_{-\infty}^{y} x^r f(x)dx$. For empirical purposes, the shape of many distributions can be usefully described by the incomplete moments. Here, we propose two methods to determine the incomplete moments of the new family. First, we can express $m_r(y)$ as

$$m_r(y) = \sum_{k=0}^{\infty} (k+1)\, b_{k+1} \int_0^{G(y;\xi)} Q_G(u)^r\, u^k\, du. \tag{21}$$

The integral in (21) can be evaluated at least numerically for most baseline distributions.

A second method for the incomplete moments of X follows from (21) using Eqs. (12) and (13). We obtain

$$m_r(y) = \sum_{k,m=0}^{\infty} \frac{(k+1)\, b_{k+1}\, c_{r,m}}{m+k+1}\, G(y;\xi)^{m+k+1}. \tag{22}$$

Equations (21) and (22) are the main results of this section.

3.6 Generating function

Here, we provide three formulae for the mgf $M(s) = E(\mathrm{e}^{sX})$ of X. A first formula for $M(s)$ comes from Eq. (10) as

$$M(s) = \sum_{k=0}^{\infty} b_{k+1}\, M_{k+1}(s), \tag{23}$$

where $M_{k+1}(s)$ is the generating function of the exp-G$(k+1)$ distribution. Hence, $M(s)$ can be determined from an infinite linear combination of the exp-G generating functions.

A second formula for $M(s)$ can be derived from Eq. (10) as

$$M(s) = \sum_{k=0}^{\infty} (k+1)\, b_{k+1}\, \rho_k(s), \tag{24}$$

where

$$\rho_k(s) = \int_0^1 \exp\left[s\, Q_G(u)\right] u^k du. \tag{25}$$

We can derive the mgfs of several EMO distributions directly from Eqs. (24) and (25). For example, the mgfs of the exponentiated Marshall-Olkin exponential (EMOE) (such that $\lambda s < 1$) and EMO-standard logistic (for $s < 1$) distributions are given by

$$M(s) = \sum_{k=0}^{\infty} (k+1)\, b_{k+1}\, B(k+1, 1-\lambda s) \quad \text{and} \quad M(s) = \sum_{k=0}^{\infty} (k+1)\, b_{k+1}\, B(s+k+1, 1-s),$$

respectively.

3.7 Mean deviations

The mean deviations about the mean ($\delta_1 = E(|X - \mu_1'|)$) and about the median ($\delta_2 = E(|X - M|)$) of X can be expressed as

$$\delta_1 = 2\mu_1' F\left(\mu_1'\right) - 2m_1\left(\mu_1'\right) \qquad \text{and} \qquad \delta_2 = \mu_1' - 2m_1(M), \tag{26}$$

respectively, $F(\mu_1')$ is easily evaluated from Eq. (2),

$$M = Q_G\left[1 - \left(\frac{1 - 2^{-1/\alpha}}{1 - p \times 2^{-1/\alpha}}\right)^{1/\lambda}\right]$$

is the median of X, $\mu_1' = \mathrm{E}(X)$ comes from (18) and $m_1(y)$ is the first incomplete moment of X determined from (22) with $r = 1$.

Next, we provide three alternative ways to compute δ_1 and δ_2. A general equation for $m_1(z)$ is given by (21). A second general formula for $m_1(z)$ can be obtained from (10) as

$$m_1(z) = \sum_{k=0}^{\infty} b_{k+1} J_{k+1}(z), \tag{27}$$

where

$$J_{k+1}(z) = \int_{-\infty}^{z} x\, h_{k+1}(x) dx. \tag{28}$$

Equation (28) is the basic quantity to compute the mean deviations for the exp-G distributions. A simple application of (27) and (28) can be conducted to the exponentiated Marshall-Olkin Weibull (EMOW) distribution. The exponentiated Weibull density function (for $x > 0$) with power parameter $k + 1$, shape parameter c and scale parameter β is given by

$$h_{k+1}(x) = c\,(k+1)\,\beta^c x^{c-1} \exp\left\{-(\beta x)^c\right\} \left[1 - \exp\left\{-(\beta x)^c\right\}\right]^k,$$

and then

$$J_{k+1}(z) = c\,(k+1)\,\beta^c \sum_{r=0}^{\infty} (-1)^r \binom{k}{r} \int_0^z x^c \exp\left\{-(r+1)(\beta x)^c\right\} dx.$$

The last integral reduces to the incomplete gamma function

$$J_{k+1}(z) = c\,(k+1)\,\beta^c \sum_{r=0}^{\infty} (-1)^r \binom{k}{r} \gamma\left(c+1, (r+1)(\beta z)^c\right),$$

where $\gamma(a,x) = \int_0^x w^{a-1}\, \mathrm{e}^{-w} \mathrm{d}w$.

A third general formula for $m_1(z)$ can be derived by setting $u = G(x)$ in (10)

$$m_1(z) = \sum_{k=0}^{\infty} (k+1)\, b_{k+1}\; T_k(z), \tag{29}$$

where $T_k(z)$ is given by

$$T_k(z) = \int_0^{G(z)} Q_G(u)\, u^k du. \tag{30}$$

Applications of these equations are straightforward to obtain Bonferroni and Lorenz curves. These curves are defined (for a given probability π) by $B(\pi) = m_1(q)/(\pi\, \mu_1')$ and $L(\pi) = m_1(q)/\mu_1'$, respectively, where $q = F^{-1}(\pi) = Q(\pi)$ comes from the qf of X for a given probability π.

4 Estimation

Several approaches for parameter estimation were proposed in the literature but the maximum likelihood method is the most commonly employed. The maximum likelihood estimators (MLEs) enjoy desirable properties and can be used when constructing confidence intervals for the model parameters. The normal approximation for these estimators in large samples can be easily handled either analytically or numerically. So, we consider the estimation of the unknown parameters for this family from complete samples only by maximum likelihood. Here, we determine the MLEs of the parameters of the new family of distributions from complete samples only.

Let $x_1, \ldots, x_n$ be the observed values from the EMOG distribution with parameters p, α, λ and $\boldsymbol{\xi}$. Let $\theta = (p, \alpha, \lambda, \boldsymbol{\xi})^\top$ be the $r \times 1$ parameter vector. The total log-likelihood function for θ is given by

$$\begin{aligned}\ell_n = \ell_n(\Theta) &= n \log\alpha + n \log\lambda + n \log(1-p) + \sum_{i=1}^{n} \log\left[g(x_i;\boldsymbol{\xi})\right] \\ &+(\lambda-1)\sum_{i=1}^{n}\log\left[1-G(x_i;\boldsymbol{\xi})\right] + (\alpha-1)\sum_{i=1}^{n}\log\left\{1-[1-G(x_i;\boldsymbol{\xi})]^{\lambda}\right\} \\ &-(\alpha+1)\sum_{i=1}^{n}\log\left\{1-p[1-G(x_i;\boldsymbol{\xi})]^{\lambda}\right\}. \end{aligned} \tag{31}$$

The maximized log-likelihood can be either directly by using the NLMIXED procedure in SAS or the sub-routine MaxBFGS in the Ox program (see Doornik 2009) or by solving the nonlinear likelihood equations obtained by differentiating (31). The components of the score function

$$U_n(\theta) = (\partial\ell_n/\partial p, \partial\ell_n/\partial\alpha, \partial\ell_n/\partial\lambda, \partial\ell_n/\partial\boldsymbol{\xi})^\top$$

are given by

$$\begin{aligned}\frac{\partial\ell_n}{\partial p} &= (\alpha+1)\sum_{i=1}^{n}\frac{[1-G(x_i;\boldsymbol{\xi})]^{\lambda}}{1-p[1-G(x_i;\boldsymbol{\xi})]^{\lambda}} - \frac{n}{1-p}, \\ \frac{\partial\ell_n}{\partial\alpha} &= \frac{n}{\alpha} + \sum_{i=1}^{n}\log\left\{1-[1-G(x_i;\boldsymbol{\xi})]^{\lambda}\right\} - \sum_{i=1}^{n}\log\left\{1-p[1-G(x_i;\boldsymbol{\xi})]^{\lambda}\right\}, \\ \frac{\partial\ell_n}{\partial\lambda} &= \frac{n}{\lambda} + \sum_{i=1}^{n}\log[1-G(x_i;\boldsymbol{\xi})] - (\alpha-1)\sum_{i=1}^{n}\frac{[1-G(x_i;\boldsymbol{\xi})]^{\lambda}\log[1-G(x_i;\boldsymbol{\xi})]}{1-[1-G(x_i;\boldsymbol{\xi})]^{\lambda}} \\ &\quad + p(\alpha+1)\sum_{i=1}^{n}\frac{[1-G(x_i;\boldsymbol{\xi})]^{\lambda}\log[1-G(x_i;\boldsymbol{\xi})]}{1-p[1-G(x_i;\boldsymbol{\xi})]^{\lambda}}\end{aligned}$$

and

$$\begin{aligned}\frac{\partial\ell_n}{\partial\boldsymbol{\xi}} &= \sum_{i=1}^{n}\frac{g^{(\xi)}(x_i;\boldsymbol{\xi})}{g(x_i;\boldsymbol{\xi})} - (\lambda-1)\sum_{i=1}^{n}\frac{G^{(\xi)}(x_i;\boldsymbol{\xi})}{1-G(x;\boldsymbol{\xi})} \\ &\quad + \lambda(\alpha-1)\sum_{i=1}^{n}\frac{g^{(\xi)}(x;\boldsymbol{\xi})[1-G(x_i;\boldsymbol{\xi})]^{\lambda-1}}{1-[1-G(x_i;\boldsymbol{\xi})]^{\lambda}} - p\lambda(\alpha+1) \\ &\quad \times\sum_{i=1}^{n}\frac{g^{(\xi)}(x_i;\boldsymbol{\xi})[1-G(x_i;\boldsymbol{\xi})]^{\lambda-1}}{1-p[1-G(x_i;\boldsymbol{\xi})]^{\lambda}},\end{aligned}$$

where $h^{(\xi)}(\cdot)$ means the derivative of the function h with respect to ξ. For interval estimation of the model parameters, we can derive the observed information matrix $J_n(\theta)$, whose elements can be obtained from the authors upon request. Let $\widehat{\theta}$ be the MLE of θ. Under standard regularity conditions (Cox and Hinkley 1974), we can approximate the distribution of $\sqrt{n}(\widehat{\theta} - \theta)$ by the multivariate normal $N_r(0, K(\theta)^{-1})$, where $K(\theta) = \lim_{n\to\infty} n^{-1} J_n(\theta)$ is the unit information matrix and r is the number of parameters of the new distribution.

Often with lifetime data and reliability studies, one encounters censoring. In a very realistic random censoring mechanism, each individual i is assumed to have a lifetime X_i and a censoring time C_i, where X_i and C_i are independent random variables. Suppose that the data consist of n independent observations $x_i = \min(X_i, C_i)$ and $\delta_i = I(X_i \leq C_i)$ is such that $\delta_i = 1$ if X_i is a time to event and $\delta_i = 0$ if it is right censored for $i = 1, \ldots, n$. The censored likelihood $L(\theta)$ for the model parameters is

$$L(\theta) \propto \prod_{i=1}^{n} [f(x_i; p, \alpha, \lambda, \boldsymbol{\xi})]^{\delta_i} \, [S(x_i; p, \alpha, \lambda, \boldsymbol{\xi})]^{1-\delta_i},$$

where $f(x; p, \alpha, \lambda, \boldsymbol{\xi})$ is given by (3) and $S(x; p, \alpha, \lambda, \boldsymbol{\xi})$ is the survival function evaluated from (2).

5 Applications

In this section, we use a real data set, collected by Prater (1956) and analyzed by Atkinson (1985), on the stress among women in Townsville, Queensland, Australia and the proportion of crude oil converted to gasoline after distillation and fractionation. This application aims to illustrate the potentiality of the EMO family. All the computations were done using the R software. For this application, we consider the following distributions: EMOB $(a, b, p, \lambda, \alpha)$ and Kw-WP$(a, b, c, \lambda, \beta)$ (Kwmaraswamy Weibull Poisson distribution) proposed by Ramos et al. (2015), both with five parameters. The density of the Kw-WP is given by

$$f(x) = \frac{\lambda abc\beta^c x^{c-1}[1 - e^{-(\beta x)^c}]^{a-1} \exp[\lambda\{1-[1 - e^{-(\beta x)^c}]^a\}^b - (\beta x)^c]}{(e^{\lambda} - 1)\{1-[1 - e^{-(\beta x)^c}]^a\}^{1-b}}.$$

A descriptive analysis of the data is presented in Table 3.

It is possible to obtain qualitative information about the hrf by means of plot analysis when we have the data censored or uncensored. We emphasize that the data sets here are uncensored. For this type of data, the total time in test (TTT) plot proposed

Table 3 Descriptive statistics

Statistics	Real data sets	
	Stress	Gasoline
Mean	0.2642	0.1966
Median	0.2500	0.1780
Mode	0.2500	0.1500
Variance	0.0376	0.0115
Skewness	0.9712	0.3867
Kurtosis	0.8272	−0.6561
Maximum	0.8500	0.4570
Minimum	0.0100	0.0280
n	166	32

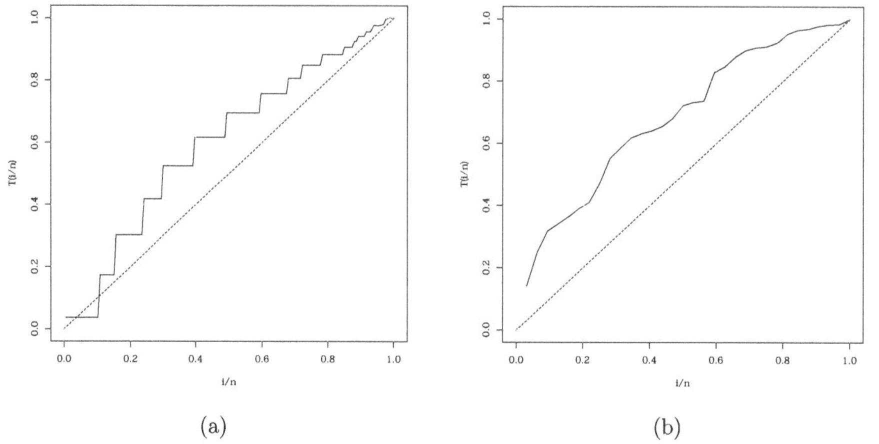

Fig. 8 The TTT plot for: (a) stress among women in Townsville, Queensland, Australia (b) proportion of crude oil converted to gasoline after distillation and fractionation

by Aarset (1987) may be used. Let T be a random variable with non-negative values that represents the survival time. The TTT curve is constructed by plotting the statistic $G(r/n) = \left[\sum_{i=1}^{r} T_{i:n} + (n+r)T_{r:n}\right] / \left(\sum_{i=1}^{n} T_{i:n}\right)$ versus r/n ($r = 1, \ldots, n$), where the values $T_{i:n}$ are the order statistics of the sample, for $i = 1, \ldots, n$. The plots can be easily obtained using the `TTT` function of the `AdequacyModel` package from the `R` software. More details about this package are available from `help(TTT)`. The TTT plots for the dataset in this application are shown in Fig. 8. For both plots, the TTT curve is concave, which, according to Aarset (1987), provides evidence that a monotonic increasing hrf is adequate.

The Fig. 9 displays the fitted densities to the current data obtained in a nonparametric manner using the gaussian kernel density estimation, defined as follows. Let $X_1, \ldots, X_n$ be a random vector of random variables independent and identically distributed where each variable follows an unknown distribution, denoted by f. The kernel density estimator is given by the following expression

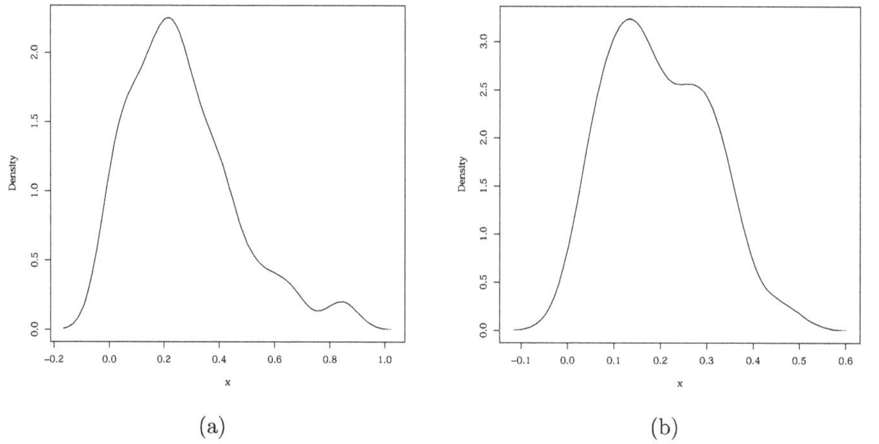

Fig. 9 Gaussian kernel density estimation for: (a) stress among women in Townsville, Queensland, Australia (b) proportion of crude oil converted to gasoline after distillation and fractionation

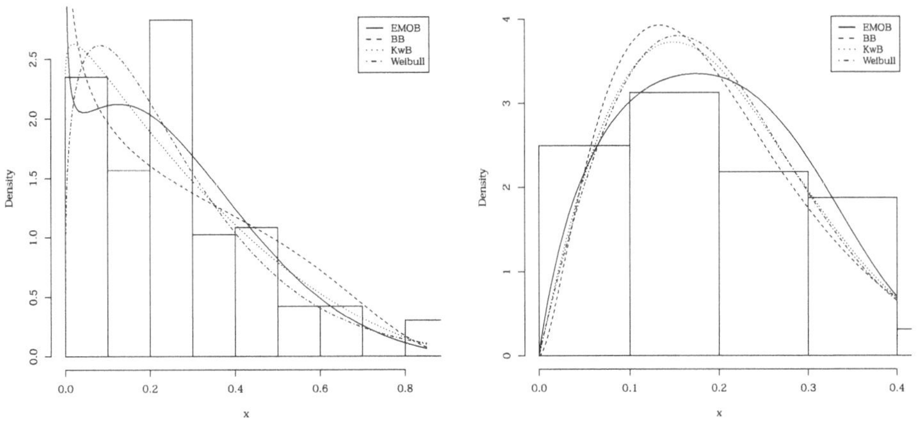

(a) (b)

Fig. 10 Estimates of the pdf for: (**a**) stress among women in Townsville, Queensland, Australia (**b**) proportion of crude oil converted to gasoline after distillation and fractionation

$$\hat{f}_h(x) = \frac{1}{n}\sum_{i=1}^{n} K_h(x - x_i) = \frac{1}{nh}\sum_{i=1}^{n} K\left(\frac{x - x_i}{h}\right), \tag{32}$$

where $K(\cdot)$ is the symmetrical kernel function and $\int_{-\infty}^{\infty} K(x)dx = 1$. Furthermore, $h > 0$ is known in literature as bandwith, which is a smoothing parameter. It is possible to find in literature numerous kernel function, as the normal standard distribution, for example. Silverman (1986) demonstrated that for the K standard normal, a reasonable bandwith is given by $h = \sqrt[5]{(4\hat{\sigma}^5/3n)} \approx 1.06\hat{\sigma}/\sqrt[5]{n}$, where $\hat{\sigma}$ is defined by the standard deviation of the sample.

The plots displayed in Fig. 10 indicates that the EMOB distribution provides the best fit compared with the other fitted distributions. We note the good adequacy of the fitted EMOB distribution in Fig. 11.

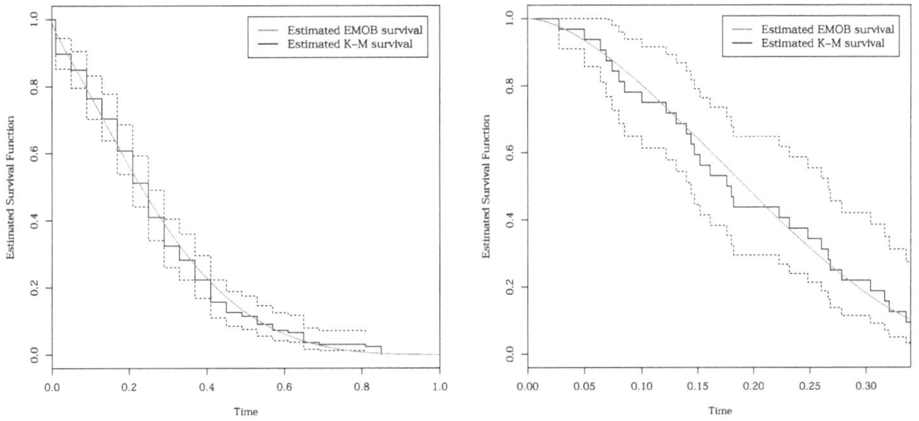

(a) (b)

Fig. 11 Estimated K-M survival (K-M estimates) compared with the EMOB survival estimates for: (**a**) tress among women in Townsville, Queensland, Australia (**b**) proportion of crude oil converted to gasoline after distillation and fractionation. The confidence intervals are 95 %

In this application, we use the package `AdequacyModel`. This package is intended to provide a computational support to work with probability distributions, mainly distributions aimed to survival analysis. This package was used to calculate some fitness statistics adjustment such as AIC (Akaike Information Criterion), CAIC (Consistent Akaikes Information Criterion), BIC (Bayesian Information Criterion), HQIC (Hannan-Quinn information criterion), KS (Test of Kolmogorov-Smirnov), A^* (statistic of Anderson-Darling) and W^* (statistic of Cramér-von Mises), which are described by Chen and Balakrishnan (1995), based on the results presented by Stephens (1986). When we want to test if one random sample, denoted by $x_1, x_2, \ldots, x_n$, with empirical distribuction function $F_n(x)$ comes from a specific distribution, we use these statistics. The Cramér-von Mises (W^*) and Anderson-Darling (A^*) statistics are given by the following expressions:

$$W^* = \left\{ n \int_{-\infty}^{+\infty} \{F_n(x) - F(x;\hat{\theta}_n)\}^2 \, dF(x;\hat{\theta}_n) \right\} \left(1 + \frac{0.5}{n}\right) \quad \text{and}$$

$$A^* = \left\{ n \int_{-\infty}^{+\infty} \frac{\{F_n(x) - F(x;\hat{\theta}_n)\}^2}{F(x;\hat{\theta}_n)(1 - F(x;\hat{\theta}_n))} dF(x;\hat{\theta}_n) \right\} \left(1 + \frac{0.75}{n} + \frac{2.25}{n^2}\right).$$

respectively. In these expressions, we have that $F_n(x)$ is the empirical distribution function, $F(x;\hat{\theta}_n)$ is the postulated distribution function evaluated at the MLE of θ, i.e. $\hat{\theta}_n$. Lower values of W^* and A^* provide evidence that $F(x;\hat{\theta}_n)$ generates the sample. More details about these statistics are given by Chen and Balakrishnan (1995). The `goodness.fit` function is used to calculate these statistics. More details can be obtained using the command `help(goodness.fit)`. The Table 4 shows the goodness-of-fit statistics (rounding to the fourth decimal place) for the dataset used in this application. The results showed that the EMOB $(a, b, p, \lambda, \alpha)$ distribution presented better results for the KS, A^* and W^* statistics when compared with the other distributions used in this application. In this study, the MLEs in Table 5 were obtained by global search heuristic method called Particle Swarm optmization - PSO proposed by Eberhart and Kennedy (1995). One of the advantages of using the PSO method in addition to being a robust optimization method is that there is no need to provide initial guesses. However, this is a computationally intensive method. The Appendix A shows the function `pso` implemented in `R`. At the end of the code there is a small example of how to use the function to minimize an objective function. The standard errors of the MLEs can be obtained by the bootstrap method. The standard errors were not obtained in these examples due to the use of the PSO method, which is computationally intensive.

Table 4 Goodness-of-fit statistics for the data: (I) stress among women in Townsville, Queensland, Australia (II) proportion of crude oil converted to gasoline after distillation and fractionation

Data set	Distribution	A*	W*
I	EMOB $(p, \alpha, \lambda, a, b)$	0.1554	0.0786
	Kw-WP$(a, b, c, \lambda, \beta)$	0.4974	0.1305
II	EMOB $(p, \alpha, \lambda, a, b)$	0.0348	0.0983
	Kw-WP$(a, b, c, \lambda, \beta)$	0.0489	0.0993

Table 5 MLEs for: (I) stress among women in Townsville, Queensland, Australia (II) proportion of crude oil converted to gasoline after distillation and fractionation

Data set	Distribution	Maximum Likelihood Estimates - MLE				
I	EMOB $(p, \alpha, \lambda, a, b)$	−19.8703	1.4276	3.4036	0.2454	0.7793
	Kw-WP$(a, b, c, \lambda, \beta)$	12.3010	20.1431	0.1647	24.6569	2.3195
II	EMOB $(p, \alpha, \lambda, a, b)$	−1.2077	0.3989	1.9915	4.6421	10.7513
	Kw-WP$(a, b, c, \lambda, \beta)$	15.2365	8.3171	0.2446	24.9571	10.5076

6 Conclusions

We derive general mathematical properties of a new continuous distributions with three extra shape parameters. We present some special models of the new EMO family of distributions. We investigate the shapes and derive explicit expressions for the ordinary and incomplete moments, quantile and generating functions and probability weighted moments, which hold for any baseline model. The estimation of the model parameters is by the method of maximum likelihood and the estimates were obtained by global search heuristic method called Particle Swarm Optmization-PSO. Ultimately, we fit some EMO-G distributions to two real data sets to demonstrate the potentiality of this family. We hope this generalization may attract wider applications in statistics.

Appendix

Code in R language for PSO method

```
pso <- function(func,S=150,lim_inf,lim_sup,e=0.0001,data=NULL,N=100){
b_lo = min(lim_inf)
b_up = max(lim_sup)
integer_max = .Machine$integer.max

if(length(lim_sup)!=length(lim_inf)){
stop("The␣vectors␣lim_inf␣and␣lim_sup␣must␣have␣the␣same␣dimension.")
}
dimension = length(lim_sup)
swarm_xi = swarm_pi = swarm_vi = matrix(NA,nrow=S,ncol=dimension)

# The best position of the particles.
g = runif(n=dimension,min=lim_inf,max=lim_sup)

# Objective function calculated in g.
f_g = func(par=as.vector(g),x=as.vector(data))

if(NaN%in%f_g==TRUE || Inf%in%abs(f_g)==TRUE{
while(NaN%in%f_g==TRUE || Inf%in%abs(f_g)==TRUE{
g = runif(n=dimension,min=lim_inf,max=lim_sup)
f_g = func(par=g,x=as.vector(data))
}
}

# Here begins initialization of the algorithm.
x_i = mapply(runif,n=S,min=lim_inf,max=lim_sup)

# Initializing the best position of particularities i to initial position.

swarm_pi = swarm_xi = x_i
f_pi = apply(X=x_i,MARGIN=1,FUN=func,x=as.vector(data))

is.integer0 <- function(x){
is.integer(x) && length(x)==0L
}

if(NaN%in%f_pi==TRUE || Inf%in%abs(f_pi)){
while(NaN%in%f_pi==TRUE || Inf%in%abs(f_pi)){
id_inf_fpi = which(abs(f_pi)==Inf)
if(is.integer0(id_inf_fpi)!=TRUE){
f_pi[id_inf_fpi] = integer_max
}
id_nan_fpi = which(f_pi==NaN)
if(is.integer0(id_nan_fpi)!=TRUE){
```

```
x_i[id_nan_fpi,] = mapply(runif,n=length(id_nan_fpi),min=lim_inf,
max=lim_sup)
swarm_pi = swarm_xi = x_i
f_pi = apply(X=x_i,MARGIN=1,FUN=func,x=as.vector(data))
}
}
}

minimo_fpi = min(f_pi)
if(minimo_fpi < f_g) g = x_i[which.min(f_pi),]

# Initializing the speeds of the particles.

swarm_vi = mapply(runif,n=S,min=-abs(rep(abs(b_up-b_lo),dimension)),
max=abs(rep(abs(b_up-b_lo),dimension)))

# Here ends the initialization of the algorithm

omega = 0.5
phi_p = 0.5
phi_g = 0.5

m=1
vector_f_g <- vector()

while(is.na(var(vector_f_g)) || m<50 ||
var(vector_f_g[length(vector_f_g):(length(vector_f_g)-10)])>e){
# r_p and r_g are randomized numbers in (0.1).
r_p = runif(n=dimension,min=0,max=1)
r_g = runif(n=dimension,min=0,max=1)

# Updating the vector speed.
swarm_vi = omega*swarm_vi+phi_p*r_p*(swarm_pi-swarm_xi)+
phi_g*r_g*(g-swarm_xi)

# Updating the position of each particle.
swarm_xi = swarm_xi+swarm_vi

myoptim = function(...) tryCatch(optim(...), error = function(e) NA)

f_xi = apply(X=swarm_xi,MARGIN=1,FUN=func,x=as.vector(data))
f_pi = apply(X=swarm_pi,MARGIN=1,FUN=func,x=as.vector(data))
f_g = func(par=g,x=as.vector(data))

if(NaN%in%f_xi==TRUE || NaN%in%f_pi==TRUE{
while(NaN%in%f_xi==TRUE){
id_comb = c(which(is.na(f_xi)==TRUE),which(is.na(f_pi)==TRUE))
if(is.integer0(id_comb)!=TRUE){
new_xi = mapply(runif,n=length(id_comb),min=lim_inf,
max=lim_sup)
swarm_pi[id_comb,]=swarm_xi[id_comb,] = new_xi
if(length(id_comb)>1){
if_xi[id_comb] = apply(X=swarm_xi[id_comb,],MARGIN=1,
FUN=func,x=as.vector(data))
f_pi[id_comb] = apply(X=swarm_pi[id_comb,],MARGIN=1,FUN=func,
x=as.vector(data))
}else{
f_xi[id_comb] = func(par=new_xi,x=as.vector(data))
}
}
}
}

if(Inf%in%abs(f_xi)==TRUE{
f_xi[which(is.infinite(f_xi))]=integer_max
}
if(InfInf%in%abs(f_pi)==TRUE{
f_pi[which(is.infinite(f_pi))]=integer_max
}

# There are values below the lower limit of restrictions?
id_test_inf=
which(apply(swarm_xi<t(matrix(rep(lim_inf,S),dimension,S)),1,sum)>=1)
id_test_sup=
which(apply(swarm_xi>t(matrix(rep(lim_sup,S),dimension,S)),1,sum)>=1)

if(is.integer0(id_test_inf)!=TRUE){
swarm_pi[id_test_inf,] = swarm_xi[id_test_inf,] =
mapply(runif,n=length(id_test_inf),
min=lim_inf,max=lim_sup)
```

```
}

if(is.integer0(id_test_sup)!=TRUE){
swarm_pi[id_test_sup,] = swarm_xi[id_test_sup,] =
mapply(runif,n=length(id_test_sup),
min=lim_inf,max=lim_sup)
}

if(is.integer0(which((f_xi<=f_pi)==TRUE))){
swarm_pi[which((f_xi<=f_pi)),] = swarm_pi[which((f_xi<=f_pi)),]
}

if(f_xi[which.min(f_xi)] <= f_pi[which.min(f_pi)]){
swarm_pi[which.min(f_pi),] = swarm_xi[which.min(f_xi),]
if(f_pi[which.min(f_pi)] < f_g) g = swarm_pi[which.min(f_pi),]
} # Here ends the block if.

vector_f_g[m] = f_g
m = m+1
if(m>N){
break
}

} # Here ends the block while.

f_x = apply(X=swarm_xi,MARGIN=1,FUN=func,x=as.vector(data))
list(par_pso=g,f_pso=vector_f_g)

} # Here ends the function.

# Example of using the PSO function. We are looking to minimize easom
# function in that -10<=x1<=10 and -10<=x2<=10.
easom <- function(par,x){
x1 = par[1]
x2 = par[2]
-cos(x1)*cos(x2)*exp(-((x1-pi)^2 + (x2-pi)^2))
}
set.seed(0)
# Using the PSO function
# S refers to the number of particles considered.
pso(func=easom,S=350,lim_inf=c(-10,-10),lim_sup=c(10,10))
```

Authors' contributions

The authors, CRBD, GMC, MA, PRDM and HFCC with the consultation of each other carried out this work and drafted the manuscript together. All authors read and approved the final manuscript.

Competing interests

The authors declare that they have no competing interests.

Author details

[1]Department of Statistics, Federal University of Pernambuco, Recife, Brazil. [2]Department of Statistics, Federal Univeristy of Paraíba, João Pessoa, Brazil. [3]Department of Statistics, Persian Gulf University, Bushehr, Iran.

References

Aarset, MV: How to identify a bathtub hazard rate. IEEE Trans. Reliability. **36**, 106–8 (1987)
Adamidis, K, Loukas, S: A lifetime distribution with decreasing failure rate. Stat. Probab. Latt. **39**, 35–42 (1998)
Alzaatreh, A, Lee, C, Famoye, F: A new method for generating families of continuous distributions. METRON. **81**, 63–79 (2013)
Atkinson, AC: An introduction to graphical methods of diagnostic regression analysis. Clarendon Press, Oxford (1985)
Barreto-Souza, W, Morais, AL, Cordeiro, GM: The Weibull-geometric distribution. J. Stat. Comput. Simulation. **81**, 645–57 (2011)
Cordeiro, GM, de Castro, M: A new family of generalized distributions. J. Stat. Comput. Simulat. **81**, 883–98 (2011)
Cordeiro, GM, Ortega, EMM, Cunha, DCC: The exponentiated generalized class of distributions. J. Data Sci. **11**, 1–27 (2013)
Chen, G, Balakrishnan, N: A general purpose approximate goodness-of-fit test. J. Qual. Technol. **27**, 154–61 (1995)
Cox, DR, Hinkley, DV: Theoretical Statistics. Chapman and Hall, London (1974)
Doornik, JA: An Object-Oriented Matrix Language Ox 6. Timberlake Consultants Press, London (2009)
Eberhart, RC, Kennedy, J: A new optimizer using particle swarm theory. In: Proceedings of the sixth international symposium on micro machine and human science, vol. 1, pp. 39–43, New York, (1995)
Eugene, N, Lee, C, Famoye, F: Beta-normal distribution and its applications. Commun. Stat. Theory Methods. **31**, 497–512 (2002)
Gradshteyn, IS, Ryzhik, IM: Table of integrals, series, and products. Academic Press, San Diego (2000)
Gupta, RC, Gupta, RD: Proportional reversed hazard rate model and its applications. J. Stat. Planning Inference. **137**, 3525–36 (2007)

Gupta, RD, Kundu, D: Exponentiated Exponential Family: An Alternative to Gamma and Weibull Distributions. Biometrical J. **43**, 117–30 (2001)

Jones, MC: Families of distributions arising from distributions of order statistics. Test. **13**, 1–43 (2004)

Kenney, JF, Keeping, ES: Mathematics of statistics, pp. 101–102, Part 1. 3rd ed, Princeton (1962)

Marshall, AW, Olkin, I: A new method for adding a parameter to a family of distributions with applications to the exponential and Weibull families. Biometrika. **84**, 641–52 (1997)

Mahmoudi, E, Shiran, M: Exponentiated weibull-geometric distribution and its applications (2012)..arXiv:1206.4008vl [stat.ME]

Moors, JJA: A quantile alternative for kurtosis. J. R. Stat. Soc. Ser. D. Stat. **37**, 25–32 (1998)

Mudholkar, GS, Hutson, AD: The exponentiated Weibull family: some properties and a flood data application. Commun. Stat. Theory Methods. **25**, 3059–83 (1996)

Nadarajah, S, Kotz, S: The Exponentiated Type Distributions. Acta Appl. Math. **92**, 97–111 (2006)

Prater, NH: Estimate gasoline yields from crudes. Petroleum Refiner. **35**, 236–8 (1956)

Ramos, MWA, Marinho, PRD, Cordeiro, GM, Silva, RV, Hamedani, G: The kumaraswamy-G Poisson family of distributions. J. Stat. Theory Appl. **14**, 222–39 (2015)

Silva, RB, Barreto-Souza, W, Cordeiro, GM: A new distribution with decreasing, increasing and upside-down bathtub failure rate. Comput. Stat. Data Anal. **54**, 935–44 (2010)

Silverman, BW: Density estimation for statistics and data analysis, Vol. 26. CRC press (1986)

Stephens, MA: Tests based on EDF statistics. Goodness-of-fit Tech. **68**, 97–193 (1986)

Zografos, K, Balakrishnan, N: On families of beta and generalized gamma-generated distributions and associated inference. Stat. Methodol. **6**, 344–362 (2009)

7

Nonparametric rank based estimation of bivariate densities given censored data conditional on marginal probabilities

Alan D. Hutson

Correspondence:
alan.hutson@roswellpark.org
Roswell Park Cancer Institute, Department of Biostatistics and Bioinformatics, Elm and Carlton Streets, Buffalo, NY 14263, USA

Abstract

In this note we develop a new Kaplan-Meier product-limit type estimator for the bivariate survival function given right censored data in one or both dimensions. Our derivation is based on extending the constrained maximum likelihood density based approach that is utilized in the univariate setting as an alternative strategy to the approach originally developed by Kaplan and Meier (1958). The key feature of our bivariate survival function is that the marginal survival functions correspond exactly to the Kaplan-Meier product limit estimators. This provides a level of consistency between the joint bivariate estimator and the marginal quantities as compared to other approaches. The approach we outline in this note may be extended to higher dimensions and different censoring mechanisms using the same techniques.

Keywords: Product-limit estimator, Bivariate survival function, Maximum likelihood, Linear programming

Mathematics Subject Classification (MSC): 62N01; 62G07

1 Introduction

In this note we develop a new Kaplan-Meier product-limit type estimator for the bivariate survival function given right censored data in one or both dimensions. Our derivation is based on extending the constrained maximum likelihood density based approach (Satten and Datta 2001; Zhou 2005) that is utilized in the univariate setting as an alternative strategy to the classical discrete nonparametric hazard function approach (Kaplan and Meier 1958). There are several methods for estimating a bivariate survival function across different censoring patterns that have been proposed in the literature based on extending the univariate hazard function approach or creating various decompositions (Akritas and Van Keilegom 2003; Gill et al. 1995; Lin and Ying 1993; Prentice et al. 2004; Wang and Wells 1997). In general, they are somewhat complex to compute and may have deficiencies such as negative mass estimates at given points (Prentice et al. 2004). The large sample theory involving these estimators is quite technical (Gill et al. 1995). To the best of our knowledge one of the key limitations of all of the completely nonparametric bivariate survival function estimators developed to-date is that they yield marginal estimators that may not be equivalent to the product-limit estimator corresponding to each dimension. Our estimator, framed as a sparse multinomial estimation problem given simplex constraints, remedies this issue. In addition, in terms of future work our method may be extended to higher dimensions and other censoring mechanisms (left and interval) using the techniques outlined in this note. We also can consider support over the entire real line.

In terms of background, we start by outlining the nonparametric maximum likelihood based density estimator in the univariate setting given right censored data. We can utilize this estimator to define a survival function estimator, which is equivalent to the product-limit estimator. We then move to the bivariate setting in the next section as a direct extension of this approach. As noted above there are two approaches towards arriving at the Kaplan-Meier product-limit estimator (Kaplan and Meier 1958). The well-known nonparametric textbook approach focuses on utilizing the discrete hazard function to define the parameters of interest. Towards this end let $X_1, X_2, \cdots, X_n$ denote i.i.d. failure times and let $C_1, C_2, \cdots, C_n$ denote the corresponding i.i.d. non-informative right censoring times, $i = 1, 2, \cdots, n$. Given right censoring we only observe $g \leq n$ of the X's. Now let $0 < x_{(1)} < x_{(2)} < \cdots x_{(g)}$ be the distinct ordered observed failure times. The classic maximum likelihood based derivation of the product-limit estimator starts by assuming the underlying distribution is discrete with probabilities $\pi_j = P(X = x_{(j)}), j = 1, 2, \cdots, g$ for $g \leq n$. Given a discrete hazard of $h_j = P(X = x_{(j)}|X \geq x_{(j)})$ for $0 \leq h_j \leq 1$ we have that $\pi_1 = h_1, \pi_2 = (1 - h_1)h_2, \cdots, \pi_g = (1 - h_1)(1 - h_2) \cdots (1 - h_{g-1})h_g$. Then the estimator of $S(x) = P(X > x) = \prod_{x_{(j)} \leq x} (1 - h_j)$ in the discrete case is given as

$$\hat{S}(x) = \prod_{x_{(j)} \leq x} \left(1 - \hat{h}_j\right). \tag{1}$$

The estimates of the discrete hazard parameters are obtained from maximizing the log-likelihood

$$\log L = \sum_{j=1}^{g} d_j \log h_j + \left(r_j - d_j\right) \log(1 - h_j), \tag{2}$$

where d_j denotes the number of events and r_j denotes the number at risk at time $x_{(j)}, j = 1, 2, \cdots, g$, e.g. See Cox and Oakes (1984) for details of this derivation. The maximization of (2) with respect to the parameters $h_j, j = 1, 2, \cdots, g$, yields the oft utilized estimates $\hat{h}_j = d_j/r_j$, where r_j denotes the number of subjects at risk at time $x_{(j)}$ and d_j denotes the number of subjects who fail at time $x_{(j)}$. For a technical treatment with respect to the behavior of the product-limit estimator and how it translates to continuous case see Chen and Lo (1997). It is well-known, but not immediately obvious, that the product-limit estimator reduces to the classic empirical estimator of the survival function $\hat{S}(x) = 1 - \sum_{i=1}^{n} I(x_{(i)} \leq x)/n$ when there are no censored observations, where $I(\cdot)$ denotes the indicator function. This is the starting point for most of the bivariate survival estimators found in the literature (Gill et al. 1995).

As an alternative approach to the Kaplan-Meier construction we start by estimating the density function first in a nonparametric fashion from which the survival function and distribution functions are then readily estimated. This approach mirrors the classic parametric maximum likelihood estimation given right censored data in terms of the likelihood containing a density function component and a survival function component whose relative contributions depends upon whether or not an observation is censored. In this framework denote the observed values as $T_i = \min(X_i, C_i)$, $i = 1, 2, \cdots, n$. Note that in our alternative derivation we allow the more general assumption that both X and C may have support over the entire real line as compared to the more common restriction in survival modeling that X and C have support only on the positive real line. Furthermore, denote the censoring indicator variable as $\delta_i = I_{(X_i \leq C_i)}$, denote the ordered observed T_i's as $t_{(1)} < t_{(2)} < \cdots < t_{(n)}$ and define the parameters of interest as $\pi_i = P(X \leq t_{(i)}|\delta_i = 1) - P(X < t_{(i)}|\delta_i = 1)$, $i = 1, 2, \cdots, n$. The parameter definition is the justification and linkage for this estimator towards underlying continuous data (Owen 1988). Note that similar to the traditional product-limit estimator $\pi_i = 0$ if $\delta_i = 0$ by definition. Now, given right-censoring we only observe $j \leq n$ of the X's, where $j = \sum_{i=1}^{n} \delta_i$.

Maximum likelihood estimation is now carried forth under the constraint that $\sum_{i=1}^{n} \pi_i = 1$ similar to classic maximum likelihood based empirical density estimation. The classic product-limit estimator is derived through a straightforward extension of the uncensored case and starts with the same assumption that the X_i's are functionally discrete, i.e. the true distributions of interest are continuous and we are discretizing the time scale with respect to are definition of the π_i's. In the case of observed ties in the data under the assumption of a truly continuous underlying distribution we can arbitrarily rank order those respective observations and combine the point masses corresponding to the respective given ties. In our alternative formulation the likelihood accounting for right-censoring now takes a form similar to the parametric setting and is given by

$$L = \prod_{i=1}^{n-1} \pi_i^{\delta_i} \left(1 - \sum_{j=1}^{i} \pi_j\right)^{1-\delta_i} \times \left(1 - \sum_{j=1}^{n-1} \pi_j\right) \tag{3}$$

The last term in the likelihood corresponds to the constraint that $\sum_{i=1}^{n} \pi_i = 1$. If the last observation is censored this implies as in the traditional approach that we still may have an improper distribution of the survival function. Hence, by definition we set this to be an uncensored observation per asymptotic consistency arguments (Chen and Lo 1997). Obviously the likelihood at (3) reduces to the likelihood for the classical empirical estimator given no censoring with $\hat{\pi}_i = 1/n$.

The form of the likelihood at (3) has been presented in other contexts such as empirical likelihood constrained maximum likelihood estimation and inference, e.g. see Zhou (2005) and the references there within. The constraint that $\sum_{i=1}^{n} \pi_i = 1$ yields $n - 1$ score equations given by $s_j = \partial \log L / \partial \pi_j$. Solving the system of equations that $s_j = 0$ for $j = 1, 2, \cdots, n-1$ yields the following nonparametric maximum likelihood estimates for the π_i's given as

$$\hat{\pi}_i = \begin{cases} \frac{\delta_1}{n}, & i = 1, \\ \frac{\prod_{j=1}^{i-1}((n-j+1)-\delta_j)\delta_i}{\prod_{j=1}^{i}(n+j-i)}, & i > 1, \end{cases} \tag{4}$$

where $\hat{\pi}_n = 1 - \sum_{j=1}^{n-1} \hat{\pi}_j$, see Satten and Datta (2001).

In addition, it follows straightforward that similar to the standard empirical distribution function estimator we have for right-censored data that

$$\hat{F}(x) = \sum_{i=1}^{n} \hat{\pi}_i I_{(t_{(i)} \leq x)} \tag{5}$$

and

$$\hat{S}(x) = 1 - \hat{F}(x) = 1 - \sum_{i=1}^{n} \hat{\pi}_i I_{(t_{(i)} \leq x)}, \tag{6}$$

where $\hat{\pi}_i$'s are given at (4) and $\hat{\pi}_n = 1 - \sum_{j=1}^{n-1} \hat{\pi}_j$. The estimator of the survival function given at (6) is equivalent to the product-limit form at (1) in terms of the actual estimated survival probabilities. This is the jumping off point of our new bivariate survival function estimator.

In Section 2 we outline the new constrained maximum likelihood procedure used to develop a bivariate estimator of the joint density $f(x, y)$ from which estimators of the bivariate distribution function $F(x, y)$ and bivariate survival function are readily calculated. In Section 3 we provide some illustrative toy examples followed by two real data examples. We finish with some basic conclusions.

2 Constrained maximum likelihood estimation

In this section we describe the process for estimating $f(x, y)$ nonparametrically conditional on marginal constraints. This in turn will lead to an estimator for the bivariate distribution function $F(x, y)$ and survival function $S(x, y)$. Towards this end let (X_i, Y_i), $i = 1, 2, \cdots, n$, be independent and identically distributed pairs of bivariate failure times with joint probability density function $f(x, y)$ and corresponding cumulative distribution function $F(x, y)$ with $S(x, y) = 1 - F(x, y)$. Furthermore, let (C_i^x, C_i^y), $i = 1, 2, \cdots, n$, be independent and identically bivariate distributed pairs of censoring variables. Under right censoring in each dimension we observe

$$(S_i, T_i) = \left((X_i \wedge C_i^x), (Y_i \wedge C_i^y)\right) \text{ and } (\delta_i^x, \delta_i^y) = \left(I\left(X_i < C_i^x\right), I\left(Y_i < C_i^y\right)\right),$$
$$i = 1, 2, \cdots, n,$$

where $\wedge$ denotes the minimum between pairs of random variables and $I(\cdot)$ denotes the indicator function. For this note we assume the (X_i, Y_i)'s and (C_i^x, C_i^y)'s are absolutely continuous and are also pairwise independent from each other. In the nonparametric setting the distribution and survival functions are defined as follows:

$$F(x, y) = \sum_{i=1}^{n} \sum_{j=1}^{n} \pi_{i,j} I_{(s_{(i)} \le x, t_{(j)} \le y)} \tag{7}$$

$$S(x, y) = 1 - F(x, y), \tag{8}$$

where the parameters, i.e. the $\pi_{i,j}$'s, $i = 1, 2, \cdots, n, j = 1, 2, \cdots, n$, are in essence weights between 0 and 1 and are defined in detail below at (17).

Now similar to the univariate case, outlined in the introduction, denote the parameters corresponding to the nonparametric estimators of the marginal densities f_x and f_y as

$$\pi_{r_{s_i},\cdot} = \Delta F_s\left(S_{r_{s_i}}\right) = F_s\left(S_{r_{s_i}}\right) - F_s\left(S_{r_{s_i}}-\right) = P\left(S_{r_{s_i}} \le s_{r_{s_i}}\right) - P\left(S_{r_{s_i}} < s_{r_{s_i}}\right) \text{ and}$$
$$\pi_{\cdot,r_{t_j}} = \Delta F_t\left(T_{r_{t_j}}\right) = F_t\left(T_{r_{t_j}}\right) - F_t\left(T_{r_{t_j}}-\right) = P\left(T_{r_{t_j}} \le t_{r_{t_j}}\right) - P\left(T_{r_{t_j}} < t_{r_{t_j}}\right), \tag{9}$$

where we denote the ranks of the observed failure or censoring times per each margin as $r_{s_i} = \text{rank}(S_i)$ and $r_{t_j} = \text{rank}(T_j)$, respectively, corresponding to the order statistics $S_{(1)} < S_{(2)} < \cdots < S_{(n)}$ and $T_{(1)} < T_{(2)} < \cdots < T_{(n)}$. The parameters at (9) are instrumental with respect to defining the simplex constraints used in our maximization procedure described below. The inter-relationships between the cell probabilities, the $\pi_{i,j}$'s, and the marginal probabilities, the $\pi_{r_{s_i},\cdot}$'s and $\pi_{\cdot,r_{t_j}}$'s, are defined in detail below at (20) and (21), respectively.

As described in Section 1, and with a simple modification of the notation, the parameter estimates derived from the form of the likelihood at (3) corresponding to the marginal density f_x were derived by Satten and Datta (2001) as

$$\hat{\pi}_{i,.} = \begin{cases} \frac{\delta^x_{(1)}}{n}, & i = 1, \\ \frac{\prod_{j=1}^{i-1}\left((n-j+1)-\delta^x_{(j)}\right)\delta^x_{(i)}}{\prod_{j=1}^{i}(n+j-i)}, & i > 1, \end{cases} \tag{10}$$

where we set $\delta^x_{(n)} = 1$. In the case of no censoring all $\hat{\pi}_{i,.}$'s are equal to $1/n$.

It follows straightforward that similar to the standard empirical distribution function estimator we have for right-censored data that

$$\hat{F}_x(x) = \sum_{i=1}^{n} \hat{\pi}_{i,.} I_{(s_{(i)} \le x)} \tag{11}$$

and

$$\hat{S}_x(x) = 1 - \hat{F}_x(x) = 1 - \sum_{i=1}^{n} \hat{\pi}_{i,.} I_{(s_{(i)} \le x)}, \tag{12}$$

where $\hat{\pi}_{i,.}$'s are given at (10) and $\hat{\pi}_{n,.} = 1 - \sum_{j=1}^{n-1} \hat{\pi}_{j,.}$. It should be obvious that $\hat{\pi}_{i,.} = 0$ from (10) when $\delta^x_{(i)} = 0$ in the case of a censored observation.

Similarly, we have the estimated parameters corresponding to the marginal density f_y, $\delta^y_{(n)} = 1$, given as

$$\hat{\pi}_{.,j} = \begin{cases} \frac{\delta^y_{(1)}}{n}, & j = 1, \\ \frac{\prod_{i=1}^{j-1}((n-i+1)-\delta^y_{(i)})\delta^y_{(j)}}{\prod_{i=1}^{j}(n+i-j)}, & j > 1, \end{cases} \tag{13}$$

which yields the marginal estimators of the distribution and survival functions as

$$\hat{F}_y(y) = \sum_{j=1}^{n} \hat{\pi}_{.,j} I_{(t_{(j)} \le y)} \tag{14}$$

and

$$\hat{S}_y(y) = 1 - \hat{F}_y(y) = 1 - \sum_{j=1}^{n} \hat{\pi}_{.,j} I_{(t_{(j)} \le y)}, \tag{15}$$

where $\hat{\pi}_{.,j}$'s are given at (13) and $\hat{\pi}_{.,n} = 1 - \sum_{i=1}^{n-1} \hat{\pi}_{i,.}$.

Let us now define the parameters associated with the nonparametric likelihood corresponding to the joint density $f(x, y)$. In terms of determining the relevant parameters for use in the nonparametric likelihood model we need to define an indicator function as a type of bookkeeping feature given censoring information from both marginal distributions. Towards this end let

$$\delta_{i,j} = \begin{cases} 1, \text{ if } \delta^x_i \delta^y_i = 1, \forall i = 1, 2, \cdots, n \\ 1, \text{ if } \left(1 - \delta^x_i\right)\delta^y_i = 1, \forall i = r_{s_j+1}, \cdots, n, j = 1, 2, \cdots, n, \\ 1, \text{ if } \delta^x_i\left(1 - \delta^y_i\right) = 1, \forall i = 1, 2, \cdots, n, j = r_{t_i+1}, \cdots, n, \\ 1, \text{ if } \left(1 - \delta^x_i\right)\left(1 - \delta^y_i\right) = 1, \forall i = r_{s_j+1}, \cdots, n, j = r_{t_i+1}, \cdots, n, \\ 0, \text{ otherwise.} \end{cases} \tag{16}$$

Then $\forall(r_{s_i}, r_{t_j})$ combinations for which $\delta_{r_{s_i},r_{t_j}} \neq 0$, $\delta^x_{r_{s_i}} \neq 0$ and $\delta^y_{r_{t_j}} \neq 0$, $i = 1, 2, \cdots, n$ and $j = 1, 2, \cdots, n$ the parameters of interest in our nonparametric model corresponding to the joint density $f(x, y)$ are given as

$$\begin{aligned}
\pi_{r_{s_i},r_{t_j}} &= \Delta F\left(S_{r_{s_i}}, T_{r_{t_j}}\right) \\
&= F\left(S_{r_{s_i}}, T_{r_{t_j}}\right) - F\left(S_{r_{s_i}}, T_{r_{t_j}}-\right) - F\left(S_{r_{s_i}}-, T_{r_{t_j}}\right) + F\left(S_{r_{s_i}}-, T_{r_{t_j}}-\right) \\
&= P\left(S_{r_{s_i}} \le s_{r_{s_i}}, T_{r_{t_j}} \le t_{r_{t_j}}\right) - P\left(S_{r_{s_i}} \le s_{r_{s_i}}, T_{r_{t_j}} < t_{r_{t_j}}\right) \\
&\quad - P\left(S_{r_{s_i}} < s_{r_{s_i}}, T_{r_{t_j}} \le t_{r_{t_j}}\right) + P\left(S_{r_{s_i}} < s_{r_{s_i}}, T_{r_{t_j}} < t_{r_{t_j}}\right), \qquad (17)
\end{aligned}$$

else we define $\pi_{r_{s_i},r_{t_j}} = 0$, i.e. $\pi_{r_{s_i},r_{t_j}} = 0$ if $\delta_{r_{s_i},r_{t_j}} = 0$ or $\delta^x_{r_{s_i}} = 0$ or $\delta^y_{r_{t_j}} = 0$. In the specific case where there is no censoring for both X and Y the number of parameters is of size n and the corresponding maximum likelihood estimator for $\pi_{r_{s_i},r_{t_j}}$ is $1/n$ as per the standard empirical density estimator, i.e. there is a point mass of $1/n$ per each set of paired observations.

It now follows similar to that of the univariate case at (3) that the likelihood function for the joint density $f(x, y)$ defined through the parameters at (17) is given as

$$\begin{aligned}
L = \prod_{i=1}^{n} \pi_{r_{s_i},r_{t_i}}^{\delta_i^x \delta_i^y} \left(\sum_{j=r_{s_i}+1}^{n} \delta_j^x \pi_{j,r_{t_i}} \right)^{(1-\delta_i^x)\delta_i^y} \left(\sum_{k=t_{s_i}+1}^{n} \delta_k^y \pi_{r_{s_i},k} \right)^{\delta_i^x(1-\delta_i^y)} \\
\times \left(\sum_{j=r_{s_i}+1}^{n} \sum_{k=t_{s_i}+1}^{n} \delta_j^x \delta_k^y \pi_{j,k} \right)^{(1-\delta_i^x)(1-\delta_i^y)}, \qquad (18)
\end{aligned}$$

where the components of the likelihood correspond to the four possible right-censoring combinations for the (δ_i^x, δ_i^y) pairs, i.e. there is a simple point mass given no censoring else probability is shifted to the right similar to the classic Kaplan-Meier estimator given the various censoring patterns. The objective is to maximize L at (18) subject to the simplex constraints

$$\sum_{i=1}^{n}\sum_{j=1}^{n} \delta_{i,j}\delta_i^x\delta_j^y \pi_{i,j} = 1, \qquad (19)$$

$$\sum_{j=1}^{n} \delta_{i,j}\delta_j^y \pi_{i,j} = \pi_{i,.}, \qquad (20)$$

$$\sum_{i=1}^{n} \delta_{i,j}\delta_i^x \pi_{i,j} = \pi_{.,j}, \qquad (21)$$

$$\text{if } \delta_{i,j} = \delta_i^x = \delta_j^y = 1 \text{ then } 0 \le \pi_{i,j} \le 1, \forall i = 1, \cdots, n, j = 1, \cdots, n. \qquad (22)$$

The constraints at (20) and (21) pertain to the marginal constraints where $\pi_{i,.}$ and $\pi_{.,j}$ are defined at (9) . Our approach is similar to the problems described for multinomial distribution parameter estimation given sparse data and a class of linear simplex constraints (Liu 2000). The argument for replacing $\pi_{i,.}$ and $\pi_{.,j}$ at (9) with their corresponding estimators $\hat{\pi}_{i,.}$ and $\hat{\pi}_{.,j}$ at (10) and (13), respectively, follows similar to the classic $R \times C$ contingency table exact inference case. The contribution of the $\hat{\pi}_{i,.}$'s and $\hat{\pi}_{.,j}$'s to the multinomial distribution given sparse data and linear simplex constraints corresponding to $\hat{\pi}_{r_{s_i},r_{t_j}}$'s of interest depend on the data through the censoring values for the δ^x's and δ^y's. The joint distribution of the $\hat{\pi}_{i,.}$'s and $\hat{\pi}_{r_{s_i},r_{t_j}}$'s is identical to the distribution of the

$\hat{\pi}_{r_{s_i},r_{t_j}}$'s since by definition the $\hat{\pi}_{i,.}$'s are determined by the $\hat{\pi}_{r_{s_i},r_{t_j}}$'s . The same holds in the other dimension relative to the $\hat{\pi}_{.,j}$'s, i.e. the $\hat{\pi}_{r_{s_i},r_{t_j}}$'s are sufficient statistics in terms of determinig the parameters that define the marginal densities.

Steps in the parameter estimation:

1. Given the observed censoring pattern utilize the constraints (19), (20) and (21) to define the parameter space.
2. Obtain the estimates of marginal probabilities $\pi_{i,.}$ and $\pi_{.,j}$ given by $\hat{\pi}_{i,.}$ and $\hat{\pi}_{.,j}$ at (10) and (13),respectively. Substitute the estimates into (20) and (21) after first processing step 1 above.
3. Utilize standard maximum likelihood technique on the likelihood defined at (18) to solve for the remaining unknown parameters given the constraints defined at (19)–(22).

It then follows that the estimators of the bivariate distribution function and survival function are given as:

$$\hat{F}(x,y) = \sum_{i=1}^{n}\sum_{j=1}^{n} \hat{\pi}_{i,j} I(s_{(i)} \le x, t_{(j)} \le y) \tag{23}$$

$$\hat{S}(x,y) = 1 - \hat{F}(x,y), \tag{24}$$

respectively. Some small sample toy examples will be provided in the next section in order to illustrate the process followed by some real data examples.

Note that if censoring occurs solely in either the x or y dimension the likelihood at (18) reduces substantially in complexity and has a form very similar to that of the univariate setting at (3).

Comment Large sample variance estimates for $\hat{F}(x,y)$ and $\hat{S}(x,y)$ are conceptually straightforward in that they follow standard methods based on obtaining the co-information matrix with dimensions that vary as a function of the proportion of censored observations. For small samples this is straightforward. However for moderate to large samples and from a programming point of view, this becomes a rather complex computational problem such that we would recommend either bootstrap or jackknife methodologies for the purpose of variance estimation.

3 Examples

In this section we provide a few straightforward small sample scenarios in order to illustrate the maximization process for the estimator of $f(x,y)$ and $S(x,y)$ given various censoring patterns. This is followed by two real data examples used by previous researchers in the past to illustrate this type of estimator. It is important to note that as we present the results we set $\delta^x_{(n)} = \delta^y_{(n)} = 1$ by definition. The rational for this was described above.

3.1 Toy data examples

Example 1. For $n = 6$ we have the following data:

$$\mathbf{s} = (6.1, 4.5, 6.2, 4.8, 5.9, 3.3), r_s = (5, 2, 6, 3, 4, 1), \delta^x = (1, 1, 0, 1, 0, 1),$$

$$\mathbf{t} = (9.6, 2.6, 7.2, 4.1, 7.7, 5.0), r_t = (6, 1, 4, 2, 5, 3), \delta^y = (0, 1, 1, 1, 0, 1).$$

The vector of parameters of interest defined by the censoring patterns as per (16) is given as $\boldsymbol{\pi} = (\pi_{1,3}, \pi_{2,1}, \pi_{3,2}, \pi_{5,6}, \pi_{6,4}, \pi_{6,6})$. The goal is to maximize the likelihood

$$L = \pi_{1,3}\pi_{2,1}\pi_{3,2}\pi_{5,6}\pi_{6,4}\left(\pi_{5,6} + \pi_{6,6}\right)$$

subject to the simplex constraints from (19), (20) and (21), respectively, and given as:

1. $\pi_{1,3} + \pi_{2,1} + \pi_{3,2} + \pi_{5,6} + \pi_{6,4} + \pi_{6,6} = 1,$
2. $\pi_{1,3} = 1/6, \pi_{2,1} = 1/6, \pi_{3,2} = 1/6, \pi_{5,6} = 1/4, \pi_{6,4} + \pi_{6,6} = 1/4,$
3. $\pi_{2,1} = 1/6, \pi_{3,2} = 1/6, \pi_{1,3} = 1/6, \pi_{6,4} = 1/6, \pi_{5,6} + \pi_{6,6} = 1/3.$

We can see that in this small sample setting that estimates for $\boldsymbol{\pi} = (\pi_{1,3}, \pi_{2,1}, \pi_{3,2}, \pi_{5,6}, \pi_{6,4}, \pi_{6,6})$ in this case are determined solely by the marginal constraints. In general, for moderate to large sample sizes this will not be the case.

The estimates given the constraints are provided in Table 1. In this specific case no maximization of the likelihood was needed. Note that the estimators of the marginal survivor functions $\hat{S}_x(x)$ and $\hat{S}_y(y)$ from (12) and (15), respectively, and based on the parameter estimates in Table 1 are exactly those corresponding to the product-limit estimator. The plot of the bivariate survival function $\hat{S}(x, y)$ from (24) is given in Fig. 1, which is clearly monotone decreasing in both dimensions for increasing values of data with positive support.

Example 2. For $n = 6$ we have the following data:

$$\mathbf{s} = (8.9, 2.6, 3.7, 5.9, 7.9, 1.2), r_s = (6, 2, 3, 4, 5, 1), \delta^x = (0, 0, 1, 1, 1, 1),$$
$$\mathbf{t} = (6.3, 9.6, 6.8, 0.1, 0.7, 4.2), r_t = (4, 6, 5, 1, 3, 2), \delta^y = (0, 0, 1, 1, 1, 1).$$

The vector of parameters of interest defined by the censoring patterns as per (16) is given as $\boldsymbol{\pi} = (\pi_{1,3}, \pi_{3,5}, \pi_{3,6}, \pi_{4,1}, \pi_{4,6}, \pi_{5,2}, \pi_{5,6}, \pi_{6,5}, \pi_{6,6})$, which is of a slightly higher dimension as compared to Example 1. The goal is to maximize the likelihood

$$L = \pi_{1,3}\pi_{3,5}\pi_{4,1}\pi_{5,2}\left(\pi_{3,6} + \pi_{4,6} + \pi_{5,6} + \pi_{6,6}\right)\left(\pi_{6,5} + \pi_{6,6}\right)$$

subject to the simplex constraints from (19), (20) and (21), respectively, and given as:

1. $\pi_{1,3} + \pi_{3,5} + \pi_{3,6} + \pi_{4,1} + \pi_{4,6} + \pi_{5,2} + \pi_{5,6} + \pi_{6,5} + \pi_{6,6} = 1$
2. $\pi_{1,3} = 1/6, \pi_{3,5} + \pi_{3,6} = 5/24, \pi_{4,1} + \pi_{4,6} = 5/24, \pi_{5,2} + \pi_{5,6} = 5/24, \pi_{6,5} + \pi_{6,6} = 5/24$
3. $\pi_{4,1} = 1/6, \pi_{5,2} = 1/6, \pi_{1,3} = 1/6, \pi_{3,5} + \pi_{6,5} = 1/4, \pi_{3,6} + \pi_{4,6} + \pi_{5,6} + \pi_{6,6} = 1/4$

The results of the maximum likelihood estimates derived from (18) are provided in Table 2. In this case note that $\hat{\pi}_{3,6}$ was at the boundary of the feasible region and set equal

Table 1 Bivariate estimates for $\hat{\pi}_{i,j}, i = 1, 2, \cdots, 6, j = 1, 2, \cdots, 6$, corresponding to example 1

i/j	$r_t = 1$	$r_t = 2$	$r_t = 3$	$r_t = 4$	$r_t = 5$	$r_t = 6$	$\hat{\pi}_{i,\cdot}$	$\delta^x_{r_{s_i}}$
$r_s = 1$	0	0	1/6	0	0	0	1/6	1
$r_s = 2$	1/6	0	0	0	0	0	1/6	1
$r_s = 3$	0	1/6	0	0	0	0	1/6	1
$r_s = 4$	0	0	0	0	0	0	0	0
$r_s = 5$	0	0	0	0	0	1/4	1/4	1
$r_s = 6$	0	0	0	1/6	0	1/12	1/4	1
$\hat{\pi}_{\cdot,j}$	1/6	1/6	1/6	1/6	0	1/3	1	
$\delta^y_{r_{t_j}}$	1	1	1	1	0	1		

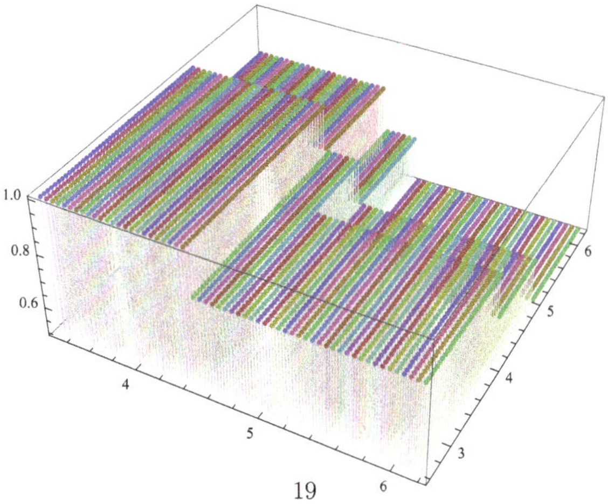

Fig. 1 Estimate of $\hat{S}(x, y)$ for example 1 data

to 0. The plot of the bivariate survival function $\hat{S}(x, y)$ from (24) is given in Fig. 2, which as before is clearly monotone decreasing in both dimensions for increasing values of the data with positive support.

Example 3. For $n = 6$ we have the following data:

$$\mathbf{s} = (6.4, 7.0, 7.2, 8.1, 6.2, 7.4), r_s = (2, 3, 4, 6, 1, 5), \delta^x = (0, 1, 0, 0, 1, 0),$$
$$\mathbf{t} = (8.6, 7.5, 8.0, 4.8, 0.8, 4.0), r_t = (6, 4, 5, 3, 1, 2), \delta^y = (0, 1, 0, 0, 1, 0).$$

The vector of parameters of interest defined by the censoring patterns as per (16) is given as $\boldsymbol{\pi} = (\pi_{1,1}, \pi_{3,4}, \pi_{3,6}\pi_{6,4}, \pi_{6,6})$. The goal is to maximize the likelihood

$$L = \pi_{1,1}\pi_{3,4}\pi_{6,6}\left(\pi_{3,6} + \pi_{6,6}\right)\left(\pi_{6,4} + \pi_{6,6}\right)^2$$

subject to the simplex constraints from (19), (20) and (21), respectively, and given as:

1. $\pi_{1,1} + \pi_{3,4} + \pi_{3,6} + \pi_{6,4} + \pi_{6,6} = 1$
2. $\pi_{1,1} = 1/6, \pi_{3,4} + \pi_{3,6} = 5/24, \pi_{6,4} + \pi_{6,6} = 5/8$
3. $\pi_{1,1} = 1/6, \pi_{3,4} + \pi_{6,4} = 5/18, \pi_{3,6} + \pi_{6,6} = 5/9$

In this example case we have heavy censoring relative to the total number of observations. Similar to example 2, $\hat{\pi}_{3,6}$ was at the boundary of the feasible region and set equal to 0. The results of the maximum likelihood estimates derived from (18) are provided in Table 3. The plot of the bivariate survival function $\hat{S}(x, y)$ from (24) is given in Fig. 3,

Table 2 Bivariate estimates for $\hat{\pi}_{i,j}, i = 1, 2, \cdots, 6, j = 1, 2, \cdots, 6$, corresponding to example 2

i/j	$r_t = 1$	$r_t = 2$	$r_t = 3$	$r_t = 4$	$r_t = 5$	$r_t = 6$	$\hat{\pi}_{i,\cdot}$	$\delta^x_{r_{s_i}}$
$r_s = 1$	0	0	1/6	0	0	0	1/6	1
$r_s = 2$	0	0	0	0	0	0	0	0
$r_s = 3$	0	0	0	0	5/24	0	5/24	1
$r_s = 4$	1/6	0	0	0	0	1/24	5/24	1
$r_s = 5$	0	1/6	0	0	0	1/24	5/24	1
$r_s = 6$	0	0	0	0	1/24	1/6	5/24	1
$\hat{\pi}_{\cdot,j}$	1/6	1/6	1/6	0	1/4	1/4	1	
$\delta^y_{r_{t_j}}$	1	1	1	0	1	1		

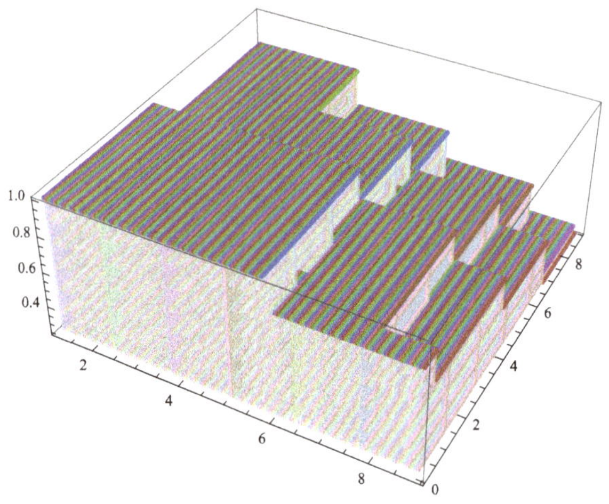

Fig. 2 Estimate of $\hat{S}(x, y)$ for example 2 data

which as before is clearly monotone decreasing in both dimensions for increasing values of data with positive support.

Note that if there was no censoring within examples 1–3 then the respective values for $\hat{\pi}_{r_{s_i}, r_{t_i}}$, $i = 1, 2, \cdots, n$, would be simply $1/n$, which corresponds to the maximum likelihood estimates for the classic empirical joint density function.

3.2 Real data examples

In this section we re-analyze two sets of real data utilized to demonstrate other approaches to bivarate survival function estimation (Akritas and Van Keilegom 2003; Wang and Wells 1997), see the references contained within relative to the source of the original data.

Survival days of skin grafts in burn patients (Wang and Wells 1997). For this data set we have $n = 11$ paired survival times and censoring indicators for skin grafts in burn patients given as:

$$\mathbf{s} = (37, 19, 57, 93, 16, 22, 20, 18, 63, 29, 60),$$

$$\mathbf{t} = (29, 13, 15, 26, 11, 17, 26, 21, 43, 15, 40),$$

$$\delta^x = (1, 1, 0, 1, 1, 1, 1, 1, 1, 1, 0),$$

$$\delta^y = (1, 1, 1, 1, 1, 1, 1, 1, 1, 1, 1).$$

Table 3 Bivariate estimates for $\hat{\pi}_{i,j}, i = 1, 2, \cdots, 6, j = 1, 2, \cdots, 6$, corresponding to example 3

i/j	$r_t = 1$	$r_t = 2$	$r_t = 3$	$r_t = 4$	$r_t = 5$	$r_t = 6$	$\hat{\pi}_{i,.}$	$\delta^x_{r_{s_i}}$
$r_s = 1$	1/6	0	0	0	0	0	1/6	1
$r_s = 2$	0	0	0	0	0	0	0	0
$r_s = 3$	0	0	0	5/24	0	0	5/24	1
$r_s = 4$	0	0	0	0	0	0	0	0
$r_s = 5$	0	0	0	0	0	0	0	0
$r_s = 6$	0	0	0	5/72	0	5/9	5/8	1
$\hat{\pi}_{.,j}$	1/6	0	0	5/18	0	5/9	1	
$\delta^y_{r_{t_j}}$	1	0	0	1	0	1		

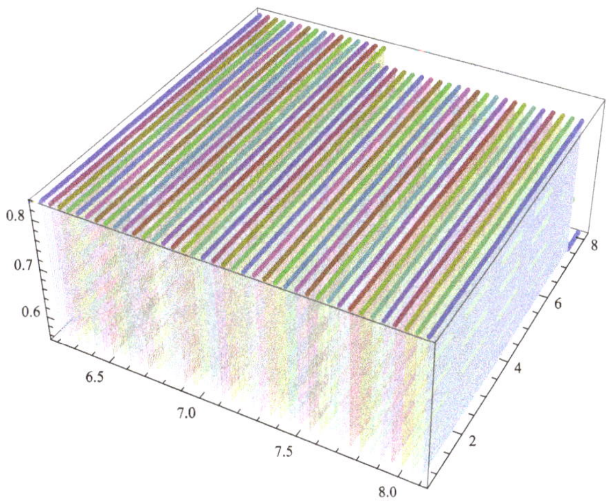

Fig. 3 Estimate of $\hat{S}(x, y)$ for example 3 data

As you can see there is no censoring in the y component and only a moderate amount of censoring in the x component. Hence in most instances the $\hat{\pi}_{r_{s_i}, r_{t_j}}$'s will be equal to $1/n$. In this case the vector of parameters with the corresponding maximum likelihood estimates are given as: $\left(\hat{\pi}_{1,1} = \frac{1}{11}, \hat{\pi}_{2,8} = \frac{1}{11}, \hat{\pi}_{3,2} = \frac{1}{11}, \hat{\pi}_{4,6} = \frac{1}{11}, \hat{\pi}_{5,5} = \frac{1}{11}, \hat{\pi}_{6,4} = \frac{1}{11}, \hat{\pi}_{7,9} = \frac{1}{11}, \hat{\pi}_{10,3} = \frac{3}{64}, \hat{\pi}_{10,10} = \frac{31}{704}, \hat{\pi}_{10,11} = \frac{1}{11}, \hat{\pi}_{11,3} = \frac{31}{704}, \hat{\pi}_{11,7} = \frac{1}{11}, \hat{\pi}_{11,10} = \frac{3}{64}\right)$. The likelihood for this example has the form

$$L = \pi_{1,1}\pi_{2,8}\pi_{3,2}\pi_{4,6}\pi_{5,5}\pi_{6,4}\pi_{7,9}\pi_{10,11}\left(\pi_{10,3} + \pi_{11,3}\right)\pi_{11,7}\left(\pi_{10,10} + \pi_{11,10}\right) \quad (25)$$

with simplex constraints from (19), (20) and (21), respectively, given as:

1. $\pi_{1,1} + \pi_{2,8} + \pi_{3,2} + \pi_{4,6} + \pi_{5,5} + \pi_{6,4} + \pi_{7,9} + \pi_{10,3} + \pi_{10,10} + \pi_{10,11} + \pi_{11,3} + \pi_{11,7} + \pi_{11,10} = 1$,
2. $\pi_{1,1} = 1/11, \pi_{2,8} = 1/11, \pi_{3,2} = 1/11, \pi_{4,6} = 1/11, \pi_{5,5} = 1/11, \pi_{6,4} = 1/11, \pi_{7,9} = 1/11, \pi_{10,3} + \pi_{10,10} + \pi_{10,11} = 2/11, \pi_{11,3} + \pi_{11,7} + \pi_{11,10} = 2/11$
3. $\pi_{1,1} = 1/11, \pi_{3,2} = 1/11, \pi_{10,3} + \pi_{11,3} = 1/11, \pi_{6,4} = 1/11, \pi_{5,5} = 1/11, \pi_{4,6} = 1/11, \pi_{11,7} = 1/11, \pi_{2,8} = 1/11, \pi_{7,9} = 1/11, \pi_{10,10} + \pi_{11,10} = 1/11, \pi_{10,11} = 1/11$.

Again, we see the most of the parameters in the likelihood are determined via the marginal constraints, which should be obvious given the low percentage of censored

Table 4 Estimated bivariate survival probabilites for skin graft data evaluated at the marginal quartiles

u_x	u_y	$\hat{Q}_x(u_x)$	$\hat{Q}_y(u_y)$	$\hat{S}\left(\hat{Q}_x(u_x), \hat{Q}_y(u_y)\right)$
0.25	0.25	19.25	15.0	0.82
0.25	0.5	19.25	21.0	0.82
0.25	0.75	19.25	28.25	0.73
0.5	0.25	29.	15.0	0.73
0.5	0.5	29.	21.0	0.55
0.5	0.75	29.	28.25	0.45
0.75	0.25	59.25	15.0	0.73
0.75	0.5	59.25	21.0	0.55
0.75	0.75	59.25	28.25	0.45

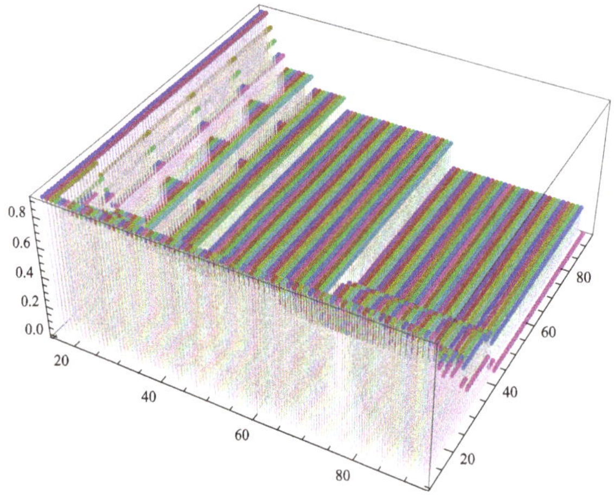

Fig. 4 Estimate of $\hat{S}(x, y)$ for survival days of skin grafts

observations. The estimated bivariate survival probabilities $\hat{S}(\hat{Q}_x(u_x), \hat{Q}_y(u_y))$ evaluated at the marginal quartiles are provided in Table 4. The joint bivariate survival function is plotted in Fig. 4. We see that our estimates provide a valid estimator of the joint survival function that is monotone decreasing in both dimensions.

Recurrence times to infection at the point of insertion of a catheter for kidney patients using portable dialysis equipment (Akritas and Van Keilegom 2003). For this data set we have $n = 38$ paired survival times corresponding to infections times at two points along with paired censoring indicators. The data given below yields 487 $\pi_{i,j}$ parameters to be estimated. The constraints from (19), (20) and (21) and likelihood are not presented for this problem. Essentially the problem is a basic symbolic linear programming problem, which in our case was readily handled within Mathematica(Mathematica 8.0 for Linux, Wolfram Research Inc.,Champaign, IL). The number of independent free parameters to be estimated after considering the constraints was 248.

$$\mathbf{s} = (8, 23, 22, 447, 30, 24, 7, 511, 53, 15, 7, 141, 96, 149, 536, 17, 185, 292, 22, 15, 152, 402, 13, 39, 12, 113, 132, 34, 2, 130, 27, 5, 152, 190, 119, 54, 6, 63),$$

Table 5 Estimated bivariate survival probabilities for recurrence times to infection at the point of insertion of a catheter for kidney patients at the marginal quantiles

u_x	u_y	$\hat{Q}_x(u_x)$	$\hat{Q}_y(u_y)$	$\hat{S}(\hat{Q}_x(u_x), \hat{Q}_y(u_y))$
0.25	0.25	15	16	0.95
0.25	0.5	15	39	0.92
0.25	0.75	15	154	0.81
0.5	0.25	46	16	0.91
0.5	0.5	46	39	0.85
0.5	0.75	46	154	0.68
0.75	0.25	149	16	0.90
0.75	0.5	149	39	0.77
0.75	0.75	149	154	0.58

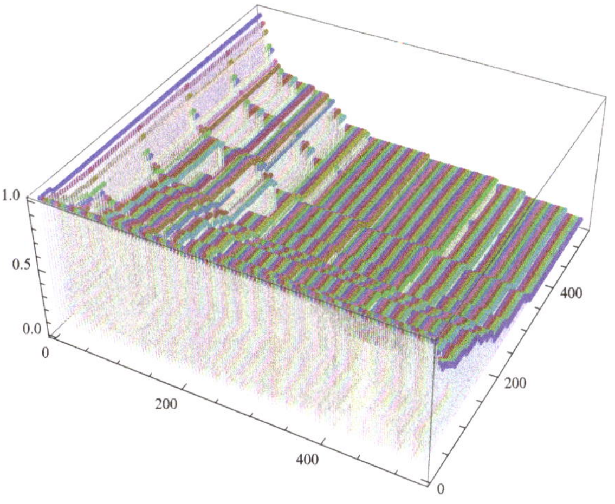

Fig. 5 Estimate of $\hat{S}(x, y)$ for survival days for recurrence times to infection

$$\mathbf{t} = (16, 13, 28, 318, 12, 245, 9, 30, 196, 154, 333, 8, 38, 70, 25, 4, 177, 114, 159, 108, 562, 24, 66, 46, 40, 201, 156, 30, 25, 26, 58, 43, 30, 5, 8, 16, 78, 8),$$

$$\delta^x = (1, 1, 1, 1, 1, 1, 1, 1, 1, 1, 1, 1, 1, 0, 1, 1, 1, 1, 0, 1, 1, 1, 1, 1, 1, 0, 1, 1, 1, 1, 1, 0, 1, 1, 1, 0, 0, 1),$$

$$\delta^y = (1, 0, 1, 1, 1, 1, 1, 1, 1, 1, 1, 0, 1, 0, 0, 0, 1, 1, 0, 0, 1, 0, 1, 0, 1, 1, 1, 1, 1, 1, 1, 1, 1, 0, 1, 0, 1, 0).$$

Due to the size of the problem we only present the estimated bivariate survival probabilities $\hat{S}(\hat{Q}_x(u_x), \hat{Q}_y(u_y))$ evaluated at the marginal quartiles as provided in Table 5. The joint bivariate survival function is plotted in Fig. 5. We see that our estimates provide a valid estimator of the joint survival function that is monotone decreasing in both dimensions. It is straightforward to evaluate all potential survival probabilities and any estimators one wishes to derive from these given the sophistication of current software packages.

4 Conclusions

In this note we provide the only method for estimating a bivariate survival function such that the marginal estimators correspond exactly to the Kaplan-Meier product limit estimators such that there is a relative consistency between the marginal estimates derived via univariate or bivariate methods. Unlike other methods developed in the literature, our approach is generalizable to higher dimensions and different censoring mechanisms(interval and left censoring). Our methodology also opens up an alternative path for kernel smoothing of both the bivariate density, distribution function and survival function via use of the estimated $\pi_{i,j}$'s over the multinomial grid of non-zero mass. Using real data we illustrated the computational approach and feasibility of this new method of simplex constraint based maximum likelihood estimation as applied to right censored data.

References

Akritas, MG, Van Keilegom, I: Estimation of bivariate and marginal distributions with censored data. J. R. Stat. Soc. Ser. B. **65**, 457–471 (2003)

Chen, K, Lo, S-H: On the rate of uniform convergence of the product-limit estimator: Strong and weak laws. Ann. Stat. **25**, 1050–1087 (1997)

Cox, DR, Oakes, D: Analysis of Survival Data. Chapman & Hall/CRC, New York (1984)

Gill, RD, van der Lann, MJ, Wellner, JA: Inefficient estimators of the bivariate survival function for three models. Annales de l'Institut Henri Poincaré. **31**, 545–597 (1995)

Kaplan, EL, Meier, P: Nonparametric estimation from incomplete observations. J. Am. Stat. Assoc. **53**, 457–481 (1958)

Lin, DY, Ying, Z: A simple nonparametric estimator of the bivariate survival function under univariate censoring. Biometrika. **80**, 573–581 (1993)

Liu, C: Estimation of discrete distributions with a class of simplex constraints. J. Am. Stat. Assoc. **95**, 109–120 (2000)

Owen, AB: Empirical likelihood ratio confidence intervals for a single functional. Biometrika. **75**, 237–249 (1988)

Prentice, RL, Zoe Moodie, F, Wu, J: Hazard-based nonparametric survivor function estimation. J. R. Stat. Soc. Ser. B. **66**, 305–319 (2004)

Satten, GA, Datta, S: The Kaplan-Meier Estimator as an Inverse-Probability-of-Censoring Weighted Average. Am. Stat. **55**, 207–210 (2001)

Wang, W, Wells, MT: Nonparametric estimators of the bivariate survival function under simplified censoring condition. Biometrika. **84**, 863–880 (1997)

Zhou, M: Empirical likelihood ratio with arbitrarily censored/truncated data by EM algorithm. J. Comput. Graph. Stat. **14**, 643–656 (2005)

8

Compounding of distributions: a survey and new generalized classes

Muhammad H Tahir[1*] and Gauss M. Cordeiro[2]

*Correspondence:
mtahir.stat@gmail.com
[1]Department Statistics, Baghdad Campus, The Islamia University of Bahawalpur, 63100 Bahawalpur, Pakistan
Full list of author information is available at the end of the article

Abstract

Generalizing distributions is an old practice and has ever been considered as precious as many other practical problems in statistics. It simply started with defining different mathematical functional forms, and then induction of location, scale or inequality parameters. The generalization through induction of shape parameter(s) started in 1997, and the last two decades were full of such practices. But to cope with the complex situations under series and parallel structures, the art of mixing discrete and continuous started in 1998. In this article, we present a survey on compounding univariate distributions, their extensions and classes. We review several available compound classes and propose some new ones. The recent trends in the construction of generalized and compounding classes are discussed, and the need for future works are addressed.

Keywords: Compound G-classes, generalized classes, G-geometric class, G-Poisson class, power series class

AMS Subject Classification: Primary 60E05; secondary 62N05; 62F10

1 Introduction

The modern era on distribution theory stresses on problem-solving faced by the practitioners and applied researchers and proposes a variety of models so that lifetime data set can be better assessed and investigated in different fields. In other words, there is strong need to introduce useful models for better exploration of the real-life phenomenons. Nowadays, the trends and practices in defining probability models totally differ in comparison to the models suggested before 1997. One main objective for proposing, extending or generalizing (models or their classes) is to explain how the lifetime phenomenon arises in fields like physics, computer science, insurance, public health, medical, engineering, biology, industry, communications, life-testing and many others. For example, the well-known and fundamental distributions such as exponential, Rayleigh, Weibull and gamma are very limited in their characteristics and are unable to show wide flexibility. The number of shape parameters, closed-form expressions of the cumulative distribution function (cdf), forms of the quantile function (qf), density and failure rate shapes, skewness and kurtosis features, entropy measures, estimation of stress reliability $R = \mathbb{P}(Y < X)$, structural properties, use of advanced mathematical functions and power series, which are available through computing platforms like Mathematica,

Maple, Matlab, Python, Ox and R, variety of estimation methods, estimation of parameters in case of censored and uncensored situations, simulation results, coping with data sets having different shapes and goodness-of-fits are some well-established characteristics that a proposed lifetime model may possess.

For complex phenomenons in reliability studies, lifetime testing, human mortality studies, engineering modeling, electronic sciences and biological surveys, the failure rate behavior can be bathtub, upside-down bathtub and other shaped but not usually monotone increasing or decreasing. Thus, in order to cope with both monotonic and non-monotonic failure rate shapes, researchers have introduced several generalizations (or G-classes) which are very flexible to study needful properties of the model and its fitness.

In the last two decades, two main generalization approaches were adopted and practiced, and have also received increased attention.

1.1 First generalization approach (G-classes)

The first approach deals with the shape parameter(s) induction in parent (or baseline) distribution to explore tail properties and to improve goodness-of-fits. Some well-known generalized (or G-) classes are: Marshall-Olkin-G (Marshall and Olkin 1997), exponentiated-G (Gupta et al. 1998), beta-G (Eugene et al. 2002), Kumaraswamy-G (Cordeiro and de-Castro 2011), McDonald-G (Alexander et al. 2012), ZBgamma-G (Zografos and Balakrishnan 2009; Amini et al. 2014), RBgamma-G (Ristić and Balakrishanan 2012; Amini et al. 2014), odd-gamma-G (Torabi and Montazari 2012), Kummer-beta-G (Pescim et al. 2012), beta extended Weibull-G (Cordeiro et al. 2012b), odd exponentiated generalized-G (Cordeiro et al. 2013a), truncated exponential-G (Barreto-Souza and Simas 2013), logistic-G (Torabi and Montazari 2014), gamma extended Weibull-G (Nascimento et al. 2014), odd Weibull-G (Bourguignon et al. 2014a), exponentiated-half-logistic-G (Cordeiro et al. 2014a), Libby-Novick beta-G (Cordeiro et al. 2014b; Ristić et al. 2015), Lomax-G (Cordeiro et al. 2014d), Harris-G (Batsidis and Lemonte 2015; Pinho et al. 2015), modified beta-G (Nadarajah et al. 2014b), odd generalized-exponential-G (Tahir et al. 2015), Kumaraswamy odd log-logistic-G (Alizadeh et al. 2015b), beta odd log-logistic-G (Cordeiro et al. 2016), Kumaraswamy-Marshall-Olkin-G (Alizadeh et al. 2015c), beta-Marshall-Olkin-G (Alizadeh et al. 2015a), Weibull-G (Tahir et al. 2016b), exponentiated-Kumaraswamy-G (da-Silva et al. 2016), ZBgamma-odd-loglogistic-G (Cordeiro et al. 2015a) and Tukey's g- and h-G (Jiménez et al. 2015). For more details on some well-established G-classes, the reader is referred to Tahir and Nadarajah (2015).

The mathematical properties of the Kumaraswamy-G family were studied by Nadarajah et al. (2012), Hussian (2013), and Cordeiro and Bager (2015). The failure rate and aging properties of the beta-G and ZBgamma-G models were addressed by Triantafyllou and Koutras (2014). The structural properties of the ZBgamma-G and RBgamma-G models were recently investigated by Nadarajah et al. (2015b), and Cordeiro and Bourguignon (2016).

The revolutionary idea in generalization begun with the work of Alzaatreh et al. (2013) who proposed transformed-transformer (T-X) (Weibull-X and gamma-X) family of distributions. This approach was further extended to the exponentiated T-X (Alzaghal et al. 2013), T-X{Y}-quantile based approach (Aljarrah et al. 2014), T-R{Y} (Alzaatreh et al. 2014), T-Weibull{Y} (Almheidat et al. 2015), T-gamma{Y} (Alzaatreh et al. 2016a),

T-Cauchy{Y}(Alzaatreh et al. 2016b), Gumbel-X (Al-Aqtash 2013; Al-Aqtash et al. 2014) and logistic-X (Tahir et al. 2016).

After the wide-spread popularity of well-established exponentiated-G, Marshall-Olkin-G, beta-G, Kumaraswamy-G and McDonald-G classes, and T-X family, the transmuted-G class of distributions has received increased attention in the last decade. This class is based on the quadratic rank transmutation map (QRTM) pioneered by Shaw and Buckley (2009) and highlighted by Aryal and Tsokos (2009, 2011).

For any baseline distribution cdf $G(x)$, the cdf of the *transmuted-G* class (for $|\eta| \leq 1$) is given by

$$F(x) = (\eta + 1)\, G(x) - \eta\, G(x)^2, \tag{1}$$

where η is the transmuted (or shape) parameter. For $\eta = 0$, the above equation reduces to the baseline distribution.

The general properties of the transmuted family were obtained by Bourguignon et al. (2016a) and Das (2015). The transmuted family was further extended as the exponentiated transmuted-G type 1 using the Lehmann alternative type 1 (LA1) class (due to Gupta et al. 1998) by Nofal et al. (2016) and Alizadeh et al. (2016a), the exponentiated transmuted-G type 2 using the Lehmann alternative type 2 (LA2) class (due to Gupta et al. 1998) by Merovci et al. (2016), and the transmuted exponentiated generalized-G by Yousof et al. (2015).

More than *fifty* transmuted distributions have been reported in the literature, which are complied in chronological order in Table 1; see also Section 8.4.

1.2 Second generalization approach (compounding)

The second approach deals with the compounding of discrete models, namely the geometric, Poisson, logarithmic, binomial, negative-binomial (NB), Conway-Maxwell-Poisson (COMP) and power-series with continuous lifetime models. The basic idea of introducing compound models or families is that a lifetime of a system with N (discrete random variable) components and the positive continuous random variable, say Y_i (the lifetime of the ith component), can be denoted by the non-negative random variable $Y = \min(Y_1, \ldots, Y_N)$ (the minimum of an unknown number of any continuous random variables) or $Z = \max(Y_1, \ldots, Y_N)$ (the maximum of an unknown number of any continuous random variables), based on whether the components are in series or in parallel structure. Some useful references for the readers are Louzada et al. (2012a), Leahu et al. (2013) and Bidram and Alavi (2014).

The objectives of our article are three-fold: Firstly, we present an up-to-date account on the compounded distributions and their generalizations for the readers of modern distribution theory. Secondly, this survey will motivate the researchers to fill the gap and to furnish their work in remaining applied areas. Thirdly, this survey will also be helpful as a tutorial to the beginners of compound modeling art.

The rest of the article is organized as follows. In Section 2, two compound G-classes are reviewed to represent the distributions of the minimum and maximum of an unknown number of continuous random variables having the same parent lifetime distribution. In Section 3, fourteen existing and new compound classes for the minimum constructed from the zero-truncated geometric (ZTG), zero-truncated Poisson (ZTP), logarithmic

Table 1 Contributed work on transmuted distributions

S.No.	Pioneer Year	Distribution	Author(s)
1.	2009	Transmuted extreme value	Aryal and Tsokos (2009)
2.	2011	Transmuted Weibull	Aryal and Tsokos (2011)
			Ahmad et al. (2015b)
			Khan et al. (2016b)
3.	2013	Transmuted log-logistic	Aryal (2013)
			Granzotto and Louzada (2015)
4.	2013	Transmuted Rayleigh	Merovci (2013a)
			Ahmad et al. (2015a)
5.	2013	Transmuted exponentiated-exponential	Merovci (2013b)
			Khan et al. (2015a)
6.	2013	Transmuted Fréchet	Mahmoud and Mandouh (2013)
7.	2013	Transmuted Lomax	Ashour and Eltehiwy (2013a)
8.	2013	Transmuted Lindley	Merovci (2013c)
9.	2013	Transmuted quasi-Lindley	Elbatal and Elgarhy (2013)
10.	2013	Transmuted exponentiated-Lomax	Ashour and Eltehiwy (2013b)
11.	2013	Transmuted modified inverse Weibull	Elbatal (2013a)
12.	2013	Transmuted generalized inverted exponential	Elbatal (2013b)
13.	2013	Transmuted exponentiated-modified Weibull	Ashour and Eltehiwy (2013c)
14.	2013	Transmuted generalized linear exponential	Elbatal et al. (2013)
15.	2013	Transmuted additive Weibull (AW)	Elbatal and Aryal (2013)
16.	2013	Transmuted modified Weibull (MW)	Khan and King (2013)
17.	2014	Transmuted Pareto	Merovci and Puka (2014)
18.	2014	Transmuted Maxwell	Iriarte and Astorga (2014)
19.	2014	Transmuted linear exponential	Tian et al. (2014)
20.	2014	Transmuted inverse Rayleigh (IR)	Sharma et al. (2014)
			Ahmad et al. (2014)
21.	2014	Transmuted generalized Rayleigh (GR)	Merovci (2014)
			Iriarte and Astorga (2015)
22.	2014	Transmuted inverted Weibull (IW)	Khan et al. (2014)
			Khan and King (2014a)
23.	2014	Transmuted generalized IW	Merovci et al. (2014)
			Khan and King (2014b)
24.	2014	Exponentiated transmuted Weibull	Ebraheim (2014)
25.	2014	Beta transmuted Weibull	Pal and Tiensuwan (2014)
26.	2014	Transmuted exponentiated-gamma	Hussian (2014)
			Lucena et al. (2015)
27.	2014	Transmuted exponentiated-Fréchet	Elbatal et al. (2014)
28.	2015	Transmuted exponential	Owoloko et al. (2015)
29.	2015	Transmuted Burr III	Abdul-Moniem (2015)
30.	2015	Transmuted Gompertz	Abdul-Moniem and Seham (2015)
			Khan et al. (2016c)
31.	2015	Transmuted modified IR	Khan and King (2015)
32.	2015	Generalized transmuted AW	Mansour et al. (2015a)
33.	2015	Kumaraswamy transmuted MW	Mansour et al. (2015b)
34.	2015	Transmuted generalized MW	Cordeiro et al. (2015c)
35.	2015	Kumaraswamy transmuted exponentiated MW	Al-Babtain et al. (2015)
36.	2015	Transmuted Marshall-Olkin (MO) Fréchet	Afify et al. (2015a)
37.	2015	Exponentiated transmuted GR	Afify et al. (2015b)
38.	2015	Transmuted Weibull-Lomax	Afify et al. (2015c)
39.	2015	Transmuted Dagum	Elbatal and Aryal (2015)
			Shahzad and Asghar (2016)
40.	2015	Transmuted exponentiated-Pareto	Luguterah and Nasiru (2015)
			Fatima and Roohi (2015)

Table 1 Contributed work on transmuted distributions *(Continued)*

S.No.	Pioneer Year	Distribution	Author(s)
41.	2015	Generalized transmuted Lindley	Mansour and Mohamed (2015)
42.	2015	Transmuted Chen	Khan et al. (2015b)
43.	2015	Beta transmuted Fréchet	da-Silva et al. (2015a)
44.	2016	Transmuted Kumaraswamy	Khan et al. (2016a)
45.	2016	Transmuted generalized Lindley	Elgarhy et al. (2016)
46.	2016	Transmuted Almalki-Yuan MW	Vardhan and Balaswamy (2016)
47.	2016	Transmuted power function	Haq et al. (2016)
48.	2016	Transmuted geometric	Chakraborty and Bhati (2016)
49.	2016	Transmuted Birnbaum-Saunders	Bourguignon et al. (2016b)
50.	2016	Kumaraswamy Transmuted MO Fréchet	Yousof et al. (2016)
51.	2016	Transmuted Weibull-Pareto	Afify et al. (2016)

(Ln), zero-truncated binomial (ZTBi) and zero-truncated negative-binomial (ZTNB) distributions are described. In Section 4, we obtain the dual models for the maximum corresponding to those models discussed in Section 3. Section 5 deals with several compounding models and their extensions, which can be derived under the construction of the minimum and maximum of random variables. Sections 7 and 8 deal with other or different types of compounded models. In Section 8, we present recent trends on compounding of distributions, their G-classes and mixing of compounded and transmuted distributions. The main purpose of Section 9 is to briefly review general inference procedure, crude rate survival models and their inference. The paper is concluded with some remarks in Section 10.

2 Construction of compound G-classes

Suppose that a system has N subsystems assumed to be independent and identically distributed (i.i.d.) at a given time, where the lifetime of the ith subsystem is denoted by Y_i, and that each subsystem is made of α parallel units, so that the system will fail if all the subsystems fail. We note that for a parallel system, the system fails only if all the subsystems fail, but for a series system, the failure of any subsystem will destroy the whole system. Further, suppose that the random variable N follows any discrete distribution with probability mass function (pmf) $\mathbb{P}(N = n)$ (for $n = 1, 2, \ldots$). We consider that the failure times of the components $Z_{i,1}, \ldots, Z_{i,\alpha}$ for the ith subsystem are i.i.d. random variables having a suitable cdf, which is a function of the baseline cdf depending on a parameter vector τ, say $T[G(x;\tau),\alpha] = G(x;\tau)^{\alpha}$ (for $x > 0$). In the following construction, although α is a positive integer called power or resilience parameter, we can consider that $\alpha > 0$.

If we define $Y = \min\{Y_1, \ldots, Y_N\}$, then the conditional cdf of Y given N is

$$\begin{aligned} F(y \mid N) &= \mathbb{P}\left[\min(Y_1, \ldots, Y_N) < y|N)\right] = 1 - \left[1 - \mathbb{P}^{\alpha}(Z_{1,1} \le y)\right]^N \\ &= 1 - \left[1 - G(y;\tau)^{\alpha}\right]^N. \end{aligned} \tag{2}$$

If we define $Z = \max\{Y_1, \ldots, Y_N\}$, then the conditional (or complementary) cdf of Z given N is

$$\begin{aligned} F(z \mid N) &= \mathbb{P}\left[\max(Y_1, \ldots, Y_N) < z|N)\right] = \left[\mathbb{P}^{\alpha}(Z_{n,n} \le z)\right]^N \\ &= \left[G(z;\tau)^{\alpha}\right]^N. \end{aligned} \tag{3}$$

The unconditional cdf of Y follows from Eq. (2) as

$$F(y) = \sum_{n=1}^{\infty} F(y|N)\,\mathbb{P}(N=n) = 1 - \sum_{n=1}^{\infty} \left[1 - G(y;\tau)^{\alpha}\right]^{n}\,\mathbb{P}(N=n). \quad (4)$$

The unconditional cdf of Z follows from Eq. (3) as

$$F(z) = \sum_{n=1}^{\infty} F(z|N)\,\mathbb{P}(N=n) = \sum_{n=1}^{\infty} \left[G(z;\tau)^{\alpha}\right]^{n}\,\mathbb{P}(N=n). \quad (5)$$

Many compound G-classes can be constructed from Eqs. (4) and (5) by choosing a discrete model with pmf $\mathbb{P}(N=n)$. The random variables $Y = \min\{Y_1,\ldots,Y_N\}$ and $Z = \max\{Y_1,\ldots,Y_N\}$ generate several models that arise in series and parallel systems with identical components and have many industrial and biological applications. For example, the time to the failure of a series system with an unknown number of protected components or the cancer recurrence of an individual after undergoing a certain treatment can be modeled by the generated distribution of Y. In a dual mechanism, the time to the failure of a parallel system with an unknown number of protected components or a disease manifestation, if it occurs only after an unknown number of factors have been active, can be modeled by the generated distribution of Z.

3 Compound G-classes

In this section, we present 14 compounded models obtained from Eq. (4). In Section 4, we present the corresponding complementary models based on Eq. (5). The list below does not include all compounded models but a large number of them and some new ones. For example, it does not describe the *exponentiated-G-Conway-Maxwell Poisson* (EGCOMP) class pioneered by Cordeiro et al. (2012a) and its special *GCOMP* model, among others. For all formulated models, we provide only their cdfs since the corresponding probability density functions (pdfs) can be determined by simple differentiation.

3.1 Exponentiated G-geometric class

If $T[\,G(y;\tau);\alpha] = G(y;\tau)^{\alpha}$ and N be a ZTG r.v. with pmf $\mathbb{P}(N=n) = q\,p^{n-1}$, $n = 1,2,\ldots$, $p \in (0,1)$, where p is the success probability, the cdf of the exponentiated G-geometric (EGG) class can follow as

$$F_{EGG}(y) = 1 - \frac{q}{p} \sum_{n=1}^{\infty} \left[p(1 - G(y;\tau)^{\alpha})\,\right]^{n} = \frac{G(y;\tau)^{\alpha}}{1 - p + p\,G(y;\tau)^{\alpha}}.$$

The EGG class has recently been introduced by Nadarajah et al. (2015a), and can also be called the generalized G-geometric class. For $\alpha = 1$, the EGG class reduces to the G-geometric (GG) class proposed by Alkarni (2012). They investigated some of its general properties.

Remark 1 *Swapping p with q in the ZTG pmf with $p + q = 1$, the cdf of an alternative EGG (denoted by EGGA) class is given by*

$$F_{EGGA}(y) = 1 - \frac{p}{q} \sum_{n=1}^{\infty} \left[q\,\left(1 - G(y;\tau)^{\alpha}\right)\right]^{n} = \frac{G(y;\tau)^{\alpha}}{p - (p-1)G(y;\tau)^{\alpha}}.$$

For $\alpha = 1$, the EGGA class reduces to the alternative G-geometric (GGA) class defined by Castellares and Lemonte (2016).

3.2 Exponentiated Kumaraswamy-G-geometric class (new)

For any baseline cdf $G(x)$ and pdf $g(x)$, Cordeiro and de-Castro (2011) defined the cdf of the Kumaraswamy-G class by $K_{a,b}(x;\tau) = 1 - [1 - G(x)^a]^b$, where $a > 0$ and $b > 0$ are both shape parameters. Let $T[G(y;\tau);\alpha] = K_{a,b}(y;\tau)^\alpha$ and N be a ZTG r.v. with pmf $\mathbb{P}(N = n) = q p^{n-1}$, $n = 1, 2, \ldots$, the cdf of the exponentiated Kumaraswamy-G-geometric (EKGG) class can follow as

$$F_{EKGG}(y) = 1 - \frac{q}{p}\sum_{n=1}^{\infty}\{p[1 - K_{a,b}(y;\tau)^\alpha]\}^n = \frac{K_{a,b}(y;\tau)^\alpha}{1 - p[1 - K_{a,b}(y;\tau)^\alpha]}.$$

For $\alpha = 1$, the EKGG class gives the new Kumaraswamy-G geometric (KGG) family.

3.3 McDonald-G-geometric class (new)

Alexander et al. (2012) defined the cdf of the McDonald-G class by $M_{a,b,c}(x;\tau) = I_{G(x;\tau)^c}(a,b)$, where $B(a,b) = \int_0^1 w^{a-1}(1-w)^{b-1}\,dw$, $B_z(a,b) = \int_0^z w^{a-1}(1-w)^{b-1}\,dw$ and $I_z(a,b) = B_z(a,b)/B(a,b)$ are the beta function, incomplete beta function and incomplete beta function ratio, respectively.

If $\alpha = 1$ and $T[G(y;\tau)] = M_{a,b,c}(y;\tau)$, and N be a ZTG r.v. with pmf $\mathbb{P}(N = n) = q p^{n-1}$, $n = 1, 2, \ldots$, the cdf of the McDonald-G geometric (MGG) class is given by

$$F_{MGG}(y) = 1 - \frac{q}{p}\sum_{n=1}^{\infty}\{p[1 - M_{a,b,c}(y;\tau)]\}^n = \frac{M_{a,b,c}(y;\tau)}{1 - p[1 - M_{a,b,c}(y;\tau)]}.$$

3.4 Beta-G-geometric class (new)

For any baseline cdf $G(x)$, Eugene et al. (2002) defined the cdf of the beta-G class by

$$B_{a,b}(x) = I_{G(x;\tau)}(a,b).$$

If $\alpha = 1$ and $T[G(y;\tau] = B_{a,b}(y;\tau)$, and N be a ZTG r.v. with pmf $\mathbb{P}(N = n) = q p^{n-1}$, $n = 1, 2, \ldots$, the cdf of the beta-G geometric (BGG) class is given by

$$F_{BGG}(y) = 1 - \frac{q}{p}\sum_{n=1}^{\infty}\{p[1 - B_{a,b}(y;\tau)]\}^n = \frac{B_{a,b}(y;\tau)}{1 - p[1 - B_{a,b}(y;\tau)]}.$$

3.5 Exponentiated G-Poisson class

If $T[G(y;\tau);\alpha] = G(y;\tau)^\alpha$ and N be a ZTP r.v. with parameter λ defined by the pmf $\mathbb{P}(N = n) = \lambda^n/[n!\,(e^\lambda - 1)]$, $n = 1, 2, \ldots$, the cdf of the exponentiated G-Poisson (EGP) class can follow as

$$F_{EGP}(y) = 1 - \sum_{n=1}^{\infty}\frac{[1 - G(y;\tau)^\alpha]^n\,\lambda^n}{n!\,(e^\lambda - 1)} = \frac{1 - e^{-\lambda\,G(y;\tau)^\alpha}}{1 - e^{-\lambda}}.$$

The EGP class has been studied by Gomes et al. (2015). For $\alpha = 1$, it becomes the G-Poisson (GP) class as defined recently by Tahir et al. (2016a) by the name of the *Poisson-X* class since it was based on the T-X family.

3.6 Exponentiated Kumaraswamy-G-Poisson class (new)

If $T[G(y;\tau);\alpha] = G(y;\tau)^\alpha$ and N be a ZTP r.v. with parameter λ, the cdf of the exponentiated Kumaraswamy-G Poisson (EKGP) class is given by

$$F_{EKGP}(y) = 1 - \sum_{n=1}^{\infty}[1 - K_{a,b}(y;\tau)^\alpha]^n\,\frac{\lambda^n}{n!\,(e^\lambda - 1)} = \frac{1 - e^{-\lambda\,K_{a,b}(y;\tau)^\alpha}}{1 - e^{-\lambda}}.$$

For $\alpha = 1$, the EKGP class reduces to the Kumaraswamy-G Poisson (KGP) family studied by Ramos et al. (2015).

3.7 McDonald-G-Poisson class (new)

Let $\alpha = 1$ and $T[G(y;\tau)] = M_{a,b,c}(y;\tau)$ and N be a ZTP r.v. with parameter λ, the cdf of the McDonald-G Poisson (MGP) class can be expressed as

$$F_{MGP}(y) = 1 - \frac{1}{e^{\lambda} - 1} \sum_{n=1}^{\infty} \frac{\{\lambda [1 - M_{a,b,c}(y;\tau)]\}^n}{n!} = \frac{1 - e^{-\lambda M_{a,b,c}(y;\tau)}}{1 - e^{-\lambda}}.$$

3.8 Beta-G-Poisson class (new)

If $\alpha = 1$ and $T[G(y);\tau] = B_{a,b}(y;\tau)$ and N be a ZTP r.v. with parameter λ, the cdf of the beta-G Poisson (BGP) class reduces to

$$F_{BGP}(y) = 1 - \frac{1}{e^{\lambda} - 1} \sum_{n=1}^{\infty} \frac{\{\lambda [1 - B_{a,b}(y;\tau)]\}^n}{n!} = \frac{1 - e^{-\lambda B_{a,b}(y;\tau)}}{1 - e^{-\lambda}}.$$

3.9 Exponentiated G-logarithmic class (new)

If $T[G(y;\tau);\alpha] = G(y;\tau)^{\alpha}$ and N be a Ln r.v. with pmf $\mathbb{P}(N = n) = (1 - \phi)^n/[-n \ln \phi]$, $\phi \in (0,1)$, $n = 1,2,\ldots$, the cdf of the exponentiated G-logarithmic (EGLn) class is given by

$$F_{EGLn}(y) = 1 - \sum_{n=1}^{\infty} \left[1 - G(y;\tau)^{\alpha}\right]^n \frac{(1-\phi)^n}{-n \ln \phi} = 1 - \frac{\ln\{1 - (1-\phi)[1 - G(y;\tau)^{\alpha}]\}}{\ln \phi},$$

by noting that $\sum_{n=1}^{\infty} Q^n/n = -\ln(1-Q)$.

For $\alpha = 1$, the EGLn class becomes the G-logarithmic (GLn) family introduced by Alkarni (2012).

3.10 Exponentiated Kumaraswamy-G-logarithmic class (new)

If $T[G(y;\tau);\alpha] = G(y;\tau)^{\alpha}$ and N be a Ln r.v. with pmf $\mathbb{P}(N = n) = (1 - \phi)^n/[-n \ln \phi]$, $\phi \in (0,1)$, $n = 1,2,\ldots$, the cdf of the exponentiated Kumaraswamy-G-logarithmic (EKGLn) is given by

$$\begin{aligned} F_{EKGLn}(y) &= 1 - \sum_{n=1}^{\infty} \left[1 - K_{a,b}(y;\tau)^{\alpha}\right]^n \frac{(1-\phi)^n}{-n \ln \phi} \\ &= 1 - \frac{\ln\{1 - (1-\phi)[1 - K_{a,b}(y;\tau)^{\alpha}]\}}{\ln \phi}. \end{aligned}$$

For $\alpha = 1$, the EKGLn cdf is identical to the cdf of the new Kumaraswamy-G-logarithmic (KGLn) class.

3.11 McDonald-G-logarithmic class (new)

If $T[G(y;\tau)] = M_{a,b,c}(y;\tau)$ and N be a Ln r.v. with pmf $\mathbb{P}(N = n) = (1 - \phi)^n/[-n \ln \phi]$, $\phi \in (0,1)$, $n = 1,2,\ldots$, the cdf of the McDonald-G-logarithmic (MGLn) class is defined by

$$F_{MGLn}(y) = 1 - \sum_{n=1}^{\infty} \left[1 - M_{a,b,c}(y;\tau)\right]^n \frac{(1-\phi)^n}{-n \ln \phi}$$
$$= 1 - \frac{\ln\left\{1 - (1-\phi)[1 - M_{a,b,c}(y;\tau)]\right\}}{\ln \phi}.$$

3.12 Beta-G-logarithmic class (new)

If $\alpha = 1$ and $T[G(y;\tau)] = B_{a,b}(y;\tau)$ and N be a Ln r.v. with pmf $\mathbb{P}(N = n) = (1 - \phi)^n/[-n \ln \phi]$, $\phi \in (0,1)$, $n = 1, 2, \ldots$, the cdf of the beta-G-logarithmic (BGLn) class is given by

$$F_{BGLn}(y) = 1 - \frac{\ln\left\{1 - (1-\phi)[1 - B_{a,b}(y;\tau)]\right\}}{\ln \phi}.$$

3.13 Exponentiated G-binomial class (new)

If $T[G(y;\tau);\alpha] = G(y;\tau)^\alpha$ and N be a ZTBi r.v. with pmf given by

$$\mathbb{P}(N = k) = \frac{\binom{n}{k}\theta^k (1-\theta)^{n-k}}{1 - (1-\theta)^n}, \quad \theta \in (0,1), \quad k = 1, 2, \ldots, n,$$

the cdf of the exponentiated G-binomial (EGBi) class reduces to

$$F_{EGBi}(y) = \frac{1 - \left[1 - \theta\, G(y;\tau)^\alpha\right]^n}{1 - (1-\theta)^n}.$$

Bakouch et al. (2012b) studied a special case of the EGBi class so-called the exponentiated-exponential binomial (EEBi) model. For $\alpha = 1$, the EGBi class becomes the G-binomial (GBi) class pioneered by Alkarni (2013).

3.14 Exponentiated G-NB class (new)

If $T[G(y;\tau);\alpha] = G(y;\tau)^\alpha$ and N be a ZTNB r.v. with parameter $\beta \in (0,1)$ as the probability of success and pmf given by

$$\mathbb{P}(N = n) = \binom{s+n-1}{n} \frac{\beta^n}{(1-\beta)^{-s} - 1}, \quad s > 0, \quad n = 1, 2, \ldots,$$

the cdf of the exponentiated G-NB (EGNB) class can follow as

$$F_{EGNB}(y) = 1 - \frac{1}{(1-\beta)^{-s} - 1} \sum_{n=1}^{\infty} \binom{s+n-1}{n} \left\{\beta\left[1 - G(y;\tau)^\alpha\right]\right\}^n.$$

Using the binomial expansion $(1-w)^{-s} - 1 = \sum_{n=1}^{\infty} \binom{s+n-1}{n} w^n$, we obtain

$$F_{EGNB}(y) = \frac{(1-\beta)^{-s} - \left\{1 - \beta\left[1 - G(y;\tau)^\alpha\right]\right\}^{-s}}{(1-\beta)^{-s} - 1}. \tag{6}$$

For $\alpha = 1$, the EGNB class leads to the G-NB (GNB) class proposed by Percontini et al. (2013b).

Remark 2 ***i*** *If we replace the probability of success β by $1 - \beta$ and the dispersion parameter s by θ in (6), the cdf of an alternate form of the GNB class (denoted by EGNB1) will be*

$$F_{EGNB1}(y) = 1 - \frac{\beta^\theta}{1 - \beta^\theta} \left\{\left[G(y;\tau)^\alpha + (1-\beta)\bar{G}(y;\tau)^\alpha\right]^{-\theta} - 1\right\}. \tag{7}$$

For $\alpha = 1$, the EGNB1 class leads to the GNB family discussed by Nadarajah et al. (2013b), who also studied, as a special case of the GNB1 class, the exponential-truncated negative-binomial (ETNB) model.

ii *If we express β in terms of the population mean of the distribution in (6), the cdf of an alternate form of the GNB class (denoted by EGNB2) will be*

$$F_{EGNB2}(y) = \frac{1 - \left[1 + \eta\theta\, G(y;\tau)^{\alpha}\right]^{-\frac{1}{\eta}}}{1 - (1 + \eta\theta)^{-\frac{1}{\eta}}}.$$

For $\alpha = 1$, the EGNB2 class leads to the GNB family. Louzada et al. (2012b) studied a special model of this class.

4 Complementary compound G-classes

Complementary compound models are constructed by considering the maximum of a sequence of i.i.d. random variables which represents the risk time of a system having components in parallel structure. In this section, we generate from Eq. (5) the complementary G-classes of those ones presented in Section 3. Some proposed complementary G-classes are really new ones.

4.1 Complementary exponentiated G-geometric class (new)

If $T[G(z;\tau);\alpha] = G(z;\tau)^{\alpha}$ and N be a ZTG r.v. with geometric pmf $\mathbb{P}(N = n) = q p^{n-1}$, $n = 1, 2, \ldots$, the cdf of the complementary exponentiated G-geometric (CEGG) class reduces to

$$F_{CEGG}(z) = \frac{q}{p}\sum_{n=1}^{\infty}\left[p\, G(z;\tau)^{\alpha}\right]^{n} = \frac{(1-p)\, G(z;\tau)^{\alpha}}{1 - p\, G(z;\tau)^{\alpha}}.$$

This equation is also called the complementary generalized G-geometric family. For $\alpha = 1$, the CEGG class becomes the complementary G-geometric (CGG) family.

Remark 3 *Swapping p with q in the ZTG pmf with $p + q = 1$, the cdf of an alternative CEGG (CEGGA) class will be*

$$F_{CEGGA}(z) = \frac{p}{q}\sum_{n=1}^{\infty}\left[q\, G(z;\tau)^{\alpha})\right]^{n} = \frac{p\, G(z;\tau)^{\alpha}}{1 - (1-p)\, G(z;\tau)^{\alpha}}.$$

For $\alpha = 1$, the CEGGA class leads to the complementary G-geometric (CGG) family proposed by Castellares and Lemonte (2016).

4.2 Complementary exponentiated Kumaraswamy-G-geometric class (new)

If $\alpha = 1$, $T[G(z;\tau)] = K_{a,b}(z;\tau)$ and N be a geometric r.v. with pmf $\mathbb{P}(N = n) = q p^{n-1}$, $n = 1, 2, \ldots$, the cdf of the CEKGG class is given by

$$F_{CEKGG}(z) = \frac{q}{p}\sum_{n=1}^{\infty}\left\{p\left[K_{a,b}(z;\tau)^{\alpha}\right]\right\}^{n} = \frac{q[K_{a,b}(z;\tau)^{\alpha}]}{1 - p\, K_{a,b}(z;\tau)^{\alpha}}.$$

For $\alpha = 1$, the CEKGG class leads to the complementary KGG (CKGG) class.

4.3 Complementary McDonald-G-geometric class (new)

If $\alpha = 1$, $T[G(z;\tau)] = M_{a,b,c}(z;\tau)$ and N be a geometric r.v. with pmf $\mathbb{P}(N = n) = qp^{n-1}$, $n = 1, 2, \ldots$, the cdf of the complementary McDonald-G-geometric (CMGG) class can follow as

$$F_{CMGG}(z) = \frac{q}{p}\sum_{n=1}^{\infty}\left\{p\left[M_{a,b,c}(z;\tau)\right]\right\}^{n} = \frac{q\left[M_{a,b,c}(z;\tau)\right]}{1 - pM_{a,b,c}(z;\tau)}.$$

4.4 Complementary beta-G-geometric class (new)

If $\alpha = 1$, $T[G(z;\tau)] = B_{a,b}(z;\tau)$, and N be a geometric r.v. with pmf $\mathbb{P}(N = n) = qp^{n-1}$, $n = 1, 2, \ldots$, the cdf of the complementary beta-G-geometric (CBGG) class is given by

$$F_{CBGG}(z) = \frac{q}{p}\sum_{n=1}^{\infty}\left[pB_{a,b}(z;\tau)\right]^{n} = \frac{qB_{a,b}(z;\tau)}{1 - pB_{a,b}(z;\tau)}.$$

4.5 Complementary exponentiated G-Poisson class (new)

If $T[G(z;\tau);\alpha] = G(z;\tau)^{\alpha}$ and N be a ZTP r.v. with parameter λ, the cdf of the complementary exponentiated G-Poisson (CEGP) can follow as

$$F_{CEGP}(z) = \sum_{n=1}^{\infty}\left[G(z;\tau)^{\alpha}\right]^{n}\frac{\lambda^{n}}{n!\,(\mathrm{e}^{\lambda} - 1)} = \frac{\mathrm{e}^{\lambda\,G(z;\tau)^{\alpha}} - 1}{\mathrm{e}^{\lambda} - 1}.$$

For $\alpha = 1$, the CEGP class leads to the complementary G-Poisson (CGP) class.

4.6 Complementary exponentiated Kumaraswamy-G Poisson class (new)

If $T[G(z;\tau);\alpha] = K_{a,b}(z;\tau)^{\alpha}$ and N be a ZTP r.v. with parameter λ, the cdf of the complementary exponentiated Kumaraswamy-G Poisson (CEKGP) class is given by

$$F_{CEKGP}(z) = \sum_{n=1}^{\infty}\left[K_{a,b}(z;\tau)^{\alpha}\right]^{n}\frac{\lambda^{n}}{n!\,(\mathrm{e}^{\lambda} - 1)} = \frac{\mathrm{e}^{\lambda\,K_{a,b}(z;\tau)^{\alpha}} - 1}{\mathrm{e}^{\lambda} - 1}.$$

For $\alpha = 1$, the CEKGP class reduces to the complementary Kumaraswamy-G Poisson (CKGP) class.

4.7 Complementary McDonald-G Poisson class (new)

If $\alpha = 1$ and $T[G(z;\tau)] = M_{a,b,c}(z;\tau)$ and N be a ZTP r.v. with parameter λ, the cdf of the complementary McDonald-G Poisson (CMGP) class reduces to

$$F_{CMGP}(z) = \sum_{n=1}^{\infty}\left[M_{a,b,c}(z;\tau)\right]^{n}\frac{\lambda^{n}}{n!\,(\mathrm{e}^{\lambda} - 1)} = \frac{\mathrm{e}^{\lambda\,M_{a,b,c}(z;\tau)} - 1}{\mathrm{e}^{\lambda} - 1}.$$

4.8 Complementary beta G-Poisson class (new)

If $\alpha = 1$, $T[G(z);\tau] = B_{a,b}(z;\tau)$ and N be a ZTP r.v. with parameter λ, the cdf of the complementary beta G-Poisson (CBGP) class is given by

$$F_{CBGP}(z) = \frac{1}{\mathrm{e}^{\lambda} - 1}\sum_{n=1}^{\infty}\frac{\left\{\lambda\left[B_{a,b}(z;\tau)\right]\right\}^{n}}{n!} = \frac{\mathrm{e}^{\lambda\,B_{a,b}(z;\tau)} - 1}{\mathrm{e}^{\lambda} - 1}.$$

4.9 Complementary exponentiated G-logarithmic class (new)

If $T[G(z;\tau);\alpha] = G(z;\tau)^{\alpha}$ and N be a Ln r.v. with pmf given by $\mathbb{P}(N = n) = (1-\phi)^n/[-n \ln \phi]$, $\phi \in (0,1)$, $n = 1,2,\ldots$, the cdf of the complementary exponentiated G-logarithmic (CEGLn) class can be expressed as

$$F_{CEGLn}(z) = \frac{1}{-\ln \phi}\sum_{n=1}^{\infty}\frac{[(1-\phi)\,G(z;\tau)^{\alpha}]^n}{n} = \frac{\ln[1-(1-\phi)\,G(z;\tau)^{\alpha}]}{\ln \phi},$$

by noting that $\sum_{n=1}^{\infty} Q^n/n = -\ln(1-Q)$. For $\alpha = 1$, the CEGLn class becomes the complementary G-logarithmic (CGLn) class.

4.10 Complementary exponentiated Kumaraswamy-G-logarithmic class (new)

If $T[G(z;\tau);\alpha] = K_{a,b}(z;\tau)^{\alpha}$ and N be a Ln r.v. with $\mathbb{P}(N = n) = (1-\phi)^n/[-n \ln \phi]$, $\phi \in (0,1)$, $n = 1,2,\ldots$, the cdf of the complementary exponentiated Kumaraswamy G-logarithmic of type 1 (CEKGLn1) class is given by

$$F_{CEKGLn}(z) = \sum_{n=1}^{\infty}\left[K_{a,b}(z;\tau)^{\alpha}\right]^n \frac{(1-\phi)^n}{-n \ln \phi} = \frac{\ln\left[1-(1-\phi)\,K_{a,b}(z;\tau)^{\alpha}\right]}{\ln \phi}.$$

For $\alpha = 1$, the complementary Kumaraswamy-G-logarithmic (CEKGLn) class becomes the complementary G-logarithmic (CGLn) class.

4.11 Complementary McDonald-G-logarithmic class (new)

If $\alpha = 1$, $T[G(z;\tau)] = M_{a,b,c}(z;\tau)$ and N be a Ln r.v. with pmf $\mathbb{P}(N = n) = (1-\phi)^n/(-n \ln \phi)$, $\phi \in (0,1)$, $n = 1,2,\ldots$, the cdf of the complementary McDonald-G-logarithmic (CMGLn) class can be expressed as

$$F_{CMGLn}(z) = \sum_{n=1}^{\infty}\left[M_{a,b,c}(z;\tau)\right]^n \frac{(1-\phi)^n}{-n \ln \phi} = \frac{\ln\left[1-(1-\phi)\,M_{a,b,c}(z;\tau)\right]}{\ln \phi}.$$

4.12 Complementary beta-G-logarithmic class (new)

If $\alpha = 1$ and $T[G(z;\tau)] = B_{a,b}(z;\tau)$ and N be a Ln r.v. with pmf $\mathbb{P}(N = n) = (1-\phi)^n/(-n \ln \phi)$, $\phi \in (0,1)$, $n = 1,2,\ldots$, the cdf of the complementary beta-G-logarithmic (CBGLn) class is given by

$$F_{CBGLn}(z) = \sum_{n=1}^{\infty}\left[B_{a,b}(z;\tau)\right]^n \frac{(1-\phi)^n}{-n \ln \phi} = \frac{\ln\left[1-(1-\phi)\,B_{a,b}(z;\tau)\right]}{\ln \phi}.$$

4.13 Complementary exponentiated G-binomial class (new)

If $T[G(z;\tau);\alpha] = G(z;\tau)^{\alpha}$ and N be a ZTBi r.v. with pmf

$$\mathbb{P}(N = k) = \frac{\binom{n}{k}\,\theta^k\,(1-\theta)^{n-k}}{1-(1-\theta)^n}, \quad \theta \in (0,1), \quad k = 1,2,\ldots,n,$$

the cdf of the complementary exponentiated G-binomial (CEGBi) class is given by

$$\begin{aligned} F_{CEGBi}(z) &= \frac{1}{1-(1-\theta)^n}\left[\sum_{k=0}^{n}\binom{n}{k}\left[\theta\,G(z;\tau)^{\alpha}\right]^k (1-\theta)^{n-k} - (1-\theta)^n\right] \\ &= \frac{\{1-\theta\,[1-G(z;\tau)^{\alpha}]\}^n - (1-\theta)^n}{1-(1-\theta)^n}. \end{aligned}$$

For $\alpha = 1$, the CEGBi class becomes the complementary G-binomial (CGBi) class.

4.14 Complementary exponentiated G-NB class (new)

If $T[\,G(z;\tau);\alpha] = G(z;\tau)^{\alpha}$ and N be a ZTNB r.v. with parameter $\beta \in (0,1)$ with pmf

$$\mathbb{P}(N=n) = \binom{s+n-1}{n} \frac{\beta^n}{(1-\beta)^{-s}-1}, \quad s>0, \quad n=1,2,\ldots$$

Using the binomial expansion $(1-w)^{-s}-1 = \sum_{n=1}^{\infty}\binom{s+n-1}{n} w^n$, the cdf of the complementary exponentiated G-NB (CEGNB) class reduces to

$$F_{CEGNB}(z) = \frac{[1-\beta\, G(z;\tau)^{\alpha}]^{-s}-1}{(1-\beta)^{-s}-1}.$$

For $\alpha = 1$, the CEGNB class leads to the new complementary GNB class.

5 Review of existing compounded models

In this section, we review some available compounded models. In the literature, several authors have reported compounding discrete distributions, namely the ZTG, ZTP, logarithmic, ZTBi, ZTNB, zero-truncated generalized Poisson and zero-truncated power-series, with continuous lifetime models.

5.1 Compounded models based on geometric distribution

For the following models, X denotes the r.v. of the baseline G model.

Exponential-geometric (EG) model. If $X \sim$ Exponential(β) have cdf $G(x;\beta) = 1 - \exp(-\beta\, x)$, then the cdf of the EG model pioneered by Adamidis and Loukas (1998) (as the first compounded model in the literature) can be determined from the GG class (Section 3.1 with $\alpha = 1$) as

$$F_{EG}(y;p,\beta) = \frac{1-\mathrm{e}^{-\beta\, y}}{1-p\,\mathrm{e}^{-\beta\, y}}, \quad y,\beta>0,\ p\in(0,1).$$

Adamidis et al. (2005) also defined an extended EG model.

Generalized EG (GEG) model. If $X \sim$ Exponential(β), Silva et al. (2010) applied the LA1 class to the EG cdf in the last equation and defined the GEG cdf by

$$F_{GEG}(y;p,\beta,\gamma) = \left(\frac{1-\mathrm{e}^{-\beta\, y}}{1-p\,\mathrm{e}^{-\beta\, y}}\right)^{\gamma}, \quad y>0,$$

where $\gamma > 0$ is the power parameter, $\beta > 0$ and $p \in (0,1)$.

Complementary EG (CEG) model (proposed). If $X \sim$ Exponential(β), then the CEG cdf can be determined from the CEGGA class (Section 4.1 with $\alpha = 1$) as

$$F_{CEG}(z;p,\beta) = \frac{(1-p)(1-\mathrm{e}^{-\beta\, z})}{1-p(1-\mathrm{e}^{-\beta\, z})}.$$

Long-term complementary EG (LCEG) model. If $X \sim$ Exponential(β), then Louzada et al. (2012d) proposed another CEG model called the LCEG model, in a latent complementary risk framework, with cdf given by

$$F_{LCEG}(z;p,\beta,\theta) = \frac{\theta(1-p)(1-\mathrm{e}^{-\beta\, z})}{\mathrm{e}^{-\beta\, z}(1-\theta)+\theta}, \quad z>0,\ \theta\in(0,1),\ p\in(0,1),$$

where $\beta > 0$ is a scale parameter, θ is a shape parameter and p is the long-term parameter.

Generalized complementary EG (GCEG) model. Bidram and Nadadrajah (2016) applied the LA1 class to the CEGA cdf (given in Section 4.1 with $\alpha = 1$) and obtained the GCEG cdf as

$$F_{GCEG}(z;p,\beta,\gamma)=\left[\frac{(1-p)(1-\mathrm{e}^{-\beta z})}{1-p\,(1-\mathrm{e}^{-\beta z})}\right]^{\gamma},\quad z>0,$$

where $\gamma>0$ is the power parameter, $\beta>0$ and $p\in(0,1)$.

Complementary exponentiated-exponential geometric (CEEG) model. If $X\sim$ ExponentiatedExponential(β,α) have cdf $G(x;\beta,\alpha)=[1-\exp(-\beta x)]^{\alpha}$, then the cdf of the CEEG model, proposed by Louzada et al. (2014a), can be determined from the alternative CEGG class (Section 4.1 with $\alpha=1$) as

$$F_{CEEG}(z;p,\beta,\alpha)=\frac{p(1-\mathrm{e}^{-\beta z})^{\alpha}}{1-(1-p)[1-\mathrm{e}^{-\beta z}]^{\alpha}},\quad z,\alpha,\beta>0,\ p\in(0,1).$$

Exponentiated-CEG (ECEG) model. Yamachi et al. (2013) generalized the CEG model $F_{CEG}(z;p,\beta,b)=\frac{p(1-\mathrm{e}^{-b\beta z})}{\mathrm{e}^{-\beta z}(1-p)+p}$ introduced by Louzada et al. (2011) by applying the LA1 class and defined the ECEG cdf as

$$F_{ECEG}(z;p,\beta,b,\gamma)=\left(\frac{p(1-\mathrm{e}^{-b\beta z})}{\mathrm{e}^{-\beta z}(1-p)+p}\right)^{\gamma},$$

where $\gamma>0$ is the power parameter.

Beta-EG (BEG) model. If $X\sim$ Exponential(β), the cdf of the BEG model proposed by Bidram (2012) and Nassar and Nada (2012) can follow from the BGG class (Section 3.4) as

$$F_{BEG}(y;p,\beta,l,m)=\frac{B_{l,m(EG)}}{1-p+p\,B_{l,m(EG)}},$$

where

$$B_{l,m(EG)}=\frac{1}{B(l,m)}\int_0^{\left[\frac{1-\mathrm{e}^{-\beta x}}{1-p\,\mathrm{e}^{-\beta x}}\right]}t^{l-1}\,(1-t)^{m-1}\,dt.$$

Exponentiated EG (EEG) model. The cdf of the EEG model defined by Louzada et al. (2014) can be determined from the EGGA class (Section 3.1) as

$$F_{EEG}(y;p,\beta,\alpha)=\frac{(1-\mathrm{e}^{-\beta y})^{\alpha}}{1-(1-p)[1-(1-\mathrm{e}^{-\beta y})^{\alpha}]},\ y>0\ \alpha,\beta>0,\ p\in(0,1).$$

Ristić and Kundu (2016) proposed the generalized geometric extreme distribution, which is identical to the EEG model given above.

Modified EG (MEG) model. Let $X\sim$ ModifiedExponential(β,π) or the MO-exponential with cdf $G(x;\beta,\pi)=1-\exp(-\beta x)/[1-(1-\pi)\exp(-\beta x)]$. The cdf of the modified EG (MEG) distribution given by Bordbar and Nematollah (2016) can follow from the GG class (Section 3.1 with $\alpha=1$) as

$$F_{MEG}(y;p,\beta,\pi)=\frac{1-\mathrm{e}^{-\beta y}}{1-(1-\pi)\mathrm{e}^{-\beta y}-p\pi\,\mathrm{e}^{-\beta y}},\quad y>0.$$

Exponentiated-LL geometric (ELLG) and complementary exponentiated-LL geometric (CELLG) models. If $X\sim$ ELL(a,b) has cdf $G(x;a,b,\alpha)=\frac{\alpha b}{a}(x/a)^{-b-1}\left[1+(x/a)^{-b}\right]^{-\alpha-1}$, then the cdfs of the ELLG and CELLG models proposed by Mendoza et al. (2016) can be determined from the GG and CGG classes described in Sections 3.1 and 4.1 (with $\alpha=1$), respectively, as

$$F_{ELLG}(y;p,a,b,\alpha)=\frac{\left[1+(y/a)^{-b}\right]^{-\alpha}}{1-p\left\{1-\left[1+(y/a)^{-b}\right]^{-\alpha}\right\}},\quad y>0,\ a,b,\alpha,\ p\in(0,1)$$

and

$$F_{CELLG}(z;p,a,b,\alpha) = \frac{(1-p)\left[1+(z/a)^{-b}\right]^{-\alpha}}{1-p\left[1+(z/a)^{-b}\right]^{-\alpha}}, \quad z>0,\ a,b,\alpha,\ p\in(0,1).$$

Generalized linear failure rate-geometric (GLFRG) model. If $X \sim$ LinearExponential(a,b) have cdf $G(x;a,b) = 1-\exp[-(ax+(b/2)x^2)]$, then the cdf of the GLFRG model proposed by Nadarajah et al. (2014a) can follow from the EGG class (Section 3.1) as

$$F_{GLFRG}(y;p,a,b,\alpha) = \frac{[1-\mathrm{e}^{-(ay+(b/2)y^2)}]^{\alpha}}{1-p\left\{1-[1-\mathrm{e}^{-(ay+(b/2)y^2)}]^{\alpha}\right\}}, \ y>0,\ a,b,\alpha>0.$$

Complementary generalized linear failure rate-geometric (CGLFRG) model. If $X \sim$ LinearExponential(a,b) have cdf $G(x;a,b) = 1-\exp[-(ax+(b/2)x^2)]$, then the CGLFRG cdf defined by Harandi and Alamatsaz (2016) can be determined from the CEGG class (Section 4.1) as

$$F_{CGLFRG}(z;p,a,b,\alpha) = \frac{(1-p)[1-\mathrm{e}^{-(az+(b/2)z^2)}]^{\alpha}}{1-p\left\{1-[1-\mathrm{e}^{-(az+(b/2)z^2)}]^{\alpha}\right\}}, \ z>0,\ a,b,\alpha>0.$$

Exponentiated power Lindley-geometric (EPLG) model. If $X \sim$ ExpPowerLindley(α,β,θ) introduced by Ashour and Eltehiwy (2015) have cdf $G(x;\alpha,\beta,\theta) = \left[1-\left(1+\frac{\theta x^{\beta}}{\theta+1}\right)\mathrm{e}^{-\theta x^{\beta}}\right]^{\alpha}$, the EPLG cdf defined by Alizadeh et al. (2016b) can follow from the GG class (Section 3.1) as

$$F_{EPLG}(y;p,\alpha,\beta,\theta) = \frac{(1-p)\left[1-\left(1+\frac{\theta y^{\beta}}{\theta+1}\right)\mathrm{e}^{-\theta y^{\beta}}\right]^{\alpha}}{1-p\left[1-\left(1+\frac{\theta y^{\beta}}{\theta+1}\right)\mathrm{e}^{-\theta y^{\beta}}\right]^{\alpha}}, \ y>0,\ \alpha,\beta,\theta>0.$$

Weibull-geometric (WG) model. If $X \sim$ Weibull(a,b) have cdf $G(x;a,b) = 1-\mathrm{e}^{-(bx)^a}$, then the cdf of the WG model proposed by Barreto-Souza et al. (2011) can be expressed from the GG class (Section 3.1 with $\alpha=1$) as

$$F_{WG}(y;p,a,b) = \frac{1-\mathrm{e}^{-(by)^a}}{1-p\,\mathrm{e}^{-(by)^a}}, \quad y,a,b>0,\ p\in(0,1).$$

Exponentiated Weibull-geometric (EWG) model. If $X \sim$ Weibull(a) have cdf $G(x;a) = 1-\mathrm{e}^{-x^a}$, then the EWG cdf given by Chung and Kang (2014) can be obtained from the EGG cdf (Section 3.1) as

$$F_{EWG}(y;p,a,\alpha) = \frac{\left(1-\mathrm{e}^{-y^a}\right)^{\alpha}}{1-p\left[1-\left(1-\mathrm{e}^{-y^a}\right)^{\alpha}\right]}, \quad y,a,\alpha>0,\ p\in(0,1).$$

Beta-Weibull-geometric (BWG) model. If $X \sim$ Weibull(a,b) have cdf $G(x;a,b) = 1-\mathrm{e}^{-(bx)^a}$, then the cdf of the BWG model defined by Bidram et al. (2013) and Cordeiro et al. (2013b) can follow from the BGG cdf (given in Section 3.4) as

$$F_{BWG}(y;p,a,b,l,m) = \frac{B_{l,m(WG)}(y)}{1-p+p\,B_{l,m(WG)}(y)},$$

where

$$B_{l,m(WG)}(y) = \frac{1}{B(l,m)}\int_0^{\frac{1-\mathrm{e}^{-(by)^a}}{1-p\,\mathrm{e}^{-(by)^a}}} t^{l-1}\,(1-t)^{m-1}\,dt.$$

Complementary exponentiated Weibull-geometric (CEWG) model. If X have Weibull cdf $G(x;a,b) = 1-\mathrm{e}^{-(bx)^a}$, the cdf of the CEWG model proposed by Mahmoudi and Shiran (2012a) follows from the CEGG class (Section 4.1) as

$$F_{CEWG}(z;p,a,b,\alpha)=\frac{(1-p)(1-e^{-(bz)^a})^\alpha}{1-p\left[1-e^{-(bz)^a}\right]^\alpha},\quad z,a,b,\alpha>0,\ p\in(0,1).$$

Complementary Weibull-geometric (CWG) model. If $X\sim$ Weibull(a,b) have cdf $G(x;a,b)=1-e^{-(bx)^a}$, the cdf of the CWG model proposed by Tojeiro et al. (2014) follows from the alternative CGG class given in Section 4.1 ($\alpha=1$) as

$$F_{CWG}(z;p,a,b)=\frac{p\left(1-e^{-(bz)^a}\right)}{p+(1-p)e^{-(bz)^a}},\quad z,a,b>0,\ p\in(0,1).$$

Modified Weibull-geometric (MWG) model. If $X\sim$ ModifiedWeibull(a,b,δ) introduced by Sarhan and Zaindin (2009) have cdf $G(x;a,b,\delta)=1-\exp\left[-(ax+bx^\delta)\right]$, then the cdf of the MWG model defined by Wang and Elbatal (2015) can be determined from the GG class (Section 3.1, $\alpha=1$ and all parameters positive) as

$$F_{MWG}(y;p,a,b,\delta)=\frac{1-e^{-(ay+by^\delta)}}{1-p\,e^{-(ay+by^\delta)}},\quad y>0.$$

Kummer-beta Weibull-geometric (KBWG) model. Ashour and Wahed (2014) extended the WG model to the KBWG model. The KBWG cdf is given by

$$F_{KBWG}(y;p,a,b,c)=K\int_0^{\frac{1-e^{-(by)^a}}{1-pe^{-(by)^a}}}y^{l-1}\,(1-y)^{m-1}\,e^{-cx}\,dx,$$

where $K=\frac{\Gamma(l)\Gamma(m)}{\Gamma(l+m)}\,{}_1F_1(l;l+m;-c)$ and ${}_1F_1$ is the Gauss hyper-geometric function.

Additive Weibull-geometric (AWG) model. If $X\sim$ AddiveWeibull(a,b,γ) introduced by Cordeiro et al. (2014c) have cdf $G(x;a,b,\gamma)=1-e^{-(ax+bx^\gamma)},\quad x,a,b>0$ and $\gamma\in(0,\infty)\setminus\{1\}$, then the cdf of the AWG model defined by Elbatal et al. (2016) can be obtained from the GG class (Section 3.1, $\alpha=1$) as

$$F_{AWG}(y;p,a,b,\gamma)=\frac{1-e^{-(ax+bx^\gamma)}}{1-p\,e^{-(ax+bx^\gamma)}},\quad y>0.$$

Lindley-geometric (LG) model. If $X\sim$ Lindley(θ) have cdf $G(x;\theta)=1-\left(1+\frac{\theta x}{\theta+1}\right)e^{-\theta x}$ (for $x>0,\ \theta>0$), then the cdf of the LG model proposed by Zakerzadeh and Mahmoudi (2013) can follow from the GG class (Section 3.1, $\alpha=1$) as

$$F_{LG}(y;p,\theta)=\frac{1-\left(1+\frac{\theta y}{\theta+1}\right)e^{-\theta y}}{1-p\left(1+\frac{\theta y}{\theta+1}\right)e^{-\theta y}},\quad y>0,\quad \theta>0,\quad p\in(0,1).$$

Exponentiated Lindley-geometric (ELG) model. If $X\sim$ Lindley(θ) have cdf $G(x;\theta)=1-\left(1+\frac{\theta x}{\theta+1}\right)e^{-\theta x}$ (for $x>0,\ \theta>0$), then the cdf of the ELG model proposed by Wang (2013) follows from the EGG class (Section 3.1) as

$$F_{ELG}(y;p,\theta)=\frac{\left[1-\left(1+\frac{\theta y}{\theta+1}\right)e^{-\theta y}\right]^\alpha}{1-p+p\left[1-\left(1+\frac{\theta y}{\theta+1}\right)e^{-\theta y}\right]^\alpha}.$$

5.2 Compounded models based on Poisson distribution

Exponential-Poisson (EP). If $X\sim$ Exponential(β), then the cdf of the EP model proposed by Kuş (2007) can be obtained from the GP class defined in Section 3.5 (with $\alpha=1$) as

$$F_{EP}(y;\lambda,\beta)=\frac{1-e^{-\lambda-\lambda e^{-\beta y}}}{1-e^{-\lambda}},\quad y,\beta>0,\ p\in(0,1).$$

For some structural properties and applications of the EP model, the reader is referred to Kuş (2007) and Cancho et al. (2011a).

Generalized EP (GEP) model. If $X \sim$ Exponential(β), Barreto-Souza and Cribari-Neto (2009) obtained the GEP cdf by applying the LA1 class to the EP cdf given in the last equation

$$F_{GEP}(y; \lambda, \beta, \gamma) = \left(\frac{1 - e^{-\lambda - \lambda e^{-\beta y}}}{1 - e^{-\lambda}} \right)^{\gamma}, \quad y > 0,$$

where $\gamma > 0$ is the power parameter, $\beta > 0$ and $p \in (0, 1)$.

Complementary EP (CEP) model. If $X \sim$ Exponential(β), the cdf of the CEP model, proposed by Cancho et al. (2011a) and Rezaei and Tahmasbi (2012), can be determined from the CGP class (Section 4.5, $\alpha = 1$) as

$$F_{CEP}(z; \lambda, \beta) = \frac{e^{-\lambda(1-e^{-\beta z})} - e^{-\lambda}}{1 - e^{-\lambda}}, \quad z, \beta \, \lambda > 0.$$

Modified EP (MEP) model. If $X \sim$ ModifiedExponential(β) or MO-exponential with cdf $G(x; \beta) = \frac{1-e^{-\beta x}}{1-(1-p)e^{-\beta x}}$, then the cdf of the MEP model, proposed by Preda et al. (2011), can follow from the GP class (Section 3.9, $\alpha = 1$) as

$$F_{MEP}(y; \lambda, p, \beta) = \frac{1 - e^{-\lambda\left(\frac{1-e^{-\beta y}}{1-(1-p)e^{-\beta y}}\right)}}{1 - e^{-\beta y}}, \quad y, \beta > 0, \; p \in (0, 1).$$

Exponentiated EP (EEP) model. If $X \sim$ Exponential(β), then the cdf of the EEP model given by Ristić and Nadarajah (2014) can be obtained from the EGP class defined in Section 3.5 as

$$F_{EEP}(y; \lambda, \beta, \alpha) = \frac{1 - e^{-\lambda(1-e^{-\beta y})^{\alpha}}}{1 - e^{-\lambda}}, \quad y > 0, \quad \alpha, \beta, \lambda > 0.$$

Weibull-Poisson (WP) model. If $X \sim$ Weibull(a) have cdf $G(x; a, b) = 1 - e^{-bx^a}$, then the cdf of the WP model proposed by Lu and Shi (2012) can follow from the GP class (Section 3.5, $\alpha = 1$) as

$$F_{WP}(y; \lambda, a, b) = \frac{e^{\lambda \, e^{-by^a}} - e^{\lambda}}{1 - e^{\lambda}}, \quad y, a, b, \lambda > 0.$$

Hemmati et al. (2011) and Bereta et al. (2011) also obtained the WP or Poisson-Weibull (PW) models from the Weibull cdf $G(y; a, b) = 1 - \exp\{-(b y)^a\}$. The cdf of their WP model is given by

$$F_{WP}(y; \lambda, a, b) = \frac{e^{\lambda \, e^{-(by)^a}} - e^{\lambda}}{1 - e^{\lambda}}, \quad y, a, b, \lambda > 0.$$

Beta-Weibull Poisson (BWP) model. If $X \sim$ Weibull(a, b) have cdf $G(x; a, b) = 1 - e^{-bx^a}$, then the cdf of the BWP model, proposed by Percontini et al. (2013a), can follow from the BGP class defined in Section 3.8 as

$$F_{BEG}(y; \lambda, a, b, l, m) = \frac{e^{-\lambda \left[1 - B_{l,m(EG)}(y)\right]} - e^{\lambda}}{1 - e^{\lambda}},$$

where

$$B_{l,m(EG)}(y) = \frac{1}{B(l, m)} \int_0^{\frac{e^{\lambda \, e^{-by^a}} - e^{\lambda}}{1-e^{\lambda}}} t^{l-1} \, (1 - t)^{m-1} \, dt.$$

Complementary modified Weibull-Poisson (CMWP) model. If $X \sim$ ModifiedWeibull(a, b, δ) introduced by Sarhan and Zaindin (2009) have cdf $G(x; a, b, \delta) = 1 - \exp\left[-(ax + b\,x^{\delta})\right]$, then the cdf of the CMWP model proposed by Ghorbani et al. (2014) can be obtained from the GP class presented in Section 3.5 (with $\alpha = 1$) as

$$F_{CMWP}(z; \lambda, a, b, \delta) = \frac{e^{\lambda[1-e^{-(az+bz^{\delta})}]} - 1}{e^{\lambda} - 1}, \quad z > 0,\ a, b, \delta, \lambda > 0.$$

Complementary exponentiated Weibull-Poisson (CEWP) model. If $X \sim$ Weibull(a, b) have cdf $G(x; a, b) = 1 - e^{-(bx)^{a}}$, then the cdf of the CEWP model defined by Mahmoudi and Sepahdar (2013) can follow from the EGP class (see Section 3.5) as

$$F_{CEWP}(z; \lambda, a, b, \alpha) = \frac{e^{\lambda(1-e^{-(bz)^{a}})^{\alpha}} - 1}{e^{\lambda} - 1}, \quad z, \lambda, a, b, \alpha > 0.$$

Log-logistic generalized Weibull Poisson (LLGWP) model. Let $X \sim$ Weibull(a, b) have cdf $G(x; a, b) = 1 - e^{-bx^{a}}$. Oluyede et al. (2016a) first defined the log-logistic generalized Weibull (LLGW) cdf by $F_{LLGW} = 1 - [1 + (x/s)^{c}]^{-1}\, e^{-bx^{a}}$ from the generalized extended Weibull family $G(x; \theta, \psi) = 1 - B(x; \psi)\, \exp\{-\alpha\, H(x; \theta)\}$, where θ and ψ are both vectors, by taking $H(x; \theta) = x^{b}$ and $B(x; \psi) = 1 - [1 + (x/s)^{c}]$. Then, the cdf of the LLGWP model proposed by Oluyede et al. (2016c) can follow from the GP class (Section 3.5, $\alpha = 1$) as

$$F_{LLGWP}(y; \lambda, b, c, s) = \frac{1 - e^{\lambda\left[1-(1+(y/s)^{c})^{-1}\, e^{-by^{a}}\right]}}{1 - e^{\lambda}}, \quad y, \lambda, b, c, s > 0.$$

Lai-modified Weibull-Poisson (LaiMWP) model. If $X \sim$ ModifiedWeibull(a, b, δ) introduced by Lai et al. (2003) have cdf $G(x; a, b, \delta) = 1 - \exp\left(-ax^{\delta}\, e^{bx}\right), a, b, \delta > 0$, then the cdf of the LaiMWP model proposed by Delgarm and Zadkarami (2015) can be obtained from the GP class (Section 3.5, $\alpha = 1$) as

$$F_{LaiMWP}(y; \lambda, a, b, \delta) = \frac{1 - e^{-\lambda\left[1-e^{-ay^{\delta}\, e^{by}}\right]}}{1 - e^{-\lambda}}, \quad y > 0,\ a, b, \delta, \lambda > 0.$$

Exponentiated Lomax-Poisson (ELoP) model. If $X \sim$ Lomax(a, b) have cdf $G(x; a, b) = 1 - (1 + b\,x)^{-a}$, then the cdf of the ELoP model proposed Ramos et al. (2013) can be determined from the EGP class (Section 3.5) as

$$F_{ELoP}(y; \lambda, a, b, \alpha) = \frac{e^{\lambda\,[1-(1+b\,y)^{-a}]^{\alpha}} - 1}{e^{\lambda} - 1}, \quad y, \lambda, a, b, \alpha > 0.$$

Complementary Poisson-Lomax (CPLo) model. Let $X \sim$ Lomax(a, b) have cdf $G(x; a, b) = 1 - (1 + b\,x)^{-a}$, then the cdf of the CLoP model proposed Al-Zahrani and Sagor (2014) can follow from the CGP class (Section 4.5, $\alpha = 1$) as

$$F_{CLoP}(z; \lambda, a, b) = \frac{1 - e^{-\lambda[1-(1+b\,z)^{-a}]}}{1 - e^{-\lambda}}, \quad z, \lambda, a, b > 0.$$

Lindley-Poisson (LP) model. Let $X \sim$ Lindley(θ) have cdf $G(x; \theta) = 1 - \left(1 + \frac{\theta\,x}{\theta+1}\right) e^{-\theta\,x}$, $x, \theta > 0$. The cdf of the LP model introduced by Gui et al. (2014) can be determined from the GP class (Section 3.5, $\alpha = 1$) as

$$F_{LP}(y; \lambda, \theta) = \frac{1 - e^{-\lambda\left[1-\left(1+\frac{\theta\,y}{\theta+1}\right)e^{-\theta\,y}\right]}}{1 - e^{-\lambda}}, \quad y > 0,\ \theta, \lambda > 0.$$

Complementary extended Lindley-Poisson (CExtLP) model. If $X \sim$ ExtLindley(θ, δ, β) proposed by Bakouch et al. (2012a) have cdf $G(x; \theta, \delta, \beta) = 1 - \left(\frac{1+\theta+\theta\, x}{\theta+1}\right)^{\delta} e^{-(\theta\, x)^{\beta}}$, the cdf of the CExtLP model introduced by Pararai et al. (2015b) can be determined from the CGP class (Section 4.5, $\alpha = 1$) as

$$F_{CExtLP}(z; \lambda, \theta, \delta, \beta) = \frac{1 - e^{\lambda\left[1-\left(\frac{1+\theta+\theta\, z}{\theta+1}\right)^{\delta} e^{-(\theta\, z)^{\beta}}\right]}}{1 - e^{\lambda}}, \quad z > 0,\ \theta, \delta, \beta, \lambda > 0.$$

Poisson Birnbaum-Saunders (PBS) model. Let $X \sim$ BS(a, b) have cdf $G(x; a, b) = \Phi\left[(1/a) + \left(\sqrt{x/b} - \sqrt{b/x}\right)\right]$, $x, a, b > 0$, then the cdf of the PBS model, proposed by Hashimoto et al. (2014) in terms of long-term survivors, can be obtained from the GP class (Section 3.5, $\alpha = 1$) as

$$F_{PBS}(y; \lambda, a, b) = \frac{1 - e^{-\lambda\, \Phi\left[\frac{1}{a}+\left(\sqrt{\frac{y}{b}}-\sqrt{\frac{b}{y}}\right)\right]}}{1 - e^{-\lambda}}, \ y > 0,\ a, b, \lambda > 0.$$

Exponentiated-Burr XII Poisson (EBP) model. Let $X \sim$ Burr(c, k, s) have cdf $G(x; c, k, s) = 1 - [1 + (x/s)^c]^{-k}$, $x, c, k, s > 0$, then the cdf of the EBP model introduced by da-Silva et al. (2015b) can be determined from the EGP class (Section 3.5) as

$$F_{EBP}(y; \lambda, c, k, s, \alpha) = \frac{1 - e^{-\lambda\left\{1+\left[1-(y/s)^c\right]^{-k}\right\}^{\alpha}}}{1 - e^{-\lambda}}, \quad y > 0,\ c, k, s, \alpha, \lambda > 0.$$

Complementary Burr III Poisson (CB3P) model Let $X \sim$ BurrIII(c, k) have cdf $G(x; c, k) = \left(1 + x^{-c}\right)^{-k}$, $x, c, k > 0$, then the cdf of the CB3P model defined by Hassan et al. (2015) can follow from the CGP class (Section 4.5, $\alpha = 1$) as

$$F_{CB3P}(z; \lambda, c, k) = \frac{e^{\lambda\,(1+z^{-c})^{-k}} - 1}{e^{\lambda} - 1}, \quad z, c, k, \lambda > 0.$$

The Burr III distribution is also known as the Dagum distribution (Dagum 1977), which is a very well-known model for studying income and wealth inequality data. Oluyede et al. (2016b) have recently introduced and studied the Dagum-Poisson distribution.

Poisson generalized linear failure rate (PGLFR) model. Let $X \sim$ GLFR(a, b, α) introduced by Sarhan and Kundu (2009) have cdf $G(x; a, b, \alpha) = \left(1 - \exp\left[-ax - (b/2)x^2\right]\right)^{\alpha}$, $x, a, b, \alpha > 0$, then the cdf of the PGLFR model defined by Cordeiro et al. (2015b) can be obtained from the EGP class (Section 3.5) as

$$F_{PGLFR}(y; \lambda, a, b, \alpha) = \frac{1 - e^{-\lambda\,\{1-\exp[-ay-(b/2)\,y^2]\}^{\alpha}}}{1 - e^{-\lambda}}, \quad y > 0,\ a, b, \alpha, \lambda > 0.$$

Complementary failure rate Poisson (CLFRP) model. Let $X \sim$ LFR(a, b, α), then the cdf of the CLFRP model defined by Gitifar et al. (2016) can follow from the CEGP class (Section 4.5, $\alpha = 1$) as

$$F_{CLFRP}(z; \lambda, a, b) = \frac{e^{\lambda\,\{1-\exp[-ay-(b/2)\,y^2]\}}}{e^{\lambda} - 1}, \quad z > 0,\ a, b, \lambda > 0.$$

Complementary exponentiated power Lindley-Poisson (CEPLP) model. If $X \sim$ ExpPowerLindley(α, β, θ) introduced by Warahena-Liyanage and Pararai (2015b) have cdf $G(x; \alpha, \beta, \theta) = \left[1 - \left(1 + \frac{\theta\, x^{\beta}}{\theta+1}\right) e^{-\theta\, x^{\beta}}\right]^{\alpha}$ (for $x > 0$, $\alpha, \beta, \theta > 0$), then the EPLG cdf defined by Pararai et al. (2016) can follow from the CGP class (Section 4.5) as

$$F_{CEPLP}(y; \lambda, \alpha, \beta, \theta) = \frac{e^{\lambda\left[1-\left(1+\frac{\theta\, x^{\beta}}{\theta+1}\right) e^{-\theta\, x^{\beta}}\right]^{\alpha}}}{e^{\lambda} - 1}, \quad z > 0,\ \alpha, \beta, \theta, \lambda > 0.$$

5.3 Compounded models based on the logarithmic distribution

The following compounded models have been reported in the literature from the logarithmic discrete model.

Exponential-logarithmic (ELn). If $X \sim$ Exponential(β) have cdf $G(x;\beta) = 1 - \exp(-\beta x)$, then the cdf of the exponential-logarithmic (ELn) model proposed by Tahmasbi and Rezaei (2008) can be obtained from the GLn class (Section 3.9, $\alpha = 1$) as

$$F_{ELn}(y;\beta,\phi) = 1 - \frac{\ln\left[1-(1-\phi)\,e^{-\beta y}\right]}{\ln\phi}. \tag{8}$$

Generalized ELn. Pappas et al. (2015) applied the LA1 class to the ELn cdf given in Eq. (8), and obtained the cdf of the generalized ELn (GELn) distribution as

$$F_{GELn}(y;\beta,\phi,\gamma) = \left(1 - \frac{\ln\left[1-(1-\phi)\,e^{-\beta y}\right]}{\ln\phi}\right)^{\gamma},$$

where $\gamma > 0$ is the power parameter.

Chen-logarithmic (ChLn). If $X \sim$ Chen(γ,β) have cdf $G(x;\gamma,\beta) = e^{\gamma(e^{x^{\beta}}-1)}$ introduced by Chen (2000), then the cdf of the Chen-logarithmic (ChLn) model proposed by Pappas et al. (2011) can be determined from the GLn class (Section 3.9 with $\alpha = 1$) as

$$F_{ChLn}(y;\gamma,\beta,\phi) = 1 - \frac{\ln\left[1-(1-\phi)e^{\gamma(e^{x^{\beta}}-1)}\right]}{\ln\phi}.$$

Complementary exponentiated Weibull-logarithmic (CEWLn) model. Let $X \sim$ EW(β,δ,α) have cdf $G(x;\beta,\delta,\gamma) = \alpha\,\delta\,\beta^{\delta}\,y^{\delta}\,e^{-(\beta y)^{\delta-1}}[1-e^{-(\beta y)^{\delta}}]^{\alpha-1}$, then the cdf of the complementary exponentiated Weibull-logarithmic (CEWLn) model introduced by Mahmoudi et al. (2014) can follow from the cdf of the CGLn class given in Section 4.9 as

$$F_{CEWLn}(z;\beta,\phi,\delta,\alpha) = \frac{\ln\left[1-\phi\left(1-e^{-(\beta z)^{\delta}}\right)^{\alpha}\right]}{\ln(1-\phi)},$$

where $\phi \in (0,1)$, $\alpha > 0$ and $\delta > 0$ are shape parameters and $\beta > 0$ is a scale parameter.

Lomax-logarithmic (LxLn) model. Let $X \sim$ Lomax(a,b) have cdf $G(x;a,b) = (1+bx)^{-a}$, then the cdf of the Lomax-logarithmic (LxLn) model given by Al-Zahrani and Sagor (2015) can be obtained from the GLn class given in Section 3.9 with $\alpha = 1$ as

$$F_{LxLn}(y;a,b,\phi) = 1 - \frac{\ln\left[1-(1-\phi)\,(1+by)^{-a}\right]}{\ln\phi}.$$

Exponentiated Poisson-logarithmic (EPoLn) model. Let $X \sim$ ExpPoisson(θ,β) have cdf $G(x;a,b) = 1 - \frac{1-\exp\{-\theta\,e^{-\beta x}\}}{1-e^{-\theta}}$, then the cdf of the exponentiated Poisson-logarithmic (EPoLn) model given by Fioruci et al. (2016) can follow from the alternative GLn cdf (Section 3.9 with $\alpha = 1$) as

$$F_{EPoLn}(y;\beta,\phi,\theta) = 1 - \frac{\ln\left[1-\phi\left\{1-\frac{e^{-\theta\,e^{-\beta y}}}{1-e^{-\theta}}\right\}\right]}{\ln(1-\phi)}.$$

6 Other compounded models

(i) *Extended Weibull power series (EWPS) models.* Let $T[\,G(y;\tau);\alpha] = 1 - e^{-\alpha H(y;\tau)}$ be the extended Weibull family, where $x > 0$, $\alpha > 0$ and $H(y;\tau)$ is a non-negative monotonically increasing function depending on a parameter vector τ. Much more than twenty known lifetime model cdfs can be expressed in this form. Further, let N be a discrete random variable following a power series (PS) distribution (truncated at zero) with pmf

$$p_n = P(N = n) = \frac{a_n \theta^n}{C(\theta)}, n = 1, 2, \ldots, \tag{9}$$

where a_n depends only on n, $C(\theta) = \sum_{n=1}^{\infty} a_n \theta^n$ and $\theta > 0$ is such that $C(\theta)$ is finite. Equation (9) summarizes some power series distributions (truncated at zero) such as the Poisson, logarithmic and geometric distributions, where $C(\theta)$ is equal to $(e^{\theta} - 1)$, $-\log(1-\theta)$ and $\theta(1-\theta)^{-1}$, respectively.

Silva et al. (2013) defined the EWPS cdf obtained under the current set-up from Eq. (4) as

$$F(y; \theta, \alpha, \tau) = 1 - \frac{C(\theta\, e^{-\alpha H(y;\tau)})}{C(\theta)}, \quad y > 0. \tag{10}$$

Equation (10) includes as special cases the Weibull power series (WPS) class, which extends the exponential power series (EPS) family. In fact, this class includes much more than 60 (20 × 3) special models, some of them given by Silva et al. (2013) and others yet not investigated. In a similar context, more recently, Silva et al. (2016) defined a family by compounding the generalized gamma (GGa) and power series distributions.

(ii) *Complementary extended Weibull power series (CEWPS) models.* The CEWPS class follows by taking $T[\,G(y;\tau);\alpha] = 1 - e^{-\alpha H(y;\tau)}$ and the PS distribution for the pmf. So, the cdf of the CEWPS class can be obtained from Eq. (5) as

$$F(z; \theta, \alpha, \tau) = \frac{C\left[\theta\,(1 - e^{-\alpha H(x;\tau)})\right]}{C(\theta)}, \quad x > 0.$$

The CEWPS class can arise in parallel systems with identical components, which appear in many industrial and biological applications.

(iii) *Exponential-geometric range (max-min) model (New).* Let N be the ZTG r.v. with pmf of type 1 $\mathbb{P}(N = n) = q p^{n-1}$, $n = 1, 2, \ldots$ and define the range $X_R = \max\{Y_1 \ldots, Y_N|p\} - \min\{Y_1 \ldots, Y_N|p\}$. The unconditional cdf of the G-geometric range ($R = \max - \min$) can be obtained as

$$F(x; p, \alpha, \tau) = \frac{(1-p)\, G(x;\tau)^{\alpha}}{1 - p\, G(x;\tau)^{\alpha}}.$$

If $Y \sim Exp(\beta)$ have density $f_{Y|\beta}(y) = \beta e^{-\beta y}$ and $Z|p \sim Geometric(p)$ have pmf $\mathbb{P}(N = n) = q p^{n-1}$, $n = 0, 1, \ldots$, Shahsanaei et al. (2012) proposed the unconditional cdf of the exponential-geometric range (EGR) model as

$$F(x; \beta, p) = \frac{(1-p)\left(1 - e^{-\beta x}\right)}{1 - p\left(1 - e^{-\beta x}\right)}.$$

The EGR distribution is useful for modeling the time between the first failure to the last failure.

(iv) *Exponential-generalized Poisson (EGP) model.* Let N have the zero truncated generalized Poisson (ZTGP) distribution with parameters λ and α and pmf

$$\mathbb{P}(N = n) = \frac{\lambda(\lambda + \alpha n)^{n-1}\, e^{-(\lambda+\alpha n)}}{(1 - e^{-\lambda})\,\Gamma(n+1)}, \quad n = 1, 2, \ldots.$$

Then, Gupta et al. (2014) proposed the EGP cdf given by

$$F(y; \alpha, \beta, \lambda) = \frac{1 - e^{-(\lambda/\alpha)[W(g)+\alpha]}}{1 - e^{-\lambda}},$$

where $g = -\alpha\, \mathrm{e}^{-(\alpha+\beta y)}$ and $W(\cdot)$ is the Lambert W function defined by $W(x)\, \mathrm{e}^{W(x)} = x$. The above cdf can be derived from the generating function of the ZTGP distribution $M_N^*(t) = \frac{M_N(t)-P(N=0)}{1-P(N=0)} = \frac{1-\mathrm{e}^{-(\lambda/\alpha)[W(g)+\alpha]}-\mathrm{e}^{-\lambda}}{1-\mathrm{e}^{-\lambda}}$, which is equivalent to the survival function $1 - F(t;\alpha,\beta,\lambda)$.

7 A different approach of compounding

The compounding of some models (continuous with continuous and discrete with discrete) are introduced such as the exponential-Weibull (Cordeiro et al. 2014c), generalized exponential-exponential (Popović et al. 2015), geometric exponential Poisson (Nadarajah et al. 2013a) and additive Weibull (Xie and Lai 1995) distributions.

(i) *Exponential-Weibull model.* Let W and T be independent random variables with exponential and Weibull distributions characterizing two series sub-system lifetimes, whose cdfs are $G_W(z) = 1 - \exp(-\alpha\, z)$ and $G_T(t) = 1 - \exp(-\beta\, t^\gamma)$, respectively. Here, $\alpha > 0$ and $\gamma > 0$ are shape parameters and $\beta > 0$ is a scale parameter. The distribution of the random variable $X = \min\{W, T\}$ is called the *exponential-Weibull* (EW) distribution, whose cdf is given by

$$F(x;\alpha,\beta,\gamma) = 1 - \mathrm{e}^{-(\alpha x+\beta x^\gamma)}, \qquad x > 0. \tag{11}$$

We consider that $\alpha > 0$, $\beta > 0$ and $\gamma \in (0,\infty) \setminus \{1\}$, which gives identifiability to the model. The mathematical properties of the EW model were investigated by Cordeiro et al. (2014c).

(ii) *Generalized exponential-exponential (GEE) model.* Gupta and Kundu (1999) introduced the two-parameter generalized exponential (GE) distribution with cdf given by $G(x;\alpha,\lambda) = (1 - \mathrm{e}^{-\lambda x})^\alpha$, where $x > 0, \alpha > 0, \lambda > 0$. For a given α, let Y be a random variable with the GE cdf with parameters $\alpha > 0$ and $\lambda > 0$. Let α be a random variable following an exponential distribution with parameter $\theta > 0$. Then, Popović et al. (2015) defined the GEE cdf (for $x > 0$) by

$$F(x;\lambda,\theta) = \int_0^\infty G(x;\alpha,\lambda)\, \theta\, \mathrm{e}^{-\theta\alpha} d\alpha = \left[1 - \frac{1}{\theta}\log(1 - \mathrm{e}^{-\lambda x})\right]^{-1}, \ \lambda,\theta > 0.$$

(iii) *Exponentiated geometric G-Poisson (EGGP) model (New).* If $N \sim ZTG(p)$ and $M \sim ZTP(\lambda)$, then the EGGP family can be determined by inserting the EGP cdf (Section 3.9) in the cdf of the EGG class (Section 3.1) as

$$F_{EGGP}(x;\alpha,\lambda,\tau,p) = \frac{\mathrm{e}^{\,-\lambda[1-G(x;\tau)^\alpha]} - \mathrm{e}^{\,-\lambda}}{1 - \mathrm{e}^{\,-\lambda} - p\left(1 - \mathrm{e}^{\,-\lambda[1-G(x;\tau)^\alpha]}\right)}.$$

If $\alpha = 1$ and $X \sim Exp(\beta)$ in the EGGP class above, then it follows the cdf of the geometric exponential Poisson (GEP) defined by Nadarajah et al. (2013a).

(iv) *Additive Weibull (AW) model.* Suppose a system composed of two interconnected independent series sub-systems that affect the system in a different way, both following the Weibull distribution with proper parameters. Xie and Lai (1995) proposed the AW model based on the simple idea of combining the failure rates of two Weibull distributions: one has a decreasing failure rate and the other one has an increasing failure rate. The cdf of the AW model is given by $F(t) = 1 - \exp(-a\, t^b - c\, t^d)$, where $a > 0$ and $c > 0$ are scale parameters and $b > d > 0$ (or $d > b > 0$) are shape parameters, which gives identifiability to the model. The interpretation of the AW model is evident. A state-of-the-art survey on the AW model can be found in Lemonte et al. (2014).

8 Recent trends in compounding

There are four very recent trends on compounding of distributions, which have received a great deal of attention.

8.1 First recent trend

Compounding a G-class with discrete model: The first recent trend deals with defining compound classes of lifetime distributions rather than studying a single compound model. In this technique, a non-compound G-family of distributions is compounded with a discrete model to generate a new flexible compounded class. Asgharzadeh et al. (2014) introduced the G-Poisson-Lindley (from discrete Poisson-Lindley) class of distributions by compounding the ZTPL distribution with any other continuous lifetime model. Four special models viz. Weibull Poisson-Lindley, Burr Poisson-Lindley, exponentiated-Weibull Poisson-Lindley and Dagum Poisson-Lindley were investigated. Nadarajah et al. (2015a) proposed the exponentiated G-geometric family, and reported two special models: exponentiated-Weibull geometric and exponentiated-log-logistic geometric. Ramos et al. (2015) introduced the Kumaraswamy-G Poisson family and showed that the special model Kumaraswamy-Weibull Poisson outperforms the competitors Kumaraswamy-Weibull and beta-Weibull models in studying real life data on bladder cancer. Gomes et al. (2015) proposed the exponentiated-G Poisson family and studied two special models, namely the exponentiated-Burr XII Poisson and exponentiated-Weibull Poisson. Two other compound G-classes are the G-Poisson (Alkarni and Oraby 2012) and Poisson-X (Tahir et al. 2016a).

8.2 Second recent trend

Combining a continuous model with compound power series class: For the second recent trend on compounding, one continuous lifetime model is compounded with the power series class of distributions truncated at zero. Chahkandi and Gangali (2009) first suggested compounding exponential and power series class, which exhibits decreasing failure rate. The power series class can be used to construct many compounding models with discrete distributions: Poisson, logarithmic, geometric, binomial and negative-binomial. Some well-known compound models defined from the power series class are: Weibull power series (WPS) (Morais and Barreto-Souza 2011), complementary generalized-exponential power series (CGEPS) (Mahmoudi and Jafari 2012), complementary exponentiated-Weibull power series (CEWPS) (Mahmoudi and Shiran 2012b), extended WPS (Silva et al. 2013), Kumaraswamy power series (KwPS) (Bidram and Nekouhou 2013), complementary exponential power series (CEPS) (Flores et al. 2013), Birnbaum-Saunders power series (BSPS) (Bourguignon et al. 2014b), complementary WPS (Munteanu 2014), complementary Erlang and Erlang power series (CErPS and ErPS) (Leahu et al. 2014), complementary extended WPS (Cordeiro and Silva 2014), exponentiated extended WPS (Tahmasebi and Jafari 2015a), Burr XII power series (BIIPS) (Silva and Cordeiro 2015), Lindley power series (LPS) (Warahena-Liyanage and Pararai 2015a), linear failure rate power series (LFRPS) (Mahmoudi and Jafari 2015), complementary normal power series (CNPS) (Mahmoudi and Mahmoodian 2015), complementary generalized Gompertz power series (CGGoPS) (Tahmasebi and Jafari 2015b), complementary inverse Weibull power series (CIWPS) (Shafiei et al. 2016), complementary generalized modified Weibull (CGMW) (Bagheri et al. 2016), complementary exponentiated inverse Weibull power series (CEIWPS) (Hassan et al. 2016), generalized gamma power series (GGPS) (Silva et al. 2016), Gompertz power series (GoPS) (Jafari and Tahmasebi 2016), complementary generalized linear failure rate power series (CGLFR) (Harandi and Alamatsaz 2016), and Dagum power series (DaPS) (Oluyede et al. 2016b).

8.3 Third recent trend

Combining compound G-class with the non-compound G-class: Here, the cdf of one compound G-class or its special model is inducted into the cdf of a non-compound G-class, thus generating a new flexible G-class. Next, we propose two such classes.

(i) If the cdf of the GP class defined in Section 3.5 with $\alpha = 1$ is inserted into the cdf of Gamma-G class, proposed by Zografos and Balakrishnan (2009), then the cdf of new class (GaGP) is obtained as

$$F_{GaGP}(y) = 1 - \frac{1}{\Gamma(a)}\,\Gamma\left(a, -\ln\bar{G}_{GP}(y)\right),$$

where $\Gamma(a) = \int_0^\infty t^{a-1}\,\mathrm{e}^{-t}\,dt$ and $\Gamma(a,z) = \int_z^\infty t^{a-1}\,\mathrm{e}^{-t}\,dt$ are the gamma and upper incomplete gamma functions, respectively, and $\bar{G}_{GP}(y) = 1 - G_{GP}(y)$.

One special model of the GaGP class, that is, the gamma-Weibull-Poisson (GaWP) was studied by Percontini et al. (2014), whose cdf is given by

$$F_{GaWP}(y) = 1 - \frac{1}{\Gamma(a)}\,\Gamma\left(a, \ln\left[\frac{1-\mathrm{e}^{\lambda}}{1-\mathrm{e}^{\lambda\bar{G}(x)}}\right]\right).$$

(ii) The cdf of the new Kumaraswamy-GP (KGP) class can be obtained by inducting the cdf of the CGP model given in Section 4.5 with $\alpha = 1$ into the cdf of the Kumaraswamy-G class, pioneered by Cordeiro and de-Castro (2011), as

$$F_{KwGP}(y) = 1 - \left[1 - G_{GP}^{a}(x)\right]^{b} = 1 - \left[1 - \left(\frac{1-\mathrm{e}^{\lambda\,G(x)}}{1-\mathrm{e}^{\lambda}}\right)^{a}\right]^{b}.$$

Pararai et al. (2015a) obtained the cdf of a special KwGP model called the Kumaraswamy Lindley-Poisson (KwLP) distribution.

8.4 Fourth recent trend

Combining transmuted G-class with well-known compound distributions: After receiving increased attention in the last decade, more than 50 transmuted distributions have been reported in the literature. Due to wide acceptability of transmuted G-class, a new trend has now begun by inserting the cdf of the compound G-class or distribution in the transmuted G-class cdf. Some models are given below:

(i) *Transmuted Weibull-geometric (TWG).* Merovci and Elbatal (2014b) extended the WG model introduced by Barreto-Souza et al. (2011) (given in Section 5) with the transmuted-G family and defined the cdf of the TWG model by

$$F_{TWG}(y;p,a,b,\eta) = \frac{1-\mathrm{e}^{-(by)^a}}{1-p\,\mathrm{e}^{-(by)^a}}\left[1+\eta-\eta\left(\frac{1-\mathrm{e}^{-(by)^a}}{1-p\,\mathrm{e}^{-(by)^a}}\right)\right].$$

(ii) *Exponentiated transmuted-Weibull geometric (ETWG) model.* If $X \sim$ Weibull(a,b) have cdf $G(x;a,b) = 1 - \mathrm{e}^{-(x/b)^a}$, Aryal and Tsokos (2011) defined the cdf of the transmuted-Weibull distribution as

$$G_{TW}(x;a,b,\eta) = \left(1-\mathrm{e}^{-(x/b)^a}\right)\left(1+\eta\,\mathrm{e}^{-(x/b)^a}\right).$$

The cdf of the ETWG model proposed by Al-Zahrani et al. (2015) can follow from the EGG cdf defined in Section 3.1 as

$$F_{ETWG}(y;p,a,b,\eta) = \frac{\left(1-\mathrm{e}^{-(y/b)^a}\right)^{\alpha}\left(1+\eta\,\mathrm{e}^{-(y/b)^a}\right)^{\alpha}}{1-p+p\left(1-\mathrm{e}^{-(y/b)^a}\right)^{\alpha}\left(1+\eta\,\mathrm{e}^{-(y/b)^a}\right)^{\alpha}}.$$

(iii) *Transmuted complementary Weibull-geometric (TCWG) model.* Afify et al. (2014) applied the LA1 class to the cdf of the Tojeiro et al.'s CWG model given in Section 5, and obtained the cdf of the TCWG model as

$$F_{TCWG}(z;p,a,b,\eta) = \frac{p\left(1-\mathrm{e}^{-(bz)^a}\right)}{p+(1-p)\mathrm{e}^{-(bz)^a}}\left[1+\eta-\eta\left(\frac{p\left(1-\mathrm{e}^{-(bz)^a}\right)}{p+(1-p)\mathrm{e}^{-(bz)^a}}\right)\right].$$

(iv) *Transmuted complementary exponentiated Weibull-geometric (TCEWG) model.* If $X \sim \text{ExpWeibull}(a,b,\alpha)$ have cdf $G(x;\beta) = \left(1-\mathrm{e}^{-(bx)^a}\right)^{\alpha}$, Saboor et al. (2016) used the CEWG cdf given in Section 5 and obtained the cdf of the TCEWG model as

$$F_{TCEWG}(z;p,a,b,\alpha,\eta) = \frac{(1-p)(1-\mathrm{e}^{-(bz)^a})^{\alpha}}{1-p\left[1-\mathrm{e}^{-(bz)^a}\right]^{\alpha}} \times\left[1+\eta-\eta\left(\frac{(1-p)(1-\mathrm{e}^{-(bz)^a})^{\alpha}}{1-p\left[1-\mathrm{e}^{-(bz)^a}\right]^{\alpha}}\right)\right].$$

(v) *Transmuted Lindley-geometric (TLG) model.* Merovci and Elbatal (2014a) extended the LG model (described in Section 5) by proposing the TLG cdf given by

$$F_{TLG}(y;p,\theta,\eta) = \frac{1-\left(1+\frac{\theta y}{\theta+1}\right)\mathrm{e}^{-\theta y}}{1-p\left(1+\frac{\theta y}{\theta+1}\right)\mathrm{e}^{-\theta y}} \times\left[1+\eta-\eta\left(\frac{1-\left(1+\frac{\theta y}{\theta+1}\right)\mathrm{e}^{-\theta y}}{1-p\left(1+\frac{\theta y}{\theta+1}\right)\mathrm{e}^{-\theta y}}\right)\right], \quad y>0,$$

where η and $\theta > 0$ are transmuted and scale parameters, respectively, and $p \in (0,1)$.

(vi) *Transmuted exponential-Weibulll (TEW) model.* Saboor et al. (2015) recently extended the exponentiated-Weibull model originally proposed by Cordeiro et al. (2014c). The TEW cdf is given by

$$F_{TLG}(y;\eta,\alpha,\beta,\gamma) = (\eta+1)\left[1-\mathrm{e}^{-(\alpha x+\beta x^{\gamma})}\right]-\eta\left[1-\mathrm{e}^{-(\alpha x+\beta x^{\gamma})}\right]^2, \quad x>0,$$

where $\alpha > 0$, $\beta > 0$ and $\gamma \in (0,\infty) \setminus \{1\}$.

9 Estimation and inference

9.1 General estimation procedure

Several approaches for parameter estimation were proposed in the literature but the maximum likelihood method is the most commonly employed. The maximum likelihood estimators (MLEs) enjoy desirable properties and can be used when constructing confidence intervals and regions and also in test statistics. The normal approximation for these estimators in large sample distribution theory is easily handled either analytically or numerically. So, we consider the estimation of the unknown parameters for each model discussed in this paper from complete or censored samples by maximum likelihood.

The log-likelihood for the model parameters can be maximized either directly by using the `R` (`optim` function), `SAS` (`PROC NLMIXED`), `Ox` program (`MaxBFGS` sub-routine) or by solving the nonlinear likelihood equations obtained by differentiating the log-likelihood. In the applications, we can also use the *AdequacyModel* package (version 2.0.0) available in the R programming language. It has been cited very frequently in papers related to new lifetime distributions. The package has been continuously updated and more information can be obtained from http://cran.rstudio.com/web/packages/AdequacyModel/index.html. It is distributed under the terms of the GNU licenses - GNU Project. An important observation is that it is not necessary to define the log-likelihood function but only the pdf and cdf of the model. The package provides some useful goodness-of-fit statistics to assess the quality of the fitted models and compare them, such as the Cramér–von Mises (W^*) and Anderson-Darling (A^*) statistics, Akaike information criterion (AIC), consistent Akaike information criterion (CAIC), Bayesian information criterion (BIC), Hannan-Quinn information criterion (HQIC) and Kolmogorov-Smirnov (K-S) test. It is important to emphasize that we can fit several competitive models to a data set and select those which yield best fits by means of the AdequacyModel package.

9.2 Cure fraction survivor models

During the last two decades, the tendency to propose probability models to deal with survival data has increased. This increased interest has lead researchers and medical practitioners to assess correct causes and information of the disease. The survival models which are receiving increased recognition in these days are *fractional survivor models*. These models have been effectively useful in some situations of clinical, medical or biological studies, where the fractional survival (or survivor fraction) models are useful to study a cure fraction of individuals. These models are also known as *cure rate models* or *long-term survival models*. In these models, it is assumed that all units under study are susceptible to an event of interest and will eventually experience it if its follow-up is sufficiently long. However, there are situations for which a fraction of individuals is not expected to experience the event of interest, that is, those individuals are cured or insusceptible. For example, researchers may be interested in analyzing the recurrence of a disease but many individuals may never have an experience or a recurrence, therefore, a cured fraction of the population exist. In other words, the cure rate models cover the situations where the sampling units insusceptible to the occurrence of the event are of interest, and also extend understanding of time-to-event data by allowing the formulation of more accurate and informative conclusions. If the cure fraction of the population is ignored, then the results will match to standard survival analysis. The cure rate models have been used for modeling time-to-event data for cardiac failure and various types of cancers including prostate, breast, leukemia, non-Hodgkin lymphoma and melanoma. That is why, the focus of the researchers is to introduce new, extended or modified distributions which accommodate cure fraction.

In the literature, two formulations of the cure rate models have received increased attention, namely, the *mixture cure rate model* (Boag 1949; Berkson and Gage 1952) and *promotion time (or accelerated) cure rate model* (Yakovlev and Tsodikov 1996; Chen et al. 1999). Perhaps, the mixture cure rate models has received increased popularity. In mixture model, it is assumed that there is a fraction $1 - \pi_0$ of susceptible individuals and, hence, a proportion π_0 of cured individuals. The survivor function for the entire population is

$$S_{pop} = \pi_0 + (1 - \pi_0) S_0,$$

where S_0 denotes the survival function for the non-cured group in the population. So, the flexibility and performance of the population model depends on the choice of S_0. Rodrigues et al. (2009a) unified both formulations considering the *negative binomial cure rate model*. The proper modeling of cure rate models reported in literature is with covariates, which have been proved very useful in getting improved results for the estimated cure rate. Some selected cure rate models are reported in Table 2. Further detail is beyond the scope of the paper.

9.3 Inference for cure fraction models

Suppose that the time-to-event is not completely observed and may be subject to right censoring. Let C_i denote the censoring time. We observe that $Y_i = \min(T_i, C_i)$ and $\delta_i = I(T_i < C_i)$ is such that $\delta_i = 1$, if T_i is a time-to-event and $\delta_i = 0$, if it is right censored, $i = 1, \ldots, n$. Let Θ denote the parameter vector of the distribution of the time-to-event. From n pairs of times and censoring indicators $Data = (y_1, \delta_1), \ldots, (y_1, \delta_1)$, the corresponding likelihood function under uninformative censoring is

$$L(Data, \boldsymbol{\Theta}) \propto \Pi_{i=1}^{n} \left[f_{CRM}(Data, \boldsymbol{\Theta})\right]^{\delta_i} \left[F_{CRM}(Data, \boldsymbol{\Theta})\right]^{1-\delta_i},$$

where $f_{CRM}(Data, \boldsymbol{\Theta})$ and $F_{CRM}(Data, \boldsymbol{\Theta})$ are the density and cdf of a cure rate model (CRM), respectively.

Maximum likelihood estimation of the parameter vector $\boldsymbol{\Theta}$ is carried out through maximization of the log-likelihood function $\ell(Data, \boldsymbol{\Theta}) = \log L(Data, \boldsymbol{\Theta})$. Details for the maximization methods are discussed in Section 9.1. Under suitable regularity conditions,

Table 2 Some selected cure rate survival models

S.No.	Investigation	Cure rate model	Author(s)
1.	Cutaneous melanoma	COM-Poisson	Rodrigues et al. (2009b)
2.	Reducing drug abuse	Generalized exponential	Kannan et al. (2010)
3.	Malignant melanoma	Negative binomial	Cancho et al. (2011b)
4.	Malignant melanoma	Geometric Birnbaum-Saunders	Cancho et al. (2012)
5.	Prostrate Cancer	Negative binomial-beta Weibull	Ortega et al. (2012)
6.	Ovarian cancer	PE-regression	Louzada et al. (2012c)
7.	Myelomatosis (bone marrow) Leukemia (autologous marrow)	EG	Roman et al. (2012)
8.	Gastric cancer	Generalized modified Weibull	Martinez et al. (2013)
9.	Cutaneous melanoma	Destructive negative-binomial	Cancho et al. (2013a)
10.	Cutaneous melanoma	Power series	Cancho et al. (2013b)
11.	Cutaneous melanoma	COM-Poisson	Balakrishnan and Pal (2013a)
12.	Cutaneous melanoma	Negative-binomial GGa	Ortega et al. (2014)
13.	Breast cancer Hemophiliacs	Poisson Birnbaum-Saunders	Hashimoto et al. (2014)
14.	Red flour beetles	log-Weibull Negative-binomial	Louzada et al. (2015)
15.	Melanoma	Weibull Negative-binomial	Yiqi et al. (2016)
16.	Congenital malformations Hemophiliacs	Negative binomial-Weibull	Hashimoto et al. (2015)
17.	Breast carcinoma	Power series BW	Ortega et al. (2015)
18.	First calving of cows	Transmuted log-logistic	Louzada and Granzotto (2015)
19.	Cutaneous melanoma	Destructive Negative-binomial	Gallardo and Bolfarine (2016)
20.	Malignant melanoma	Negative binomial Birnbaum-Saunders	Cordeiro et al. (2016)

it can be shown that the asymptotic distribution of the maximum likelihood estimator (MLE) $\widehat{\boldsymbol{\Theta}}$ is normal with mean vector $\boldsymbol{\Theta}$ and covariance matrix $\sum(\widehat{\boldsymbol{\Theta}})$ given by

$$\widehat{\sum}(\widehat{\boldsymbol{\Theta}}) = \left[-\frac{\partial^2 \ell(Data, \boldsymbol{\Theta})}{\partial \boldsymbol{\Theta}\, \partial \boldsymbol{\Theta}^T} \right]^{-1},$$

evaluated at $\boldsymbol{\Theta} = \widehat{\boldsymbol{\Theta}}$. The required second derivatives are computed numerically.

The EM algorithm (Dempster et al. 1977) is also a very popular maximization alternative used to obtain the estimates when the model has missing data. In the literature, Balakrishnan and Pal (2012, 2013b, 2015a, 2015b), Gallardo et al. (2016), Gallardo and Bolfarine (2016) and some others have considered estimation of parameters of cure fraction survival models using the EM algorithm.

10 Final remarks

The need of compounding was first felt in actuarial science and later researchers of many other fields adopted this approach for lifetime and reliability modeling. We follow the two basic principles (the minimum and the maximum) used in series and parallel structure, and report more than 30 compound G-classes. In this way, the possible available compound G-classes are surveyed and using these basic principles nearly 25 new G-classes are proposed. The purpose of providing a variety of G-classes is to test flexibility of the proposed compound models to cope with the data available in complex situations. The parameters inducted in this way might be helpful in exploring phenomenon generated from real-lifetime data sets. We expect that these G-classes or generated compounded models from them will be an addition to the art of constructing useful probability models. One can imagine its motivation and usefulness in the fields which are not touched earlier. We have also briefly described the latest trends in the development of compounding technique, which portray better exposition of the strategies adopted for the researchers and practitioners. We hope to produce many more new compound G-classes from the function $T[\,G(x;\tau);\alpha]$ but due to space problem we did our best to explore and present the elusive task in most tenable way by mentioning only the cumulative distributions of the classes. The remaining G-classes will appear in another article under the same series. Lastly, we offer more choices to the learners and practitioners of modeling to compare different models and to study pros and cons of old and new G-classes. The possible future projects are: (i) to propose more new compound G-classes of distributions; (ii) to review and develop bivariate compound G-classes; and (iii) to prepare a review and new developments on cure rate survival models.

Acknowledgements

Both authors are very grateful to the Editor-in-Chief Felix Famoye, an associate editor and two referees for their constructive comments which greatly improved the earlier version of our manuscript.

Authors' contributions

The authors MHT and GMC with the consultation of each other carried out this work and drafted the manuscript together. Both authors read and approved the final manuscript.

Competing interests

The authors declare that they have no competing interests.

Author details
[1]Department Statistics, Baghdad Campus, The Islamia University of Bahawalpur, 63100 Bahawalpur, Pakistan.
[2]Departamento de Estatística, Universidade Federal de Pernambuco, PE 50740-540 Recife, Brazil.

References

Abdul-Moniem, IB: Transmuted Burr III distribution. J. Statist.-Adv. Theory Applic. **14**, 37–47 (2015)
Abdul-Moniem, IB, Seham, M: Transmuted Gompertz distribution. Comput. Appl. Math. J. **1**, 88–96 (2015)
Adamidis, K, Dimitrakopoulou, T, Loukas, S: On an extension of the exponential-geometric distribution. Statist. Probab. Lett. **73**, 259–269 (2005)
Adamidis, K, Loukas, S: A lifetime distribution with decreasing failure rate. Statist. Probab. Lett. **39**, 35–42 (1998)
Afify, AZ, Nofal, ZM, Butt, NS: Transmuted complementary Weibull geometric distribution. Pak. J. Statist. Oper. Res. **10**, 435–454 (2014)
Afify, AZ, Hamedani, GG, Ghosh, I, Mead, ME: The transmuted Marshall-Olkin Fréchet distribution: Properties and applications. Int. J. Statist. Probab. **4**, 132–148 (2015a)
Afify, AZ, Nofal, ZM, Butt, NS: Exponentiated transmuted generalized Rayleigh distribution: A new four-parameter Rayleigh distribution. Pak. J. Statist. Oper. Res. **11**, 115–134 (2015b)
Afify, AZ, Nofal, ZM, Yousof, HM, El-Gebaly, YM, Butt, NS: The transmuted Weibull-Lomax distribution: Properties and application. Pak. J. Statist. Oper. Res. **11**, 135–152 (2015c)
Afify, AZ, Yousof, HM, Butt, NS, Hamedani, GG: The transmuted Weibull-Pareto distribution. Pak. J. Statist. **32**, 183–206 (2016)
Ahmad, A, Ahmad, SP, Ahmad, A: Transmuted inverse Rayleigh distribution: A generalization of the inverse Rayleigh distribution. Math. Theory Model. **4**, 177–185 (2014)
Ahmad, A, Ahmad, SP, Ahmad, A: Characterization and estimation of transmuted Rayleigh distribution. J. Statist. Applic. Probab. **4**, 315–321 (2015a)
Ahmad, K, Ahmad, SP, Ahmad, A: Structural properties of transmuted Weibull distribution. J. Mod. Appl. Stat. Methods. **14**, 141–158 (2015b)
Al-Aqtash, R: On Generating New Families of Distributions Using the Logit Function. Central Michigan University. Department of Mathematics, Ph.D. thesis (2013)
Al-Aqtash, R, Lee, C, Famoye, F: Gumbel-Weibull distribution: Properties and applications. J. Mod. Appl. Stat. Methods. **13**, 201–225 (2014)
Al-Babtain, A, Fattah, AA, Ahmed, AHN: The Kumaraswamy-transmuted exponentiated modified Weibull distribution. Commun. Stat. Simul. Comput (2015). forthcoming
Alexander, C, Cordeiro, GM, Ortega, EMM, Sarabia, JM: Generalized beta-generated distributions. Comput. Stat. Data Anal. **56**, 1880–1897 (2012)
Alizadeh, M, Cordeiro, GM, de-Brito, E, Demétrio, CGB: The beta Marshall-Olkin family of distributions. J. Stat. Dist. Applic. 2, Art, 4 (2015a)
Alizadeh, M, Emadi, M, Doostparast, M, Cordeiro, GM, Ortega, EMM, Pescim, RR: A new family of distributions: the Kumaraswamy odd log-logistic, properties and applications. Hacet. J. Math. Stat. **44**, 1491–1512 (2015b)
Alizadeh, M, Tahir, MH, Cordeiro, GM, Mansoor, M, Zubair, M, Hamedani, GG: The Kumaraswamy Marshal-Olkin family of distributions. J. Egyptian Math. Soc. **23**, 546–557 (2015c)
Alizadeh, M, Bagheri, SF, Alizadeh, M, Nadarajah, S: A new four-parameter lifetime distribution. J. Appl. Statist. (2016a). doi:10.1080/02664763.2016.1182137. forthcoming
Alizadeh, M, Merovci, F, Hamedani, GG: Generalized transmuted family of distributions. Properties and applications. Hacet. J. Math. Stat. (2016b). doi:10.15672/HJMS.201610915478. forthcoming
Aljarrah, MA, Lee, C, Famoye, F: On generating T-X family of distributions using quantile functions. J. Stat. Dist. Applic. 1, Art, 2 (2014)
Alkarni, SH: A compound class of geometric and lifetime disributions. Open Statist. Probab. J. **5**, 1–5 (2012)
Alkarni, S, Oraby, A: A compound class of Poisson and lifetime disributions. J. Statist. Applic. Probab. **1**, 45–51 (2012)
Alkarni, SH: A compound class of truncated binomial lifetime disributions. Open. J. Statist. **3**, 305–311 (2013)
Almheidat, M, Famoye, F, Lee, C: Some generalized families of Weibull distribution: Properties and applications. Int. J. Statist. Probab. **4**, 18–35 (2015)
Alzaatreh, A, Lee, C, Famoye, F: A new method for generating families of continuous distributions. Metron. **71**, 63–79 (2013)
Alzaatreh, A, Lee, C, Famoye, F: T-normal family of distributions: a new approach to generalize the normal distribution. J. Stat. Dist. Applic. 1, Art, 16 (2014)
Alzaatreh, A, Lee, C, Famoye, F: Family of generalized gamma distributions: Properties and applications. Hacet. J. Math. Stat. **45**, 869–886 (2016a)
Alzaatreh, A, Lee, C, Famoye, F, Ghosh, I: The generalized Cauchy family of dstributions with applications. J. Stat. Dist. Applic. 3, Art, 12 (2016b)
Alzaghal, A, Famoye, F, Lee, C: Exponentiated T-X family of distributions with some applications. Int. J. Probab. Statist. **2**, 31–49 (2013)
Al-Zahrani, B, Fattah, AA, Nadarajah, S, Ahmed, AHN: The exponentiated transmuted Weibull geometric distribution with application in survival analysis. Commun. Stat. Simul. Comput. (2015). forthcoming
Al-Zahrani, B, Sagor, H: The Poisson-Lomax distribution. Rev. Colombiana Estadíst. **37**, 225–245 (2014)
Al-Zahrani, B, Sagor, H: Statistical analysis of Lomax-logarithmic distribution. J. Stat. Comput. Simul. **85**, 1883–1901 (2015)
Amini, M, MirMostafaee, SMTK, Ahmadi, J: Log-gamma-generated families of distributions. Statistics. **48**, 913–932 (2014)
Aryal, GR: Transmuted log-logistic distribution. J. Statist. Applic. Probab. **2**, 11–20 (2013)
Aryal, GR, Tsokos, CP: On the transmuted extreme value distribution with application. Nonlinear Anal. **71**, 1401–1407 (2009)
Aryal, GR, Tosokos, CP: Transmuted Weibull distribution: A generalization of Weibull probability distribution. Eur. J. Pure Appl. Math. **4**, 89–102 (2011)
Asgharzadeh, A, Bakouch, HS, Nadarajah, S, Esmaeili, L: A new family of compound lifetime distributions. Kybernetika. **50**, 142–169 (2014)

Ashour, SK, Eltehiwy, MA: Transmuted Lomax distribution. Amer. J. Appl. Math. Statist. **1**, 121–127 (2013a)

Ashour, SK, Eltehiwy, MA: Transmuted exponentiated Lomax distribution. Austral. J. Basic Appl. Sci. **7**, 658–667 (2013b)

Ashour, SK, Eltehiwy, MA: Transmuted exponentiated modified Weibull distribution. Int. J. Basic Appl. Sci. **2**, 258–269 (2013c)

Ashour, SK, Eltehiwy, MA: Exponentiated power Lindley distribution. J. Adv. Res. **6**, 895–905 (2015)

Ashour, SK, Wahed, MLA: Kummer beta-Weibull geometric distribution: A new generalizations of beta-Weibull geometric distribution. Int. J. Sci. Basic Appl. Res. **16**, 258–273 (2014)

Bagheri, SF, Samani, EB, Ganjali, M: The generalized modified Weibull power series distribution: Theory and applications. Comput. Stat. Data Anal. **94**, 136–160 (2016)

Bakouch, HS, Al-Zahrani, BM, Al-Shomrani, AA, Marchi, VAA, Louzada, F: An extended Lindley distribution. J. Korean Stat. Soc. **41**, 75–85 (2012a)

Bakouch, HS, Ristić, MM, Asgharzadeh, A, Esmaily, L, Al-Zahrani, BM: An exponentiated exponential binomial distribution with application. Statist. Probab. Lett. **82**, 1067–1081 (2012b)

Balakrishnan, N, Pal, S: EM algorithm-based likelihood estimation for some cure rate models. J. Stat. Theory Prac. **6**, 698–724 (2012)

Balakrishnan, N, Pal, S: Lognormal lifetimes and likelihood-based inference for flexible cure rate models based on COM-Poisson family. Comput. Stat. Data Anal. **67**, 41–67 (2013a)

Balakrishnan, N, Pal, S: Expectation maximization-based likelihood inference for flexible cure rate models with Weibull lifetimes (2013b). doi:10.1177/0962280213

Balakrishnan, N, Pal, S: An EM algorithm for the estimation of parameters of a flexible cure rate model with generalized gamma lifetime and model discrimination using likelihood-and information-based methods. Comput. Statist. **30**, 151–189 (2015a)

Balakrishnan, N, Pal, S: Likelihood inference for flexible cure rate models with gamma lifetimes. Commun. Stat. Theory Methods. **44**, 4007–4048 (2015b)

Barreto-Souza, W, de-Morais, AL, Cordeiro, GM: The Weibull-geometric distribution. J. Stat. Comput. Simul. **81**, 645–657 (2011)

Barreto-Souza, B, Cribari-Neto, F: A generalization of the exponential-Poisson distribution. Statist. Probab. Lett. **79**, 2493–2500 (2009)

Barreto-Souza, W, Simas, AB: The exp-G family of distributions. Braz. J. Probab. Statist. **27**, 84–109 (2013)

Batsidis, A, Lemonte, AJ: On the Harris extended family of distributions. Statistics. **49**, 1400–1421 (2015)

Bereta, EMP, Louzanda, F, Franco, MAP: The Poisson-Weibull distribution. Adv. Applic. Statist. **22**, 107–118 (2011)

Berkson, J, Gage, RP: Survival curve for cancer patients following treatment. J. Amer. Statist. Assoc. **47**, 501–515 (1952)

Bidram, H: The beta exponential-geometric distribution. Commun. Stat. Simul. Comput. **41**, 1606–1622 (2012)

Bidram, H, Behboodian, J, Towhidi, M: The beta Weibull-geometric distribution. J. Stat. Comput. Simul. **83**, 52–67 (2013)

Bidram, H, Alavi, SM: A note on exponentiated F-geometric distributions. J. Mod. Math. Front. **3**, 18–23 (2014)

Bidram, H, Nadadrajah, S: A new lifetime model with decreasing, increasing, bathtub-shaped, and upside-down bathtub-shaped hazard rate function. Statistics. **50**, 139–156 (2016)

Bidram, H, Nekouhou, V: Double bounded Kumaraswamy-power series class of distributions. Statist. Oper. Res. Trans. **37**, 211–230 (2013)

Boag, JW: Maximum likelihood estimates of the proportion of patients cured by cancer therapy. J. R. Stat. Soc. B 11, 15–53 (1949)

Bordbar, F, Nematollah, AR: The modified exponential-geometric distribution. Commun. Stat. Theory Methods. **45**, 173–181 (2016)

Bourguignon, M, Silva, RB, Cordeiro, GM: The Weibull-G family of probability distributions. J. Data Sci. **12**, 53–68 (2014a)

Bourguignon, M, Silva, RB, Cordeiro, GM: A new class of fatigue life distributions. J. Stat. Comput. Simul. **84**, 2619–2635 (2014b)

Bourguignon, M, Ghosh, I, Cordeiro, GM: General results for the transmuted family of distributions and new models. J. Probab. Statist. Art.ID. 7208425, 21 (2016a)

Bourguignon, M, Leao, J, Leiva, V, Santos-Neto, M: The transmuted Birnbaum-Saunders distribution. REVSTAT (2016b). forthcoming

Cancho, VG, Louzanda-Neto, F, Barriga, GDC: The Poisson-exponential lifetime distribution. Comput. Stat. Data Anal. **55**, 677–686 (2011a)

Cancho, VG, Rodrigues, J, de-Castro, M: A flexible model for survival data with a cure rate: a Bayesian approach. J. Appl. Statist. **38**, 57–70 (2011b)

Cancho, VG, Louzada, F, Barriga, GDC: The geometric Birnbaum-Saunders regression model with cure rate. J. Stat. Plann. Infer. **142**, 993–1000 (2012)

Cancho, VG, Bandyopadhyay, D, Louzada, F, Yiqi, B: The destructive negarive binomial cure rate model with a latent activation scheme. Stat. Methodol. **13**, 48–68 (2013a)

Cancho, VG, Louzada, F, Ortega, EMM: The power series cure rate model: An application to a cutaneous melanoma data. Commun. Stat. Simul. Comput. **42**, 586–602 (2013b)

Castellares, F, Lemonte, AJ: On the Marshall-Olkin extended distributions. Commun. Stat. Theory Methods. **45**, 4537–4555 (2016)

Chahkandi, M, Gangali, M: On some lifetime distributions with decrasing failure rate. Comput. Stat. Data Anal. **53**, 4433–4440 (2009)

Chakraborty, S, Bhati, D: Transmuted geometric distribution with applications in modelling and regression analysis of count data. Statist. Oper. Res. Trans. **40**, 153–176 (2016)

Chen, MH, Ibrahim, JG, Sinha, D: A new Bayesian model for survival data with a surviving fraction. J. Amer. Statist. Assoc. **94**, 909–919 (1999)

Chen, Z: A new two-parameter lifetime distribution with bathtub shape or increasing failure rate function. Statist. Probab. Lett. **49**, 155–161 (2000)

Chung, Y, Kang, Y: The exponentiated Weibull-geometric distribution: Properties and estimations. Commun. Stat. Appl. Methods (Korean). **21**, 147–160 (2014)

Cordeiro, GM, Alizadeh, M, Ortega, EMM: The exponentiated half-logistic family of distributions: properties and applications. J. Probab. Statist. Art.ID. **864396**, 21 (2014a)
Cordeiro, GM, de-Santana, LH, Ortega, EMM, Pescim, RR: A new family of distributions: Libby-Novick beta. Int. J. Statist. Probab. **3**, 63–80 (2014b)
Cordeiro, GM, Ortega, EMM, Lemonte, A: The exponential-Weibull lifetime distribution. J. Stat. Comput. Simul. **84**, 2592–2606 (2014c)
Cordeiro, GM, Ortega, EMM, Popović, BV, Pescim, RR: The Lomax generator of distributions: Properties, minification process and regression model. Appl. Math. Comput. **247**, 465–486 (2014d)
Cordeiro, GM, Alizadeh, M, Ortega, EMM, Serrano, LHV: The Zografos-Balakrishnan odd log-logistic family of distributions: Properties and applications. Hacet. J. Math. Stat. **45** (2015a). doi:10.15672/HJMS.20159714145. forthcoming
Cordeiro, GM, Ortega, EMM, Lemonte, A: The Poisson generalized linear failure rate model. Commun. Stat. Theory Methods. **44**, 2037–2058 (2015b)
Cordeiro, GM, Saboor, A, Khan, MN: The transmuted generalized modified Weibull distribution. Filomat (2015c). forthcoming
Cordeiro, GM, Alizadeh, M, Tahir, MH, Mansoor, M, Bourguignon, M, Hamedani, GG: The beta odd log-logistic generalized family of distributions. Hacet. J. Math. Stat. **45**, 1175–1202 (2016). forthcoming
Cordeiro, GM, Bager, RDSB: Moments for some Kumaraswamy generalized distributions. Commun. Stat. Theory Methods. **44**, 2720–2737 (2015)
Cordeiro, GM, Bourguignon, M: New results on the Ristić-Balakrishnan family of distributions. Commun. Stat. Theory Methods. **54**, 13–53 (2016)
Cordeiro, GM, de-Castro, M: A new family of generalized distributions. J. Stat. Comput. Simul. **81**, 883–893 (2011)
Cordeiro, GM, Cancho, VG, Ortega, EMM, Barriga, GDC: A model with long-term survivors: negative binomial Birnbaum-Saunders. Commun. Stat. Theory Methods. **45**, 1370–1387 (2016)
Cordeiro, GM, Ortega, EMM, da-Cunha, DCC: The exponentiated generalized class of distributions. J. Data Sci. **11**, 777–803 (2013a)
Cordeiro, GM, Silva, GO, Ortega, EMM: The beta Weibull geometric distribution. Statistics. **47**, 817–834 (2013b)
Cordeiro, GM, Rodrigues, J, de-Castro, M: The exponential COM-Poisson distribution. Stat. Papers. **53**, 653–664 (2012a)
Cordeiro, GM, Silva, GO, Ortega, EMM: The beta extended Weibull family. J. Probab. Stat. Sci. **10**, 15–40 (2012b)
Cordeiro, GM, Silva, RB: The complementary extended Weibull power series class of distributions. Ciênc. Nat. **36**, 1–13 (2014)
Dagum, C: A new model of personal income distribution: specification and estimation. Econ. Appl. **30**, 413–437 (1977)
Das, KK: On some generalized transmuted distributions. Int. J. Sci. Eng. Res. **6**, 1686–1691 (2015)
da-Silva, AL, Rodrigues, J, Silva, G: The beta transmuted Fréchet distribution: Properties and application to survival data (in Portuguese). ForSci. **3**, 57–69 (2015a)
da-Silva, RV, Gomes-Silva, F, Ramos, MWA, Cordeiro, GM: The exponentiated Burr XII Poisson distribution with application to lifetime data. Int. J. Statist. Probab. **4**, 112–131 (2015b)
da-Silva, RV, Ramos, MWA, Gomes-Silva, F, Cordeiro, GM: The exponentiated Kumaraswamy-G class. J. Egyptian Math. Soc. (2016). forthcoming
Delgarm, L, Zadkarami, MR: A new generalization of lifetime distributions. Comput. Statist. **30**, 1185–1198 (2015)
Dempster, AP, Laird, NM, Rubin, DB: Maximum likelihood from incomplete data via the EM algorithm. J. Roy. Stat. Soc. B 39, 1–38 (1977)
Ebraheim, AEHN: Exponentiated transmuted Weibull distribution: A generalization of the Weibull distribution. Int. J. Math. Comput. Phys. Quantum Eng. **8**, 897–905 (2014)
Elbatal, I: Transmuted modified inverse Weibull distribution: A generalization of the modified inverse Weibull probability distribution. Int. J. Math. Arch. **4**, 117–129 (2013a)
Elbatal, I: Transmuted generalized inverted exponential distribution. Econ. Qual. Control. **28**, 125–133 (2013b)
Elbatal, I, Aryal, G: On the transmuted additive Weibull distribution. Aust. J. Statist. **42**, 117–132 (2013)
Elbatal, I, Aryal, G: Transmuted Dagum distribution with applications. Chil. J. Statist. **6**, 31–45 (2015)
Elbatal, I, Diab, LS, Alim, NAA: Transmuted generalized linear exponential distribution. Int. J. Computer Appl. **83**, 29–37 (2013)
Elbatal, I, Asha, G, Raja, AV: Transmuted exponentiated Fréchet distribution: Properties and applications. J. Statist. Applic. Probab. **3**, 379–394 (2014)
Elbatal, I, Elgarhy, M: Transmuted quasi-Lindley distribution: A generalization of the quasi-Lindley distribution. Int. J. Pure Appl. Sci. Technol. **18**, 59–70 (2013)
Elbatal, I, Mansour, MM, Ahsanullah, M: The additive Weibull-geometric distribution: Theory and applications. J. Stat. Theory Applic. **15**, 125–141 (2016)
Elgarhy, M, Rashed, M, Shawki, AW: Transmuted generalized Lindley distribution. Int. J. Math.– Trends Technol. **29**, 145–154 (2016)
Eugene, N, Lee, C, Famoye, F: Beta-normal distribution and its applications. Commun. Stat. Theory Methods. **31**, 497–512 (2002)
Fatima, A, Roohi, A: Transmuted exponentiated Pareto-I distribution. Pak. J. Statist. **32**, 63–80 (2015)
Fioruci, JA, Yiqi, B, Louzada, F, Cancho, VG: The exponential Poisson logarithmic distribution. Commun. Stat. Theory Methods. **45**, 2556–2575 (2016)
Flores, JD, Borges, P, Cancho, VG, Louzada, F: The complementary exponential power series distribution. Braz. J. Probab. Statist. **27**, 565–584 (2013)
Gallardo, DI, Bolfarine, H: Two alternative estimation procedures for the negative binomial cure rate model with a latent activation scheme. Statist. Oper. Res. Trans. **40**, 31–54 (2016)
Gallardo, DI, Bolfarine, H, Pedroso-de-Lima, AC: An EM algorithm for estimating the destructive weighted Poisson cure rate model. J. Stat. Comput. Simul. **86**, 1497–1515 (2016)
Gitifar, N, Rezaei, S, Nadarajah, S: Compound distributions motivated by linear failure rate. Statist. Oper. Res. Trans. **40**, 177–200 (2016)
Ghorbani, M, Bagheri, SF, Alizadeh, M: A new lifetime distribution: The modified Weibull Poisson distribution. Int. J. Oper. Res. Dec. Sci. Stud. **1**, 28–47 (2014)

Gomes, AE, da-Silva, CQ, Cordeiro, GM: The exponentiated-G Poisson model. Commun. Stat. Theory Methods. **44**, 4217–4240 (2015)
Granzotto, DCT, Louzada, F: The transmuted log-logistic distribution: Modeling, inference, and an application to a polled tabapua race time up to first calving data. Commun. Stat. Theory Methods. **43**, 3387–3402 (2015)
Gui, W, Zhang, S, Lu, X: The Lindley-Poisson distribution in lifetime analysis and its properties. Hacet. J. Math. Stat. **43**, 1063–1077 (2014)
Gupta, RC, Gupta, PI, Gupta, RD: Modeling failure time data by Lehmann alternatives. Commun. Stat. Theory Methods. **27**, 887–904 (1998)
Gupta, RD, Kundu, D: Generalized exponential distribution. Austral. N. Z. J. Stat. **41**, 173–188 (1999)
Gupta, RC, Wu, Q, Huang, J: Analysis of survival data by an exponential-generalized Poisson distribution. J. Stat. Comput. Simul. **84**, 2495–2505 (2014)
Haq, MA, Butt, NS, Usman, RM, Fattah, AA: Transmuted power function distribution. Ghazi Uni. J. Sci. **29**, 177–185 (2016)
Harandi, SS, Alamatsaz, MA: A complementary generalized linear failure rate geometric distribution. Commun. Stat. Theory Methods. **45**, 2204–2227 (2016)
Hashimoto, EM, Ortega, EMM, Cordeiro, GM, Cancho, VG: The Poisson Birnbaum-Saunders model with long-term survivors. Statistics. **48**, 1394–1413 (2014)
Hashimoto, EM, Ortega, EMM, Cordeiro, GM, Cancho, VG: A new long-term survival model with interval-censored data. Sankhyā B 77, 207–239 (2015)
Hassan, AS, Abd-Elfattah, AM, Mokhtar, AH: The complementary Burr III Poisson distribution. Austral. J. Basic Appl. Sci. **9**, 219–228 (2015)
Hassan, AS, Abd-Elfattah, AM, Mokhtar, AH: The complementary exponentiated inverted Weibull power series family of distributions and its application. British J. Math. Comput. Sci. **13**, 1–20 (2016)
Hemmati, F, Khorram, E, Rezakhah, S: A new three-parameter ageing distribution. J. Stat. Plann. Infer. **141**, 2266–2275 (2011)
Hussian, MA: Estimation of $P(Y < X)$ for the class of Kumaraswamy-G distributions. Austral. J. Basic Appl. Sci. **7**, 158–169 (2013)
Hussian, MA: Transmuted exponentiated gamma distribution: A generalization of exponentiated gamma probability distribution. Appl. Math. Sci. **27**, 1297–1310 (2014)
Iriarte, YA, Astorga, JM: Transmuted Maxwell probability distribution (in Portuguese). Rev. Integr. **32**, 211–221 (2014)
Iriarte, YA, Astorga, JM: A version of transmuted generalized Rayleigh distribution (in Portuguese). Rev. Integr. **33**, 83–95 (2015)
Jafari, AA, Tahmasebi, S: Gompertz-power series distributions. Commun. Stat. Theory Methods. **45**, 3761–3781 (2016)
Jiménez, JA, Arunachalam, V, Serna, GM: A generalization of Tukey's g-h family of distributions. J. Stat. Theory Applic. **14**, 28–44 (2015)
Kannan, N, Kundu, D, Nair, P, Tripathi, RC: The generalized exponential cure rate model with covariates. J. Appl. Statist. **37**, 1625–1636 (2010)
Khan, MS, King, R: Transmuted modified Weibull distribution: A generalization of the modified Weibull probability distribution. Eur. J. Pure Appl. Math. **6**, 66–88 (2013)
Khan, MS, King, R: A new class of transmuted inverse Weibull distribution for reliability analysis. Amer. J. Math. Manag. Sci. **33**, 261–286 (2014a)
Khan, MS, King, R: Transmuted generalized inverse Weibull distribution. J. Appl. Stat. Sci. **20**, 213–230 (2014b)
Khan, MS, King, R: Transmuted modified inverse Rayleigh distribution. Aust. J. Statist. **44**, 17–29 (2015)
Khan, MS, King, R, Hudson, IL: Characterizations of the transmuted inverse Weibull distribution. ANZIAM J. **55**, C197—C217 (2014)
Khan, MS, King, R, Hudson, IL: Transmuted generalized exponential distribution: A generalization of the exponential distribution with applications to survival data. Commun. Stat. Theory Methods (2015a). forthcoming
Khan, MS, King, R, Hudson, IL: A new three parameter transmuted Chen lifetime distribution with application. J. Appl. Stat. Sci. **21**, 239–259 (2015b)
Khan, MS, King, R, Hudson, IL: Transmuted Kumaraswamy distribution. Statist. Transition. **17**, 1–28 (2016a)
Khan, MS, King, R, Hudson, IL: Transmuted Weibull distribution: Properties and estimation. Commun. Stat. Theory Methods (2016b). doi:10.1080/03610926.2015.1100744. forthcoming
Khan, MS, King, R, Hudson, IL: Transmuted Gompertz distribution: Properties and estimation. Pak. J. Statist. **32**, 161–182 (2016c)
Kuş, C: A new lifetime distribution. Comput. Stat. Data Anal. **51**, 4497–4509 (2007)
Lai, CD, Xie, M, Murthy, DNP: A modified Weibull distribution. IEEE Trans. Reliab. **52**, 33–37 (2003)
Leahu, A, Munteanu, BG, Cataranciuc, S: On the lifetime as the maximum or minimum of the sample with power series distributed size. ROMAI J. **9**, 119–128 (2013)
Leahu, A, Munteanu, BG, Cataranciuc, S: Max-Erlang and Min-Erlang power series distributions as two new families of lifetime distribution. Bull. Acad. Ştiinţe. **2**, 60–73 (2014)
Lemonte, AJ, Cordeiro, GM, Ortega, EMM: On the Additive Weibull Distribution. Commun. Stat. Theory Methods. **43**, 2066–2080 (2014)
Louzada, F, Bereta, EMP, Franco, MAP: On the distribution of the minimum or maximum of a random number of i.i.d. lifetime random variable. Appl. Math. **3**, 350–353 (2012a)
Louzada, F, Borges, P, Cancho, V: The exponential negative-binomial distribution: A continuous bridge between under and over dispersion on a lifetime modelling structure. J. Statist. Adv. Theory Applic. **7**, 67–83 (2012b)
Louzada, F, Cancho, VG, Barriga, GDC: The Poisson-exponential regression model under different latent activation schemes. Comput. Appl. Math. **31**, 617–632 (2012c)
Louzada, F, Cancho, VG, Roman, M, Leite, JG: A new long-term lifetime distribution induced by a latent complementary risk framework. J. Appl. Statist. **39**, 2209–2222 (2012d)
Louzada, F, Cancho, VG, Yiqi, B: The log-Weibull-negative-binomial regression model under latent failure causes and presence of randomized schemes. Statistics. **49**, 930–949 (2015)

Louzada, F, Granzotto, DCT: The transmuted log-logistic regression model: a new model for time up to first calving of cows. Stat. Papers (2015). doi:10.1007/s00362-015-0671-5. forthcoming

Louzada, F, Marchi, V, Carpenter, J: The complementary exponentiated exponential geometric lifetime distribution. J Probab Statist. Art.ID. 502159, 12 (2014a)

Louzada, F, Marchi, V, Romana, M: The exponentiated exponential geometric distribution: a distribution with decreasing, increasing and unimodel failure rate. Statistics. **48**, 167–181 (2014)

Louzada, F, Roman, M, Cancho, VG: The complementary exponential geometric distribution: Model, properties, and a comparison with its counterpart. Comput. Stat. Data Anal. **55**, 2516–2524 (2011)

Lu, W, Shi, D: A new compounding life distribution: the Weibull Poisson distribution. J. Appl. Statist. **39**, 21–38 (2012)

Lucena, SEF, Silva, AHA, Cordeiro, GM: The transmuted generalized gamma distribution: Properties and application. J. Data Sci. **13**, 409–420 (2015)

Luguterah, A, Nasiru, S: Transmuted exponential Pareto distribution. Far East. J. Theor. Statist. **50**, 31–49 (2015)

Mahmoud, MR, Mandouh, RM: On the transmuted Fréchet distribution. J. Appl. Sci. Res. **9**, 5553–5561 (2013)

Mahmoudi, E, Jafari, AA: Generalized exponential–power series distributions. Comput. Stat. Data Anal. **56**, 4047–4066 (2012)

Mahmoudi, E, Jafari, AA: The compound class of linear failure rate-power series distributions: Model, properties and applications. Commun. Stat. Simul. Comput. (2015). doi:10.1080/03610918.2015.1005232. forthcoming

Mahmoudi, E, Mahmoodian, H: Normal power series class of distributions: Model, properties and applications (2015). arXiv:1510.07180v1 [stat.CO]

Mahmoudi, E, Sepahdar, A: Exponentiated Weibull-Poisson distribution: Model, properties and applications. Math. Comput. Simul. **92**, 76–97 (2013)

Mahmoudi, E, Sepahdar, A, Lemonte, AJ: Exponentiated Weibull-logarithmic distribution: Model, properties and applications (2014). arXiv:1402.2564v1 [stat.ME]

Mahmoudi, E, Shiran, M: Exponentiated Weibull-geometric distribution and its applications (2012a). arXiv:1206.4008v1 [stat.CO]

Mahmoudi, E, Shiran, M: Exponentiated Weibull-power series distributions and its applications (2012b). arXiv:1212.5613v1 [stat.CO]

Mansour, MM, Elrazik, EMB, Hamed, MS, Mohamed, SM: A new transmuted additive Weibull distribution: Based on a new method for adding a parameter to a family of distribution. Int. J. Appl. Math. Sci. **8**, 31–51 (2015a)

Mansour, MM, Hamed, MS, Mohamed, SM: A new Kumaraswamy transmuted modified Weibull distribution with application. J. Stat. Adv. Theory Applic. **13**, 101–133 (2015b)

Mansour, MM, Mohamed, SM: A new generalized of transmuted Lindley distribution. Appl. Math. Sci. **9**, 2729–2748 (2015)

Marshall, AW, Olkin, I: A new method for adding a parameter to a family of distributions with application to the exponential and Weibull families. Biometrika. **84**, 641–652 (1997)

Martinez, EZ, Achcar, JA, Jácome, AAA, Santos, JS: Mixture and non-mixture cure fraction models based on the generalized modified Weibull distribution with an application to gastric cancer data. Comput. Methods Prog. Biomed. **112**, 343–355 (2013)

Mendoza, NVR, Ortega, EMM, Cordeiro, GM: The exponentiated-log-logistic geometric distribution: Dual activation. Commun. Stat. Theory Methods. **45**, 3838–3859 (2016)

Merovci, F: Transmuted Rayleigh distribution. Aust. J. Statist. **42**, 21–31 (2013a)

Merovci, F: Transmuted exponentiated exponential distribution. Math. Sci. Applic. E-Notes. **1**, 112–122 (2013b)

Merovci, F: Transmuted Lindley distribution. Int. J. Open Problems Comput. Math. **6**, 63–72 (2013c)

Merovci, F: Transmuted generalized Rayleigh distribution. J. Stat. Applic. Probab. **3**, 9–20 (2014)

Merovci, F, Elbatal, I, Ahmed, A: The transmuted generalized inverse Weibull distribution. Aust. J. Statist. **43**, 119–131 (2014)

Merovci, F, Alizadeh, M, Hamedani, GG: Another generalized transmuted family of distributions: Properties and applications. Aust. J. Statist. **45**, 71–93 (2016)

Merovci, F, Elbatal, I: Transmuted Lindley-geometric distribution and its aplications. J. Stat. Applic. Probab. **3**, 77–91 (2014a)

Merovci, F, Elbatal, I: Transmuted Weibull-geometric distribution and its aplications. Sci. Magna. **10**, 68–82 (2014b)

Merovci, F, Puka, L: Transmuted Pareto distribution. ProbStat Forum. **7**, 1–11 (2014)

Morais, AL, Barreto-Souza, W: A compound class of Weibull and power series distributions. Comput. Stat. Data Anal. **55**, 1410–1425 (2011)

Munteanu, BG: The max-Weibull power series distribution. An. Univ. Oradea Fasc. Mat. **21**, 133–139 (2014)

Nadarajah, S, Cancho, VG, Ortega, EMM: The geometric exponential Poisson distribution. Stat. Methods Applic. **22**, 355–380 (2013a)

Nadarajah, S, Jayakumar, K, Ristić, MM: A new family of lifetime models. J. Stat. Comput. Simul. **83**, 1389–1404 (2013b)

Nadarajah, S, Cordeiro, GM, Ortega, EMM: General results for the Kumaraswamy-G distribution. J. Stat. Comput. Simul. **82**, 951–979 (2012)

Nadarajah, S, Cordeiro, GM, Ortega, EMM: The exponentiated-G-geometric family of distributions. J. Stat. Comput. Simul. **85**, 1634–1650 (2015a)

Nadarajah, S, Cordeiro, GM, Ortega, EMM: The Zografos-Balakrishnan-G family of distributions: Mathematical properties and application. Commun. Stat. Theory Methods. **44**, 186–215 (2015b)

Nadarajah, S, Shahsanaei, F, Rezaei, S: A new four-parameter lifetime distribution. J. Stat. Comput. Simul. **84**, 248–263 (2014a)

Nadarajah, S, Teimouri, M, Shih, SH: Modified beta distributions. Sankhyā B 76, 19–48 (2014b)

Nascimento, ADC, Bourguignon, M, Zea, LM, Santos-Neto, M, Silva, RB, Cordeiro, GM: The gamma extended Weibull family of distributions. J. Stat. Theory Applic. **13**, 1–16 (2014)

Nassar, M, Nada, N: A new generalization of the exponential-geometric distribution. J. Statist. Adv. Theory Applic. **7**, 25–48 (2012)

Nofal, Z, Afify, A, Yousof, H, Cordeiro, GM: The generalized transmuted-G family of distributions. Commun. Stat. Theory Methods (2016). doi:10.1080/03610926.2015.1078478. forthcoming

Oluyede, BO, Foya, S, Warahena-Liyanage, G, Huang, S: The log-logistic Weibull distribution with applications to lifetime data. Aust. J. Statist. **45**, 43–69 (2016a)

Oluyede, BO, Motsewabagale, G, Huang, S, Warahena-Liyanage, G, Rararai, M: The Dagum-Poisson distribution: model, properties and application. Electron. J. Appl. Stat. Anal. **9**, 169–197 (2016b)
Oluyede, BO, Warahena-Liyanage, G, Rararai, M: A new compund class of log-logistic Weibull-Poisson distribution: model, properties and application. J. Stat. Comput. Simul. **86**, 1363–1391 (2016c)
Ortega, EMM, Cordeiro, GM, Kattan, MW: The negative binomial-beta Weibull regresion model to predict the cure rate of prostrate cancer. J. Appl. Statist. **39**, 1191–1210 (2012)
Ortega, EMM, Barriga, GDC, Hashimoto, EM, Cancho, VG, Cordeiro, GM: A new class of survival regression models with cure fraction. J. Data Sci. **12**, 107–136 (2014)
Ortega, EMM, Cordeiro, GM, Campelo, AK, Kattan, MW, Cancho, VG: A power series beta Weibull regression model for predicing breast carcinoma. Statist. Med. **34**, 1366–1388 (2015)
Owoloko, EA, Oguntunde, PE, Adejumo, AO: Performance rating of the transmuted exponential distribution: an analytical approach. SpringerPlus. **4**, 8–18 (2015)
Pal, M, Tiensuwan, M: The beta transmuted Weibull distribution. Aust. J. Statist. **43**, 133–149 (2014)
Pappas, V, Adamidis, K, Loukas, S: A three-parameter lifetime distribution. Adv. Applic. Statist. **20**, 159–167 (2011)
Pappas, V, Adamidis, K, Loukas, S: A generalization of the exponential-logarithmic distribution. J. Stat. Theory Prac. **9**, 122–133 (2015)
Pararai, M, Oluyede, BO, Warahena-Liyanage, G: Kumaraswamy Lindley-Poisson distribution: Theory and applications. Asian. J. Math. Appl. Art. ID. ama0261, 30 (2015a)
Pararai, M, Warahena-Liyanage, G, Oluyede, BO: An extended Lindley Poisson distribution with applications. J. Math. Stat. Sci. **1**, 167–198 (2015b)
Pararai, M, Warahena-Liyanage, G, Oluyede, BO: Exponentiated power Lindley Poisson distribution: properties and applications. Commun. Stat. Theory Methods (2016). doi:10.1080/03610926.2015.1076473. forthcoming
Pescim, RR, Cordeiro, GM, Demétrio, CGB, Ortega, EMM, Nadarajah, S: The new class of Kummer beta generalized distributions. Statist. Oper. Res. Trans. **36**, 153–180 (2012)
Pinho, LGB, Cordeiro, GM, Nobre, JS: On the Harris-G class of distributions: general results and application. Braz. J. Probab. Statist. **29**, 813–832 (2015)
Preda, V, Panaitescu, E, Ciumara, R: The modified exponential-Poisson distribution. Proc. Rom. Acad. **A12**, 22–29 (2011)
Percontini, A, Blas, B, Cordeiro, GM: The beta Weibull Poisson distribution. Chil. J. Statist. **4**, 3–26 (2013a)
Percontini, A, Cordeiro, GM, Bourguignon, M: The G-negative binomail family: General properties and applications. Adv. Applic. Statist. **35**, 127–160 (2013b)
Percontini, A, Gomes-Silva, FS, Ramos, MWA, Venancio, R, Cordeiro, GM: The gamma Weibull Poisson distribution applied to survival data. Trends Appl. Comput. Math. (TEMA). **15**, 165–176 (2014)
Popović, BV, Ristić, MM, Cordeiro, GM: A two parameter distribution obtained by compounding the generalized exponential and exponential distributions. Mediterr. J. Math (2015). doi:10.1007/s00009-015-0665-5. forthcoming
Ramos, MWA, Marinho, PRD, Cordeiro, GM, da-Silva, RV, Hamedani, GG: The Kumaraswamy-G Poisson family of distributions. J. Stat. Theory Applic. **14**, 222–239 (2015)
Ramos, MWA, Marinho, PRD, da-Silva, RV, Cordeiro, GM: The exponentiated Lomax Poisson distribution with an application to lifetime data. Adv. Applic. Statist. **34**, 107–135 (2013)
Rezaei, S, Tahmasbi, R: A new lifetime distribution with increasing failure rate: Exponential truncated Poisson. J. Basic Appl. Sci. Res. **2**, 1749–1762 (2012)
Ristić, MM, Balakrishnan, N: The gamma-exponentiated exponential distribution. J. Stat. Comput. Simul. **82**, 1191–1206 (2012)
Ristić, MM, Kundu, D: Generalized exponential geometric extreme distribution. J. Stat. Theory Prac. **10**, 179–201 (2016)
Ristić, MM, Nadarajah, S: A new lifetime distribution. J. Stat. Comput. Simul. **84**, 135–150 (2014)
Ristić, MM, Popvić, BV, Nadarajah, S: Libby and Novick's generalized beta exponential distribution. J. Stat. Comput. Simul. **85**, 740–761 (2015)
Rodrigues, J, Cancho, VG, de-Castro, M, Louzada-Neto, F: On the unification of long-term survival models. Statist. Probab. Lett. **79**, 753–759 (2009a)
Rodrigues, J, de-Castro, M, Cancho, VG, Balakrishnan, N: COM-Poisson cure rate survival models and an application to a cutaneous melanoma data. J. Stat. Plann. Infer. **139**, 3605–3611 (2009b)
Roman, M, Louzada, F, Cancho, VG, Leite, JG: A new long-term survival distribution for cancer data. J. Data Sci. **10**, 241–258 (2012)
Saboor, A, Kamal, M, Ahmad, M: The transmuted exponentiated-Weibull distribution with applications. Pak. J. Statist. **31**, 229–250 (2015)
Saboor, A, Elbatal, I, Cordeiro, GM: The transmuted exponentiated Weibull geometric distribution: Theory and applications. Hacet. J. Math. Stat. **45**, 973–987 (2016)
Sarhan, AM, Kundu, D: Generalized linear failure rate distribution. Commun. Stat. Theory Methods. **38**, 642–660 (2009)
Sarhan, AM, Zaindin, M: Modified Weibull distribution. Appl. Sci. **11**, 123–136 (2009)
Shahsanaei, F, Rezaei, S, Pak, A: A new two-parameter lifetime distribution with increasing failure rate. Econ. Qual. Control. **27**, 1–17 (2012)
Shafiei, S, Darijani, S, Saboori, H: Inverse Weibull power series distributions: properties and applications. J. Stat. Comput. Simul. **86**, 1069–1094 (2016)
Sharma, VK, Singh, SK, Singh, U: A new upside-down bathtub shaped hazard rate model for survival data analysis. Appl. Math. Comput. **239**, 242–253 (2014)
Shaw, WT, Buckley, IR: The alchemy of probability distributions: Beyond Gram-Charlier expansions, and a skew-kurtotic-normal distribution from a rank transmutation map (2009). arXiv:0901.0434 [q-fin.ST]
Shahzad, MN, Asghar, Z: Transmuted Dagum distribution: A more flexible and broad shaped hazard function model. Hacet. J. Math. Stat. **45**, 227–224 (2016)
Silva, RB, Barreto-Souza, W, Cordeiro, GM: A new distribution with decreasing, increasing and upside-down bathtub failure rate. Comput. Stat. Data Anal. **54**, 935–944 (2010)

Silva, RB, Bourguignon, M, Dias, CRB, Cordeiro, GM: The compound class of extended Weibull power series distributions. Comput. Stat. Data Anal. **58**, 352–367 (2013)

Silva, RB, Bourguignon, M, Cordeiro, GM: A new compounding family of distributions: The generalized gamma power series distributions. J. Comput. Appl. Math. **303**, 119–139 (2016)

Silva, RB, Cordeiro, GM: The Burr XII power series distributions: A new compounding family. Braz. J. Probab. Statist. **29**, 565–589 (2015)

Tahir, MH, Cordeiro, GM, Alizadeh, M, Mansoor, M, Zubair, M, Hamedani, GG: The odd generalized exponential family of distributions with applications. J. Stat. Dist. Applic. 2, Art, 1 (2015)

Tahir, MH, Cordeiro, GM, Alzaatreh, A, Mansoor, M, Zubair, M: The Logistic-X family of distributions and its applications. Commun. Stat. Theory Methods. **45**, 732–7349 (2016). forthcoming

Tahir, MH, Nadarajah, S: Parameter induction in continuous univariate distribution: Well-established G families. Ann. Braz. Acad. Sci. **87**, 539–568 (2015)

Tahir, MH, Zubair, M, Cordeiro, GM, Alzaatreh, A, Mansoor, M: The Poisson-X family of distributions. J. Stat. Comput. Simul. **86**, 2901–2921 (2016a)

Tahir, MH, Zubair, M, Mansoor, M, Cordeiro, GM, Alizadeh, M, Hamedani, GG: A new Weibull-G family of distributions. Hacet. J. Math. Stat. **45**, 629–647 (2016b)

Tahmasbi, R, Rezaei, S: A two-parameter lifetime distribution with decreasing failure rate. Comput. Stat. Data Anal. **52**, 3889–3901 (2008)

Tahmasebi, S, Jafari, AA: Exponentiated extended Weibull-power series class of distributions. Ciênc. Nat. **37**, 183–193 (2015a)

Tahmasebi, S, Jafari, AA: Generalized Gompertz-power series distributions. Hacet. J. Math. Stat. (2015b). doi:10.15672/HJMS.20158312681. forthcoming

Tian, Y, Tian, M, Zhu, Q: Transmuted linear exponential distribution: A new generalization of the linear exponential distribution. Commun. Stat. Simul. Comput. **43**, 2661–2671 (2014)

Tojeiro, C, Louzada, F, Roman, M, Borges, P: The complementary Weibull geometric distribution. J. Stat. Comput. Simul. **84**, 1345–1362 (2014)

Torabi, H, Montazari, NH: The gamma-uniform distribution and its application. Kybernetika. **48**, 16–30 (2012)

Torabi, H, Montazari, NH: The logistic-uniform distribution and its application. Commun. Stat. Simul. Comput. **43**, 2551–2569 (2014)

Triantafyllou, IS, Koutras, MV: Failure rate and aging properties of generalized beta- and gamma-generated distributions. Commun. Stat. Theory Methods. **43**, 4046–4061 (2014)

Vardhan, RV, Balaswamy, S: Transmuted new modified Weibull distribution. Math. Sci. Applic. E-Notes. **4**, 125–135 (2016)

Wang, M: A new three-parameter lifetime distribution and associated inference (2013). arXiv:1308.4128v1 [stat.ME]

Wang, M, Elbatal, I: The modified Weibull geometric distribution. Metron. **73**, 303–315 (2015)

Warahena-Liyanage, G, Pararai, M: The Lindley power series class of distributions: Model, properties and applications. J. Comput. Model. **5**, 35–80 (2015a)

Warahena-Liyanage, G, Pararai, M: A generalized power Lindley Poisson distribution with applications. Asian J. Math. Appl. Art.ID. ama0169, 23 (2015b)

Xie, M, Lai, CD: Reliability analysis using an additive Weibull model with bathtub shaped failure rate function. Reliab. Eng. Syst. Safe. **52**, 87–93 (1995)

Yakovlev, AY, Tsodikov, AD: Stochastic Models of Tumor Latency and Their Biostatistical Applications. World Scientific, Singapore (1996)

Yamachi, CY, Romana, M, Louzada, F, Franco, MAP, Cancho, VG: The exponentiated complementary exponential geometric distribution. J. Mod. Math. Front. **2**, 78–83 (2013)

Yiqi, B, Russo, CM, Cancho, VG, Louzada, F: Influence diagnostics for the Weibull-negative-binomial regression model with cure rate under latent failure causes. J. Appl. Statist. **43**, 1027–1060 (2016)

Yousof, HM, Afify, AZ, Alizadeh, M, Butt, NS, Hamedani, GG, Ali, MM: The transmuted exponentiated generalized-G family of distributions. Pak. J. Statist. Oper. Res. **11**, 441–464 (2015)

Yousof, HM, Afify, AZ, Ebraheim, AEHN, Hamedani, GG, Butt, NS: On six-parameter Fréchet distribution: Properties and applications. Pak. J. Statist. Oper. Res. **12**, 281–299 (2016)

Zakerzadeh, H, Mahmoudi, E: A new two parameter lifetime distribution: Model and properties (2013). arXiv:1204.4248v1 [stat.CO]

Zografos, K, Balakrishnan, N: On families of beta- and generalized gamma-generated distributions and associated inference. Stat. Methodol. **6**, 344–362 (2009)

9

An extended-G geometric family

Gauss M. Cordeiro[1], Giovana O. Silva[2*] and Edwin M. M. Ortega[3]

*Correspondence: giovana@ufba.br
[2]Departamento de Estatística, Universidade Federal da Bahia, Av. Adhemar de Barros, s/n°, 40170-110 Salvador, Brazil
Full list of author information is available at the end of the article

Abstract

We introduce and study the extended-G geometric family of distributions, which contains as special models some important distributions such as the XTG (Xie et al. 2002) geometric, Weibull geometric, Chen (Chen 2000) geometric, Gompertz geometric, among others. This family not only includes distributions with bathtub and unimodal failure rate functions but provides a broader class of monotone failure rates. Its density function can be expressed as a linear mixture of extended-G densities. We derive explicit expansions for the ordinary and incomplete moments, generating function, mean deviations and Rényy entropy. The density of the order statistics can also be given as a linear mixture of extended-G densities. The model parameters are estimated by maximum likelihood. The potentiality of the new family is illustrated by means of an application to real data.

MSC: 60E05, 62P10, 62P30

Keywords: Expected information, Extended-G distribution, Geometric distribution, Lifetime distribution, Maximum likelihood estimation

1 Introduction

In the last few years, new classes of distributions were proposed by extending the Weibull distribution to cope with bathtub shaped failure rates. A good review of some of these models is addressed by Pham and Lai (2007). Among these models, we point out the exponentiated Weibull (Mudholkar et al. 1995, 1996), additive Weibull (AW) (Xie and Lai 1995), XTG (Xie et al. 2002), modified Weibull (MW) (Lai et al. 2003), beta exponential (Nadarajah and Kotz 2006), BLZ (Bebbington et al. 2007), generalized modified Weibull (GMW) (Carrasco et al. 2008), beta modified Weibull (BMW) (Silva et al. 2010a) and beta Weibull geometric (Cordeiro et al. 2013) distributions.

Alternatively, various works introduced more flexible distributions in modeling monotone or unimodal failure rates but it is not useful for modeling the bathtub shaped failure rates. Adamidis and Loukas (1998) defined the exponential geometric distribution to model lifetime data with decreasing hazard rate function (hrf). Gupta and Kundu 1999, 2001a, b proposed the generalized exponential (GE) (also called the exponentiated exponential) distribution, and investigated some of its mathematical properties. This distribution has only increasing or decreasing hrf. Following the same idea of the GE distribution, Silva et al. (2010b) proposed the generalized exponential geometric (GEG) model and demonstrated that its hrf can be increasing, decreasing or unimodal. Another generalization of the Weibull for modeling monotone or unimodal failure rates, referred to as the Weibull geometric (WG) distribution, was investigated by Barreto-Souza et al. (2011). The Kumaraswamy log-logistic distribution was studied by de Santana et al. (2012). This generalization contains distributions with unimodal and bathtub shaped hrfs.

In this paper, we propose the new *extended-G geometric* (EGG) family of lifetime models by compounding the *extended-G* (EG) (Nadarajah and Kotz 2005) class of models and the geometric distribution. This family has two important aspects: it involves one additional shape parameter to the baseline model and the new parameter has a clear physical interpretation. We expect that it will attract wider applications in biology, medicine and reliability, and other areas of research.

Further, several distributions are obtained as special cases of this family including the extended exponential geometric (EEG), modified Weibull geometric (MWG), exponential power geometric (EPG), log-Weibull geometric (LWG), generalized power Weibull geometric (GPWG), among several others. Besides these distributions, the EGG family contains other promising new distributions as, for example, the additive Weibull geometric (AWG) and XTG geometric (XTGG) distributions (see "Appendix"). Due to its flexibility in accommodating different forms of the risk function, the new family is an important tool to be used in a variety of problems in modeling survival data. Various EGG distributions are not only convenient for modeling comfortable bathtub-shaped failure rates but they are also suitable for testing goodness-of-fit of some special models.

The paper is organized as follows. In Section 2, we define the EGG family and demonstrate that its probability density function (pdf) is given as a linear mixture of EG densities. In Sections 3 and 4, we derive the moments and moment generating function (mgf), respectively. The mean deviations and Bonferroni and Lorenz curves are obtained in Section 5. In Section 6, we demonstrate that the Rényi entropy of the EGG family is a linear combination of EG entropies with different scale parameters. In Section 7, the density function of the EGG order statistics is expressed as a linear mixture of EG densities. Maximum likelihood estimation of the model parameters is addressed in Section 8. In Section 9, we give an application to ozone level data to prove empirically the potentiality of the proposed family. Concluding remarks are provided in Section 10. Some special models are presented in the "Appendix".

2 The EGG family

The cumulative distribution function (cdf) of the EG class is defined by

$$G_{\alpha,\boldsymbol{\tau}}(y) = 1 - \exp\{-\alpha\, H(y)\},\ \ y > 0, \tag{1}$$

where $\alpha > 0$ is a rate parameter, $H(y)$ is a monotonically increasing function of y with the only limitation $H(y) \geq 0$ and $\boldsymbol{\tau}$ denotes the vector of unknown parameters in $H(y)$. If $H(y)$ is a power function, Eq. (1) reduces to the Weibull distribution. The EG class pdf is given by

$$g_{\alpha,\boldsymbol{\tau}}(y) = \alpha h(y)\ \exp\{-\alpha H(y)\},\ \ y > 0, \tag{2}$$

where $h(y) = \partial H(y)/\partial y$. We denote by $Y_\alpha \sim EG(\alpha, \boldsymbol{\tau})$ a random variable Y_α having density function (2). The MW distribution is a special case of (2) when $H(y) = y^\gamma \exp(\lambda y)$, where $\gamma > 0$ and $\lambda \geq 0$. Clearly, the Weibull distribution is obtained as a basic exemplar when $\lambda = 0$.

The EG quantile function (qf) is given by

$$Q_{\alpha,\boldsymbol{\tau}}(u) = H^{-1}\left\{-\frac{1}{\alpha}\log(1-u)\right\}, \tag{3}$$

which plays an important role in the algebraic developments in the paper. We only need the inverse function of $H(y)$ to determine (3). Table 1 lists closed-form inverses of $H(\cdot)$ for some EG sub-models.

Let Z be a geometric random variable with probability mass function given by $P(z;p) = (1-p)\,p^{z-1}$ for $z \in N$ and $p \in (0,1)$. Suppose that $\{Y_i\}_{i=1}^{Z}$ are independent and identically distributed (iid) random variables having the same density function (2), where the unknown number Z of random variables follows this geometric distribution. We assume that the random variables Z and Y_i's are independent. The EGG family is defined by $X = \min\left(\{Y_i\}_{i=1}^{Z}\right)$, where Z has this geometric distribution. We provide three interesting situations for applying this family:

- *Reliability* - the new distribution can arise in series and parallel systems with identical components, which appear in many industrial applications and biological organisms;
- *Times to the first and last failure* - suppose the failure of a device occurs due to the presence of an unknown number Z of initial defects of same kind, which can be identifiable only after causing failure and are repaired perfectly. Let Y_i be the time to the failure of the device due to the ith defect, for $i \geq 1$. Under the assumptions that the Y_i's are iid EG random variables independent of Z, the EGG distribution is appropriate for modeling the times to the first and last failures, respectively;
- *Time to relapse of cancer under the first-activation scheme* - suppose that an individual in the population is susceptible to a certain type of cancer. Let Z be the number of carcinogenic cells for that individual left active after the initial treatment and denote by Y_i the time spent for the ith carcinogenic cell to produce a detectable cancer mass, for $i \geq 1$. Under the assumptions that $\{Y_i\}_{i\geq 1}$ is a sequence of iid EG random variables independent of Z, we conclude that the time to relapse of cancer of a susceptible individual has the EGG distribution.

The conditional density function of X given $Z = z$ is given by

$$f(x|z;\alpha,\boldsymbol{\tau}) = z\,\alpha\,h(x)\,\exp\{-\alpha H(x)\}\left[1-\exp\{-\alpha H(x)\}\right],$$

Table 1 Inverse function $x = H^{-1}(y)$ for some EG models

Distribution	$x = H^{-1}(y)$
Exponential power	$\frac{[\log(y+1)]^{1/\beta}}{\lambda}$
Chen	$\left[\log(y+1)\right]^{1/\beta}$
XTG	$\lambda\left[\log\left(y/\lambda+1\right)\right]^{\frac{1}{\beta}}$
Log-Weibull	$\sigma\log(y)+\mu$
Kies	$\frac{y^{1/\beta}\sigma+\mu}{y^{1/\beta}+1}$
Generalized power Weibull	$\beta\left[(y+1)^{1/\theta}-1\right]^{1/\alpha_1}$
BLZ	$\frac{\log(t)\pm\sqrt{[\log(y)]^2+4\alpha_1\beta}}{2\alpha_1}$
Gompertz	$\frac{\log(\alpha_1 y+1)}{\alpha_1}$
Pham	$\left[\frac{\log(1+y)}{\log(a_1)}\right]^{1/\alpha_1}$

and then the EGG density function reduces to

$$f(x;p,\alpha,\boldsymbol{\tau}) = \alpha\,(1-p)\,h(x)\,\exp\{-\alpha H(x)\}[\,1-p\exp\{-\alpha H(x)\}]^{-2}, \quad x>0. \tag{4}$$

The MWG distribution is obtained from (4) when $H(x) = x^{\gamma}\exp(\lambda x)$, where $\gamma > 0$ and $\lambda \geq 0$. Further, the WG distribution is also obtained as a special case when $\lambda = 0$. The EG distribution follows as the limiting distribution (the limit is defined in terms of the convergence in distribution) of the EGG distribution when $p \to 0^{+}$. On the other hand, if $p \to 1^{-}$, we obtain the distribution of a random variable Y such that $P(Y=0)=1$.

Hereafter, a random variable X having density function (4) is denoted by $X \sim EGG(p,\alpha,\boldsymbol{\tau})$. The EGG cdf is given by

$$F(x;p,\alpha,\boldsymbol{\tau}) = \frac{1-\exp\{-\alpha H(x)\}}{1-p\exp\{-\alpha H(x)\}}, \quad x>0. \tag{5}$$

The hrf corresponding to (4) becomes

$$h(x;p,\alpha,\boldsymbol{\tau}) = \frac{\alpha h(x)}{1-p\exp\{-\alpha H(x)\}}, \quad x>0. \tag{6}$$

The EGG family contains as special models some well-known distributions. Several new models can also be easily generated as those listed in the "Appendix". Plots of the density functions (24)-(25) (given in the "Appendix") and the corresponding hrfs for the XTGG and MWG distributions are displayed in Figs. 1 and 2, respectively.

For $|z| < 1$ and $\rho > 0$, we can write

$$(1-z)^{-\rho} = \sum_{j=0}^{\infty} \frac{\Gamma(\rho+j)}{\Gamma(\rho)\,j!}\,z^{j}. \tag{7}$$

A useful representation for the EGG pdf ($x > 0$) follows by applying (7) in Eq. (4)

$$f(x;p,\alpha,\boldsymbol{\tau}) = \sum_{j=0}^{\infty} w_j\, g_{(j+1)\alpha,\boldsymbol{\tau}}(x), \tag{8}$$

where $w_j = (1-p)\,p^{j}$ for $j = 0,1,\ldots$ and $g_{(j+1)\alpha,\boldsymbol{\tau}}(x)$ denotes the density function of the random variable $Y_{(j+1)\alpha}$. Evidently, $\sum_{j=0}^{\infty} w_j = 1$. Note that the right-hand side of (8) is mathematically more tractable than the density of X. So, the EGG density function can be expressed as a linear mixture of EG densities. Equation (8) is very useful to derive

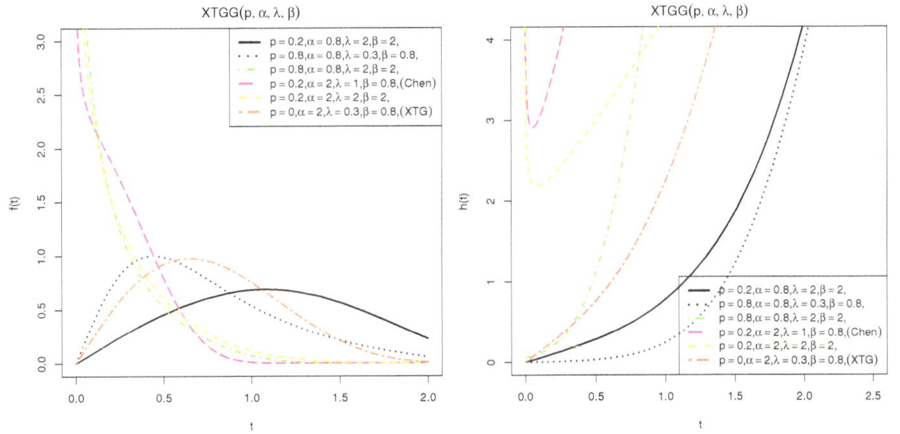

Fig. 1 Plots of the XTGG density and hazard rate functions for some parameter values

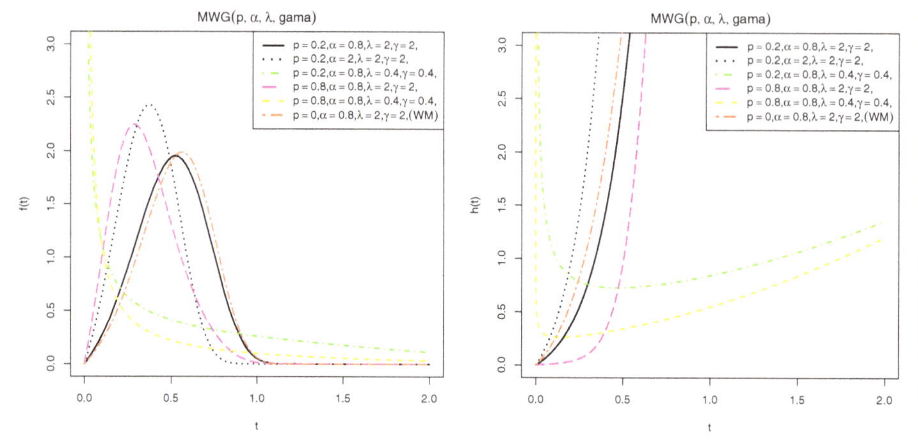

Fig. 2 Plots of the MWG density and hazard rate functions for some parameter values

some mathematical quantities (for example, ordinary, incomplete and factorial moments and mgf) for the EGG family from an infinite weighted linear combination of those EG quantities.

3 Moments

In this section, we derive an explicit expression for the EGG moments. Let $\beta_s(\alpha, \boldsymbol{\tau}) = E\left(Y_\alpha^s\right)$ be the sth moment of the EG class (2) with parameters α and $\boldsymbol{\tau}$. The sth moment of X, say μ_s', can be expressed in terms of the corresponding moments of EG distributions. In fact, we obtain a very simple expression from (8)

$$\mu_s' = E(X^s) = \sum_{j=0}^{\infty} w_j \, \beta_s((j+1)\alpha, \boldsymbol{\tau}). \tag{9}$$

Equation (9) reveals that the moments of X are linear combinations of the corresponding moments of $Y_{(j+1)\alpha}$ for $j \geq 0$. So, the moments of the EGG family are expressed as infinite linear combinations of those EG moments.

Next, we provide three alternative expressions for $\beta_s(\alpha, \boldsymbol{\tau})$. First, by substituting $u = Q_{\alpha,\boldsymbol{\tau}}(y)$ given by (3) in the sth moment of Y_α determined from (2), we obtain

$$\beta_s(\alpha, \boldsymbol{\tau}) = \int_0^1 Q_{\alpha,\boldsymbol{\tau}}(u)^s du. \tag{10}$$

A second formula follows from (2) by changing variable $\nu = H(y)$

$$\beta_s(\alpha, \boldsymbol{\tau}) = s \int_0^{\infty} p_{s,\boldsymbol{\tau}}(\nu)\, \mathrm{e}^{-\alpha \mathrm{v}} \mathrm{dv} = \mathrm{P}_{s,\boldsymbol{\tau}}(-\alpha), \tag{11}$$

where $p_{s,\boldsymbol{\tau}}(\nu) = \frac{H^{-1}(\nu)^{s-1}}{h(H^{-1}(\nu))}$ and $P_{s,\boldsymbol{\tau}}(-\alpha) = \mathfrak{L}(p_{s,\boldsymbol{\tau}}(\nu); \alpha)$ is the Laplace transform of $p_{s,\boldsymbol{\tau}}(\nu)$. A third formula for $\beta_s(\alpha, \boldsymbol{\tau})$ comes from the Mellin transform of $h(y)\, \mathrm{e}^{-\alpha \mathrm{H(y)}}$, namely

$$\beta_s(\alpha, \boldsymbol{\tau}) = \alpha \, \mathrm{M}\left(\mathrm{h(y)}\, \mathrm{e}^{-\alpha \mathrm{H(y)}}; \mathrm{s}+1\right).$$

Several tables provide Mellin transforms for common functions. The Laplace and Mellin transforms are defined in Prudnikov et al. (1986). Equations (9)–(11) are the main results of this section. They can be used to obtain analytically or numerically the moments of several EGG special models.

Further, we provide an application of (9) in conjunction with (10) and (11). Substituting $H(y)$ of the XTG distribution given in the "Appendix" into Eq. (11) and assuming that $m = s/\beta$ is an integer, and after some algebra, we have

$$\begin{aligned}\beta_s(\alpha,\beta,\lambda) &= s\int_0^\infty v^{s-1}\,\exp(\lambda\alpha)\,\exp\left\{-\lambda\alpha\exp\left[(v/\lambda)^\beta\right]\right\}dv\\ &= m\,\lambda^s\,\exp(\lambda\alpha)\left.\frac{\partial^{m-1}(\lambda\alpha)^{-t}\Gamma(t,\lambda\alpha)}{\partial t^{m-1}}\right|_{t=0},\end{aligned} \qquad (12)$$

where $\Gamma(t,\alpha) = \int_t^\infty w^{\alpha-1}\,\mathrm{e}^{-\mathrm{w}}\mathrm{dw}$ is the upper incomplete gamma function. Equation (12) gives the moments of the XTG distribution. Hence, we can express the XTGG moments by combining Eqs. (9) and (12).

The central moments (μ_s) and cumulants (κ_s) of X can be obtained from (9) as

$$\mu_s = \sum_{j=0}^{s}(-1)^j\binom{s}{j}\mu_1'^s\,\mu_{s-j}' \quad \text{and} \quad \kappa_s = \mu_s' - \sum_{j=1}^{s-1}\binom{s-1}{j-1}\kappa_j\,\mu_{s-j}',$$

respectively, where $\kappa_1 = \mu_1'$. The skewness and kurtosis measures can be determined from the central moments using well-known relationships. Plots of the skewness and kurtosis for some parameter values as functions of α for the XTGG and MWG distributions are displayed in Figs. 3 and 4, respectively.

Finally, for empirical purposes, the shape of many distributions can be usefully described by the incomplete moments. These types of moments play an important role for measuring inequality, for example, income quantiles and Lorenz and Bonferroni curves. The sth incomplete moment of X is determined from (8) as

$$m_s(y) = \int_0^y x^s f(x;p,\alpha,\boldsymbol{\tau})dx = \sum_{j=0}^\infty w_j\int_0^{Q_{(j+1)\alpha,\boldsymbol{\tau}}(y)} Q_{(j+1)\alpha,\boldsymbol{\tau}}(u)^s du. \qquad (13)$$

Both integrals in (13) can be evaluated numerically for most EGG distributions.

4 Generating function

An expression for the mgf of X can be determined from Eq. (8) using the EG generating function. Initially, we can derive a simple linear combination for the mgf of X, say

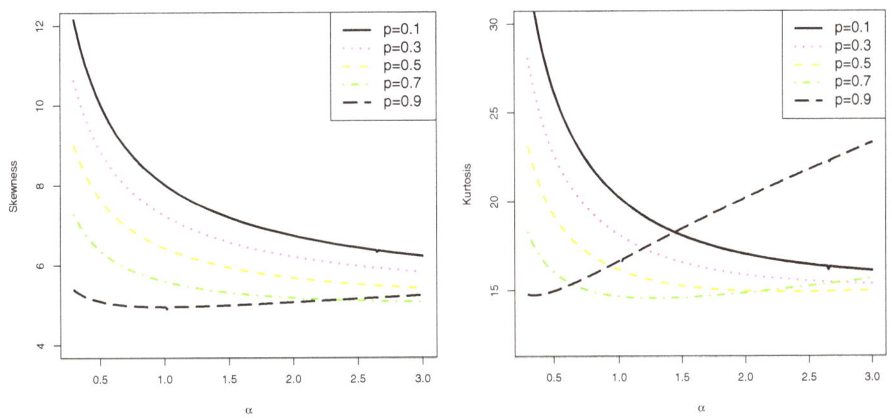

Fig. 3 Skewness and kurtosis of the XTGG distribution as functions of α for some values of p, with $\lambda = 0.3$ and $\beta = 0.8$

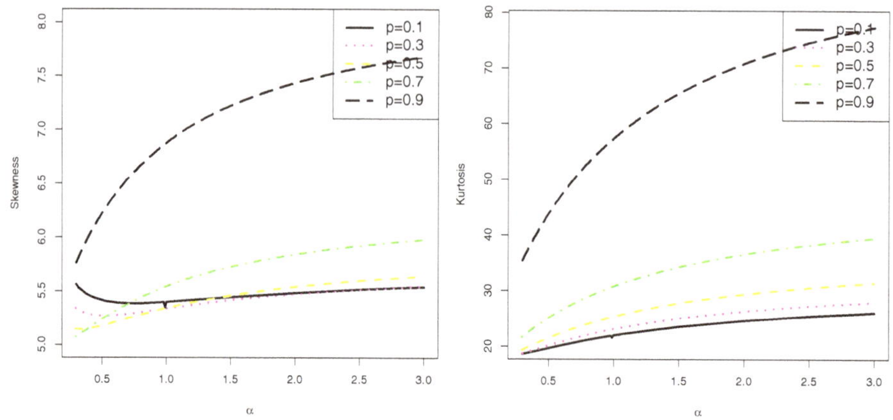

Fig. 4 Skewness and kurtosis of the MWG distribution as functions of α for some values of p, with $\lambda = 0.04$ and $\gamma = 0.8$

$M(t; p, \alpha, \boldsymbol{\tau})$, from Eq. (8) given by

$$M(t; p, \alpha, \boldsymbol{\tau}) = \sum_{j=0}^{\infty} w_j M_{(j+1)\alpha,\boldsymbol{\tau}}(t), \tag{14}$$

where $M_{\alpha,\boldsymbol{\tau}}(t) = \int_0^{\infty} \mathrm{e}^{\mathrm{t}\mathrm{y}}\, \mathrm{g}_{\alpha,\boldsymbol{\tau}}(\mathrm{y})\mathrm{dy}$ is the generating function of Y_α. A simple formula for $M_{\alpha,\boldsymbol{\tau}}(t)$ follows from Eqs. (2) and (3)

$$M_{\alpha,\boldsymbol{\tau}}(t) = \int_0^1 \exp\{t\, Q_{\alpha,\boldsymbol{\tau}}(u)\}du. \tag{15}$$

Equations (14) and (15) are the main results of this section. Combining these two equations, we can derive the generating functions of some EGG special models.

5 Mean deviations

The amount of scatter in a population is measured to some extent by the totality of deviations from the mean and median. We can obtain the mean deviations about the ordinary mean $\mu_1' = E(X)$ and about the median M of X using the relationships

$$\delta_1 = \int_0^{\infty} |t - \mu_1'|\, f(t)dt \quad \text{and} \quad \delta_2 = \int_0^{\infty} |t - M|\, f(t)dt,$$

and then

$$\delta_1 = 2\big[\mu_1' F(\mu_1') - m_1(\mu_1')\big] \text{ and } \delta_2 = \mu_1' - 2m_1(M), \tag{16}$$

where μ_1' comes from (9) with $s = 1$ and M is given by $M = H^{-1}\left(\alpha^{-1}\log[(2-p)]\right)$. Further, we obtain from (13) with $s = 1$

$$m_1(y) = \sum_{j=0}^{\infty} w_j \int_0^{Q_{(j+1)\alpha,\boldsymbol{\tau}}(y)} Q_{(j+1)\alpha,\boldsymbol{\tau}}(u)du.$$

We define the quantity $A_{\alpha,\tau}(z) = \int_0^z y g_{\alpha,\tau}(y)dy = \int_0^{Q_{\alpha,\tau}(z)} Q_{\alpha,\tau}(u)du$. A simple application of $A_{\alpha,\tau}(z)$ to the log-Weibull (LW) distribution gives

$$\begin{aligned} A_{\alpha,\tau}(z) &= \mu\, G_{\alpha,\tau}(z) + \sigma \int_0^{G_{\alpha,\tau}(z)} \log\Big\{ -\frac{1}{\alpha}\log(1-u)\Big\} du \\ &= \big[\mu - \sigma\, \log(\alpha)\big]\, G_{\alpha,\tau}(z) + \sigma \int_0^{G_{\alpha,\tau}(z)} \log\big\{-\log(1-u)\big\}\, du, \end{aligned}$$

where $G_{\alpha,\tau}(z) = 1 - \exp\big\{ - \exp\big(\frac{t-\mu}{\sigma}\big)\big\}$.

Setting $v = 1 - u$, the integral can be determined using MAPLE as

$$\int_x^1 \log\big\{-\log(v)\big\}\, dv = -x\log\big[-\log(x)\big] - E_i(1, -\log(x)) - \gamma,$$

where $E_i(a,z) = \int_1^\infty x^{-a}\, e^{zx} dx$ is the exponential integral and γ is the Euler constant. Then,

$$\begin{aligned} A_{\alpha,\tau}(z) = \big[\mu - \sigma\, \log(\alpha)\big]\, G_{\alpha,\tau}(z) + \sigma \big\{-\big[1 - G_{\alpha,\tau}(z)\big]\log\big\{-\log\big[1 - G_{\alpha,\tau}(z)\big]\big\} \\ -E_i\left(1, -\log\big[1 - G_{\alpha,\tau}(z)\big]\right) - \gamma\big\}. \end{aligned}$$

Finally, a practical application of δ_1 and δ_2 refers to the Bonferroni and Lorenz curves of X defined by $B(\pi) = m_1(q)/\left(\pi\, \mu_1'\right)$ and $L(\pi) = m_1(q)/\mu_1'$, respectively, where π is a specified probability and $q = F^{-1}(\pi)$ is given by

$$q = H^{-1}\left(\alpha^{-1}\log\left[\frac{(1-p\pi)}{(1-\pi)}\right]\right).$$

These curves have applications in economics to study income and poverty, reliability, demography, insurance and medicine. Plots of the Bonferroni and Lorenz curves for the XTGG distribution for some parameter values are displayed in Fig. 5.

6 Rényi entropy

The entropy of a random variable X is a measure of uncertainty variation. The Rényi entropy is defined by $I_{EGG}(\gamma) = \frac{1}{1-\gamma}\log\left\{\int_{\mathbb{R}} f^\gamma(x)dx\right\}$, where $\gamma > 0$ and $\gamma \neq 1$. Based on the EGG density function (4) and using the power series (7), we obtain

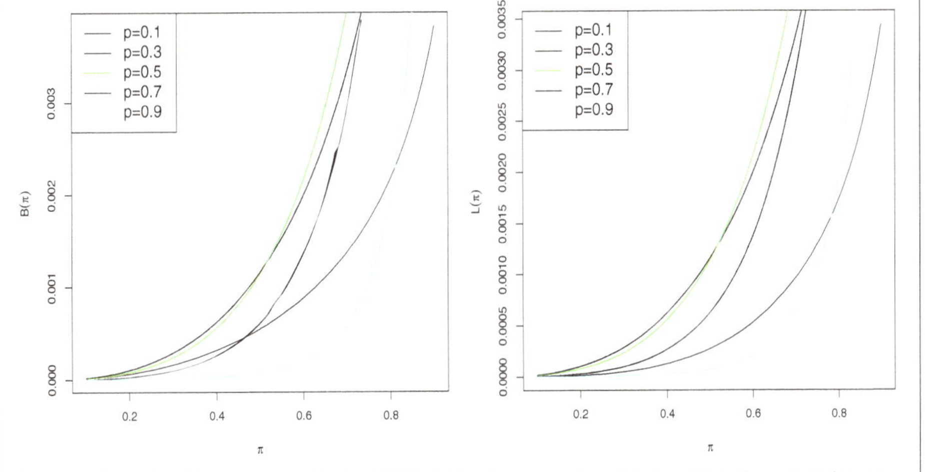

Fig. 5 Bonferroni and Lorenz curves for the XTGG distribution when $\lambda = 0.3$, $\beta = 0.8$ and $\alpha = 2$ and some values of p

$$f(x;p,\alpha,\boldsymbol{\tau})^{\gamma} = \alpha^{\gamma}\,(1-p)^{\gamma}\,h(x)^{\gamma}\sum_{j=0}^{\infty}\frac{\Gamma(2\gamma+j)\,p^{j}}{\Gamma(2\gamma)\,j!}\,\exp\left\{-(\gamma+j)\,\alpha\,H(x)\right\},\quad x>0. \tag{17}$$

Setting $\alpha_j^{\star} = (\gamma + j)\alpha/\gamma$, we can write $I_{EGG}(\gamma)$ as a linear combination of EG entropies. We obtain

$$I_{EGG}(\gamma) = \frac{\left[\gamma(1-p)\right]^{\gamma}}{(1-\gamma)}\sum_{j=0}^{\infty}\frac{\Gamma\,(2\gamma+j)\;p^{j}}{(\gamma+j)^{\gamma}\;\Gamma\,(2\gamma)\;j!}\,I_j(\gamma),$$

where $I_j(\gamma)$ denotes the EG Rényi entropy with scale parameter $\alpha_j^{\star}$. Hence, the EGG Rényi entropy is a linear combination of EG Rényi entropies with scale parameters given by $\alpha_j^{\star}$ for $j = 0, 1, \ldots$ The EG entropy depends on the form of $H(\cdot)$ and can be evaluated (at least numerically) using most statistical software programs.

7 Order statistics

Order statistics make their appearance in many areas of statistical theory and practice. The density function $f_{i:n}(x)$ of the ith order statistic, say $X_{i:n}$, for $i = 1, \ldots, n$, in a random sample of size n from the EGG family can be expressed as

$$f_{i:n}(x) = \frac{f(x)}{B(i,n-i+1)}\sum_{j=0}^{n-i}(-1)^{j}\binom{n-i}{j}F(x)^{i+j-1}, \tag{18}$$

where $B(\cdot,\cdot)$ is the beta function. Let $0 < u = \exp\{-\alpha H(y)\} < 1$. From Eqs. (4), (5) and (18), we can write

$$f_{i:n}(x) = \frac{\alpha\,(1-p)\,h(x)\,u}{B(i,n-i+1)}\sum_{j=0}^{n-i}(-1)^{j}\binom{n-i}{j}\frac{(1-u)^{i+j-1}}{(1-p\,u)^{i+j+1}}.$$

Using the binomial expansion and (7), we obtain after some algebra

$$f_{i:n}(x) = \sum_{s=0}^{\infty}\sum_{j=0}^{n-i}\sum_{r=0}^{i+j-1} v_{i:n}(s,j,r)\;g_{\widetilde{\alpha},\boldsymbol{\tau}}(x), \tag{19}$$

where $\widetilde{\alpha} = (r + s + 1)\alpha$ and

$$v_{i:n}(s,j,r) = (1-p)\sum_{s=0}^{\infty}\sum_{j=0}^{n-i}\sum_{r=0}^{i+j-1}\frac{(-1)^{j+r}\,n!\;(i+j+s)!\;p^{s}}{(i+j)\;(r+s+1)\;(i-1)!\;(n-i-j)!\;(i+j-1-r)!\;j!\;r!\;s!}.$$

Equation (19) reveals that the density function of the EGG order statistics can be expressed as a triple linear mixture of EG densities. We have checked using MAPLE and MATHEMATICA that

$$\sum_{s=0}^{\infty}\sum_{j=0}^{n-i}\sum_{r=0}^{i+j-1} v_{i:n}(s,j,r) = 1.$$

The triple sums in (19) can be computed using most packages with a large number (say 20) of terms in place of infinity. So, we can derive several mathematical properties (ordinary and incomplete moments, mgf, etc) for the EGG order statistics directly from those quantities of the EG distribution. For example, the pth ordinary moment of $X_{i:n}$ can be obtained from (19) and the quantity $\beta_s(\alpha,\boldsymbol{\tau})$ defined at the beginning of Section 3 as

$$E\left(X_{i:n}^{p}\right) = \sum_{s=0}^{\infty}\sum_{j=0}^{n-i}\sum_{r=0}^{i+j-1} v_{i:n}(s,j,r)\;\beta_p(\widetilde{\alpha},\boldsymbol{\tau}). \tag{20}$$

8 Maximum likelihood estimation

We determine the maximum likelihood estimates (MLEs) of the parameters of the EGG family from complete samples only. Let $x_1, \ldots, x_n$ be a random sample of size n from the EGG $(p, \alpha, \boldsymbol{\tau})$ distribution. The log-likelihood function for the vector of parameters $\boldsymbol{\theta} = (p, \alpha, \boldsymbol{\tau})^T$ becomes

$$l(\boldsymbol{\theta}) = n\left[\log(\alpha) + \log(1-p)\right] + \sum_{i=1}^{n} \log\left[h(x_i)\right] - \alpha \sum_{i=1}^{n} H(x_i) \qquad (21)$$
$$- 2\sum_{i=1}^{n} \log\left[1 - p\exp\{-\alpha H(x_i)\}\right].$$

The components of the score vector $U(\boldsymbol{\theta})$ are obtained by differentiating (21)

$$U_p(\boldsymbol{\theta}) = -\frac{n}{(1-p)} + 2\sum_{i=1}^{n} \frac{\exp\{-\alpha H(x_i)\}}{1 - p\exp\{-\alpha H(x_i)\}},$$
$$U_\alpha(\boldsymbol{\theta}) = \frac{n}{\alpha} - \sum_{i=1}^{n} H(x_i) + 2\sum_{i=1}^{n} \frac{pH(x_i)\exp\{-\alpha H(x_i)\}}{1 - p\exp\{-\alpha H(x_i)\}},$$
$$U_{\boldsymbol{\tau}}(\boldsymbol{\theta}) = \sum_{i=1}^{n} \frac{\dot{h}(x_i)_{\boldsymbol{\tau}}}{h(x_i)} - \alpha\sum_{i=1}^{n} \dot{H}(x_i)_{\boldsymbol{\tau}} - 2\sum_{i=1}^{n} \frac{\alpha p\dot{H}(x_i)_{\boldsymbol{\tau}}\exp\{-\alpha H(x_i)\}}{1 - p\exp\{-\alpha H(x_i)\}},$$

where $\dot{h}(x_i)_{\boldsymbol{\tau}} = \partial h(x_i)/\partial\boldsymbol{\tau}$ and $\dot{H}(x_i)_{\boldsymbol{\tau}} = \partial H(x_i)/\partial\boldsymbol{\tau}$ are $p \times 1$ vectors. Because the equations $U_p(\boldsymbol{\theta}) = 0$, $U_\alpha(\boldsymbol{\theta}) = 0$ and $U_{\boldsymbol{\tau}}(\boldsymbol{\theta}) = 0$ are nonlinear, the MLE $\widehat{\boldsymbol{\theta}}$ have to be evaluated numerically. The log-likelihood can be maximized either directly by using the MaxBFGS routine in the matrix programming language Ox (see, Doornik 2007).

For interval estimation and hypothesis tests on the model parameters, we require the observed information matrix $J = J(\boldsymbol{\theta})$, whose elements can be evaluated numerically. Under general regularity conditions, the asymptotic distribution of $(\widehat{\boldsymbol{\theta}} - \boldsymbol{\theta})$ is $N_{p+2}\left(0, I(\boldsymbol{\theta})^{-1}\right)$, where $I(\boldsymbol{\theta})$ is the expected information matrix. We can substitute $I(\boldsymbol{\theta})$ by the observed information matrix $J(\widehat{\boldsymbol{\theta}})$ evaluated at $\widehat{\boldsymbol{\theta}}$. The multivariate normal $N_{p+2}\left(0, J(\widehat{\boldsymbol{\theta}})^{-1}\right)$ distribution can be used to obtain approximate confidence intervals for the individual parameters.

We can compute the maximum values of the unrestricted and restricted log-likelihoods to construct likelihood ratio (LR) statistics for testing some sub-models of the EGG family. For example, we may use the LR statistic to check if the fit using the MWG distribution is statistically "superior" to a fit using the MW distribution for a given data set. In any case, hypothesis tests of the type $H_0 : \boldsymbol{\theta} = \boldsymbol{\theta}_0$ versus $H : \boldsymbol{\theta} \neq \boldsymbol{\theta}_0$, can be performed using LR statistics. For example, the comparison of the MWG and MW distributions, where $\boldsymbol{\theta} = (p, \alpha, \lambda, \gamma)^T$, is equivalent to test $H_0 : p = 0$ versus $H : H_0$ *is not true*. The LR statistic reduces to

$$w = 2\left\{\ell\left(\widehat{p}, 1, \widehat{\alpha}, \widehat{\lambda}, \widehat{\gamma}\right) - \ell\left(0, \widetilde{\alpha}, \widetilde{\lambda}, \widetilde{\gamma}\right)\right\},$$

where $\widehat{p}$, $\widehat{\alpha}$, $\widehat{\lambda}$ and $\widehat{\gamma}$ are the MLEs under H and $\widetilde{\alpha}$, $\widetilde{\lambda}$ and $\widetilde{\gamma}$ are the estimates under H_0.

We apply formal goodness-of-fit tests in order to verify which distribution gives the best fit. We consider the Cramér-Von Mises (W^*) and Anderson-Darling (A^*) statistics. In general, the smaller the values of these statistics, the better the fit to the data. Let $F(x; \boldsymbol{\theta})$ be the cdf, where the form of F is known but $\boldsymbol{\theta}$ (a k-dimensional parameter vector, say) is unknown. To obtain the W^* and A^* statistics, we can proceed as follows: (i) compute $v_i = F\left(x_i; \widehat{\theta}\right)$, where the x_i's are in ascending order, $y_i = \Phi^{-1}(v_i)$ is the standard normal qf and $u_i = \Phi\{(y_i - \bar{y})/s_y\}$, where $\bar{y} = n^{-1}\sum_{i=1}^{n} y_i$ and $s_y^2 = (n-1)^{-1}\sum_{i=1}^{n}(y_i - \bar{y})^2$; (ii) calculate $W^2 = \sum_{i=1}^{n}\{u_i - (2i-1)/(2n)\}^2 + 1/(12n)$ and $A^2 = -n - n^{-1}\sum_{i=1}^{n}\{(2i-1)\log(u_i) + (2n+1-2i)\log(1-u_i)\}$ and (iii) modify W^2 into $W^* = W^2(1+0.5/n)$ and A^* into $A^* = A^2\left(1 + 0.75/n + 2.25/n^2\right)$. For further details, the reader is referred to Chen and Balakrishnan (1995).

9 Application: ozone level data

We illustrate the superiority of some EGG distributions over their sub-models. We consider the data set from the New York State Department of Conservation corresponding to the daily ozone level measurements in New York in May-September, 1973. The numerical evaluations are performed using the MaxBFGS sub-routine of the Ox program (Doornik 2007).

We consider the reparametrizations, $\alpha = \alpha_1^{-\gamma}$ for the MWG model, $\alpha = \alpha_1^{-\beta}$ for the XTGG model and $\alpha_1 = \alpha_3^{-\beta_1}$ and $\alpha_2 = \alpha_4^{-\beta_2}$ for the AWG model. The density functions of the XTGG and MWG models are presented in the "Appendix".

Alternative models to fit these data can be based on the Burr XII (BXII) distribution. There are various extensions of this distribution. Recently, Lanjoni et al. (2015) introduced two extended BXII distributions by compounding the BXII and geometric distributions. These models are called the BXII geometric type I (BXIIGI) and BXII geometric type II (BXIIGII) distributions, whose density functions are given below:

- **BXIIGI distribution**

$$f(x; s, k, c, p) = \frac{(1-p)c\,k\,s^{-c}x^{c-1}\left[1 + \left(\frac{x}{s}\right)^c\right]^{-k-1}}{\left\{1 - p\left[1 + \left(\frac{x}{s}\right)^c\right]^{-k}\right\}^2}, \quad x > 0, \tag{22}$$

  where $s > 0$ is a scale parameter and $k > 0$, $c > 0$ and $p \in (0, 1)$ are shape parameters.

- **BXIIGII distribution**

$$f(x; s, k, c, p) = \frac{(1-p)c\,k\,s^{-c}x^{c-1}\left[1 + \left(\frac{x}{s}\right)^c\right]^{-k-1}}{\left\{1 - p + p\left[1 + \left(\frac{x}{s}\right)^c\right]^{-k}\right\}^2} \quad x > 0. \tag{23}$$

We fit the MWG, XTGG and AWG models, their parent distributions and the BXIIGI and BXIIGII models to the current data. Table 2 lists the MLEs of the model parameters, their standard errors (in parentheses) and the values of the Akaike information criterion (AIC). These results indicate that the special MW, XTG and AW models have larger AIC values when compared with those values of the new MWG, XTGG and AWG models. Since the value of the AIC statistic is smaller for the XTGG distribution compared to the values of the other five models, this distribution can be chosen as the best model to fit these data among these models.

The values of the W^* and A^* statistics for all eight fitted models are given in Table 3. Thus, according to these statistics, the MWG, XTGG and AWG models fit the current data better than their sub-models. The figures in this table also reveal that the XTGG model has the lowest values for the W^* and A^* statistics among all eight fitted models. Hence, it could be chosen as the best model to explain these data. Figure 6 displays the histogram of the data and some plots comparing the fitted densities. We can also conclude

Table 2 MLEs of the model parameters for the daily ozone level data, the corresponding SEs (given in parentheses) and the AIC statistics

Model	p	α_1	λ	γ		AIC
MWG	0.83032	104.81	0.0030932	1.6764		1089.21
	(0.18959)	(78.852)	(0.0068905)	(0.23498)		
MW	–	46.080	2.9046e-013	1.3402		1091.22
		(3.3755)	(0.00053284)	(0.095420)		
	p	α_1	λ	β		AIC
XTGG	0.92198	19.384	119.79	1.7396	–	1088.39
	(0.16726)	(28.978)	(92.031)	(0.19738)		
XTG	–	3.7020	20682.	1.3401	–	1091.22
		(1.8864)	(85.071)	(0.095512)		
	p	α_3	β_1	α_4	β_2	AIC
AWG	0.75545	0.99528	116.54	0.99528	116.54	1091.45
	(0.17410)	(0.0393)	(982.7309)	(0.0393)	(0.0393)	
AW	–	0.99622	77.29	0.99622	77.29	1093.22
		(0.0251070)	(512.43)	(0.025107)	(512.43)	
	p	s	c	k		AIC
BXIIGI	0.7553	16102	1.7363	10405	–	1089.5
	(0.1760)	(4316.05)	(0.2089)	(3936.62)		
BXIIGII	1E-8	134.00	1.4860	5.5678	–	1092.7
	(1E-9)	(17.35)	(0.2235)	(2.4753)		

that the distributions in the new family are very competitive models for describing the daily ozone level data. Finally, these results prove empirically the potentiality of the EGG family.

10 Concluding remarks

In this paper, we introduce and study a new class of distributions called the extended-G geometric (EGG) family that generalizes the Weibull geometric and modified Weibull geometric distributions proposed by Barreto-Souza et al. (2011) and Wang and Elbatal (2015), respectively, among other distributions. This is achieved by compounding the class of extended-G (EG) distributions (Nadarajah and Kotz 2005) with the geometric distribution. The EGG family is quite flexible in analyzing positive data instead of some other special models.

Table 3 Formal tests for some models

Model	Statistics	
	W^*	A^*
MWG	0.091	0.545
MW	0.170	0.966
XTGG	0.078	0.474
XTG	0.170	0.966
AWG	0.088	0.539
AW	0.170	0.966
BXIIGI	0.088	0.539
BXIIGII	0.139	0.820

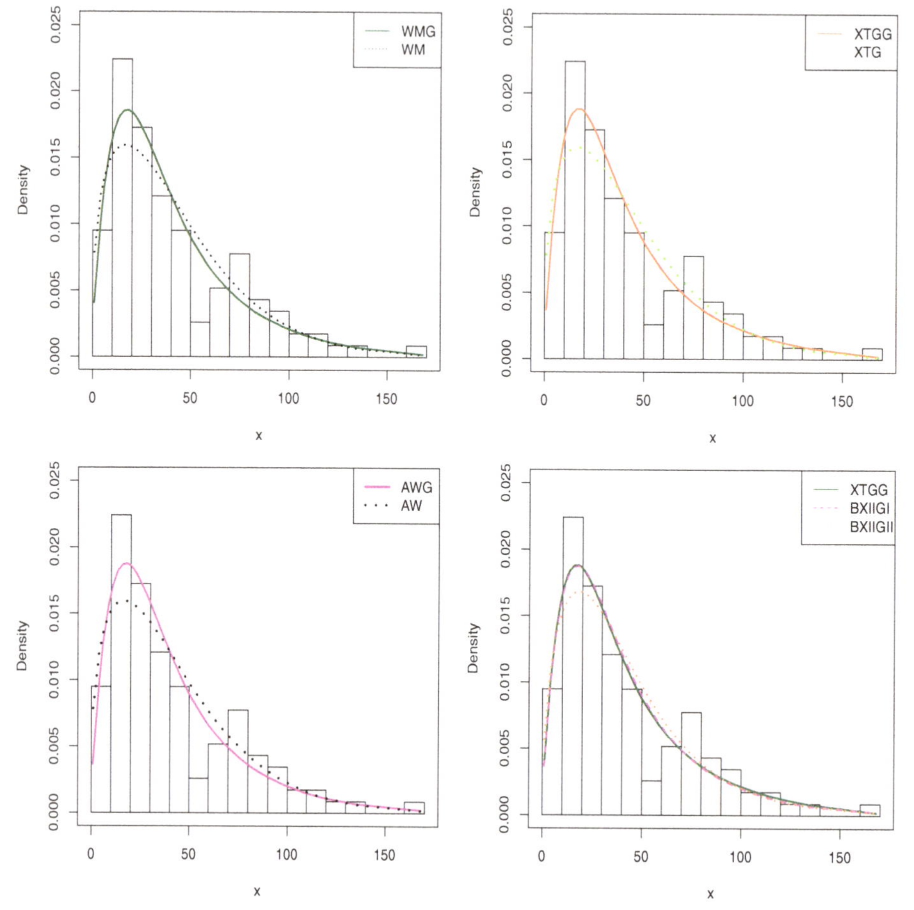

Fig. 6 Fitted densities of the more adequate models for daily ozone level data

The density function of the EGG family can be expressed as a linear mixture of EG densities. We provide a mathematical treatment of the new family including explicit expressions for the ordinary and incomplete moments, generating function, mean deviations, Bonferroni and Lorenz curves, order statistics and their moments. The estimation of the parameters is approached by the method of maximum likelihood. An application to real lifetime data indicate that the EGG family could provide better fits than other well-known lifetime models. We expect that the new family of models will attract wider applications in Statistics.

Appendix: new special EGG models

- **Modified Weibull geometric distribution**
  The case $H(x) = x^{\gamma} \exp(\lambda x)$ and $h(x) = x^{\gamma-1} \exp(\lambda x)(\gamma + \lambda x)$, where $\gamma > 0$ and $\lambda \geq 0$, in Eq. (4), gives the modified Weibull geometric (MWG) distribution. If $p \to 0^{+}$, it reduces to the MW distribution. The MWG distribution is very flexible to accommodate a hrf that has increasing, decreasing, bathtub and unimodal shapes. It is also suitable for testing goodness of fit of some special models such as the modified Weibull (MW), Weibull geometric (WG) and Weibull distributions.
- **Exponential power geometric distribution**

For the case $H(x) = \exp\left[(\lambda x)^{\beta}\right] - 1$, $h(x) = \beta\lambda \exp\left[(\lambda x)^{\beta}\right](\lambda x)^{\beta-1}$ and $\alpha = 1$, where $\beta, \lambda > 0$, we obtain the exponential power geometric (EPG) distribution. If $p \to 0^+$ in addition to $\alpha = 1$, it becomes the exponential power (EP) distribution (Smith and Bain 1975). This distribution has the property that its hrf may take a U-shaped form. Smith and Bain (1975) presented some general properties of least squares estimators and discussed the case of location-scale parameter distributions.

- **Chen geometric distribution**
  The case $H(x) = \exp\left(x^{\beta}\right) - 1$, $h(x) = \beta x^{\beta-1} \exp\left(x^{\beta}\right)$, where $\beta > 0$, corresponds to the Chen geometric (CG) distribution. If $p \to 0^+$, it becomes the Chen distribution (Chen, 2000), which has increasing or bathtub-shaped hrf. Chen (2000) discussed exact confidence intervals and exact joint confidence regions for the parameters based on type-II censored samples.
- **XTG geometric distribution**
  For $H(x) = \lambda\left[\exp\left\{(x/\lambda)^{\beta}\right\} - 1\right]$ and $h(x) = \beta \exp\left\{(x/\lambda)^{\beta}\right\}(x/\lambda)^{\beta-1}$, where $\beta > 0$ and $\lambda > 0$, we obtain the XTG geometric (XTGG) distribution. If $p \to 0^+$, it reduces to Xie et al.'s (2002) model. They studied parameter estimation methods and showed its applicability. If $\lambda = 1$, this model reduces to the CG distribution above.
- **Log-Weibull geometric distribution**
  For $H(x) = \exp\left[(x - \mu)/\sigma\right]$, $h(x) = (1/\sigma)\exp\left[(x - \mu)/\sigma\right]$ and $\alpha = 1$, where $-\infty < \mu < \infty$ and $\sigma > 0$, we have the log-Weibull geometric (LWG) distribution. If $p \to 0^+$, it gives as special case the log-Weibull (LW) distribution (White 1969; Lawless 2003). White (1969) obtained the means and variances of the order statistics of the LW distribution and listed these values in special tables. Examples of these tables to obtain weighted least squares estimates from censored samples from a Weibull distribution are also presented. The LW distribution is a very popular distribution for modeling lifetime data and phenomenon with monotone failure rates.
- **Kies geometric distribution**
  For $H(x) = \left[(x-\mu)/(\sigma - x)\right]^{\beta}$ and $h(x) = \beta\left[(x-\mu)/(\sigma - x)\right]^{\beta-1}\left[(\sigma - \mu)/(\sigma - x)^2\right]$, where $0 < \mu < t < \sigma < \infty$, we obtain the Kies geometric (KG) distribution. If $p \to 0^+$, it yields as a special model the Kies distribution (Kies 1958), which is an extension of the Weibull distribution for strength modeling.
- **Phani geometric distribution**
  The case $H(x) = (x - \mu)^{\beta_1}/(\sigma - x)^{\beta_2}$ and $h(x) = (x - \mu)^{\beta_1-1}(\sigma - x)^{-(\beta_2-1)}$ $\left[\beta_1(\sigma - x) + \beta_2(x - \mu)\right]$, where $0 < \mu < t < \sigma < \infty$ and $\beta_1, \beta_2 > 0$, leads to the Phani geometric (PG) distribution. If $p \to 0^+$, it reduces to the Phani distribution (Phani 1987). If $\beta_1 = \beta_2 = \beta$, it gives the KG distribution discussed before. Phani (1987) presented statistical justification for modifying the Weibull distribution for the analysis of fibre strength data.
- **Additive Weibull geometric distribution**
  For $H(x) = (x/\beta_1)^{\alpha_1} + (x/\beta_2)^{\alpha_2}$, $h(x) = (\alpha_1/\beta_1)(x/\beta_1)^{\alpha_1-1} + (\alpha_2/\beta_2)(x/\beta_2)^{\alpha_2-1}$ and $\alpha = 1$, where $\alpha_1, \alpha_2, \beta_1, \beta_2 > 0$, we obtain the additive Weibull geometric (AWG) distribution. If $a = b = 1$ in addition to $\alpha = 1$, it gives the additive Weibull (AW) distribution (Xie and Lai 1995). This distribution has a bathtub-shaped failure rate function. Xie and Lai (1995) studied a simple model based on adding two Weibull survival functions, presented some simplifications of the model and analyzed the graphical estimation technique based on the conventional Weibull plot.

- **Generalized power Weibull geometric distribution**
For $H(x) = \left[1 + (x/\beta)^{\alpha_1}\right]^{\theta} - 1$, $h(x) = (\theta\alpha_1/\beta)\left[1 + (x/\beta)^{\alpha_1}\right]^{\theta-1}(x/\beta)^{\alpha_1}$ and $\alpha = 1$, where $\alpha_1, \beta > 0$ and $\theta \geq 0$, we obtain the generalized power Weibull geometric (GPWG) distribution. If $p \to 0^+$, it becomes the generalized power Weibull (GPW) distribution (Nikulin and Haghighi 2006). They proposed a chi-squared type statistic to test the validity of the GPW family based on the head-and-neck cancer censored data.
- **BLZ geometric distribution**
The case $H(x) = \exp(\alpha_1 x - \beta/x)$, $h(x) = \exp(\alpha_1 x - \beta/x)\left(\alpha_1 + \beta x^{-2}\right)$ and $\alpha = 1$, where $\alpha_1, \beta > 0$, leads to the BLZ geometric (BLZG) distribution. It extends the Bebbington et al.'s (2007) distribution, which corresponds to $p \to 0^+$ and $\alpha = 1$. Its hrf has four different forms: bathtub shape, increasing, decreasing and upside-down bathtub. They presented the BLZ distribution as an extension of the Weibull, studied its properties and derived explicit formulas for the turning points of the failure rate function in terms of its parameters.
- **Gompertz geometric distribution**
For $H(x) = \alpha_1^{-1}\left[\exp(\alpha_1 x) - 1\right]$ and $h(x) = \exp(\alpha_1 x)$, where $-\infty < \alpha_1 < \infty$, we obtain the Gompertz geometric (GG) distribution. If $p \to 0^+$, it gives the Gompertz distribution (Pham and Lai 2007).
- **Pham geometric distribution**
For $H(x) = a_1^{x^{\alpha_1}} - 1$, $h(x) = \alpha_1 x^{\alpha_1 - 1}\log(a_1)\, a_1^{x^{\alpha_1}}$ and $\alpha = 1$, where $\alpha_1, a_1 > 0$, we obtain the Pham geometric (PG) distribution. If $p \to 0^+$, it yields the Pham distribution (Pham 2002). Pham (2002) proposed this distribution, also referred to as the log-log distribution, and showed that it has a U-shaped hrf.
- **Nadarajah-Kotz geometric distribution**
For $H(x) = x^{b_1}[\exp\left(cx^d\right) - 1]$ and
$h(x) = b_1 x^{b_1 - 1}[\exp\left(cx^d\right) - 1] + cdx^{(b_1+d-1)}\exp\left(cx^d\right)$, where $b_1, c \geq 0$ and $d > 0$, we obtain the Nadarajah-Kotz geometric (NKG) distribution. If $p \to 0^+$, it gives the distribution due to Nadarajah and Kotz (2005). They presented some modifications of the Weibull distribution and also discussed some modifications suggested by Gurvich et al. (1997). For $b_1 = 0$ and $c = 1$, this model reduces to the CG distribution discussed before.
- **Slymen-Lachenbruch geometric distribution**
For $H(x) = \exp\left[\alpha_1 + (\beta/2\theta)\left(x^{\theta} - x^{-\theta}\right)\right]$,
$h(x) = (\beta/2x)\exp\left[\alpha_1 + (\beta/2\theta)\left(x^{\theta} - x^{-\theta}\right)\right]\left(x^{\theta} + x^{-\theta}\right)$ and $\alpha = 1$, where $\alpha_1, \theta > 0$, we obtain the Slymen-Lachenbruch geometric (SLG) distribution. If $p \to 0^+$, it becomes the distribution proposed by Slymen and Lachenbruch (1984). They introduced and studied two classes of distributions within the framework of parametric survival analysis. These classes are derived from a general linear form by specifying a function of the survival function under certain restrictions.

We now give the density functions for two distributions in the EGG family.

1. The XTGG density function is given by

$$f(x) = \alpha\beta\exp\left\{(x/\lambda)^{\beta}\right\}(x/\lambda)^{\beta-1}\exp\left\{-\alpha\lambda\left[\exp\left\{(x/\lambda)^{\beta}\right\} - 1\right]\right\}(1-p)$$
$$\left[1 - p\exp\left\{-\alpha\lambda\left[\exp\left\{(x/\lambda)^{\beta}\right\} - 1\right]\right\}\right]^{-2}. \quad (24)$$

If X is a random variable with density function (24), we write $X \sim \mathrm{XTGG}(p, \alpha, \lambda, \beta)$.

2. The MWG density function is given by

$$f(x) = \frac{\alpha x^{\gamma-1} \exp(\lambda x)(\gamma + \lambda x) \exp\{-\alpha x^{\gamma} \exp(\lambda x)\} (1-p)^2}{\{1 - p[\exp\{-\alpha x^{\gamma-1} \exp(\lambda x)\}]\}}. \tag{25}$$

If X is a random variable with density function (25), we write $X \sim \mathrm{MWG}(p, \alpha, \lambda, \gamma)$.

Acknowledgments
The financial support from CNPq is gratefully acknowledged.

Author details
[1]Departamento de Estatística, Universidade Federal de Pernambuco, Cidade Universitária, 50740-540 Recife, PE, Brazil.
[2]Departamento de Estatística, Universidade Federal da Bahia, Av. Adhemar de Barros, s/n°, 40170-110 Salvador, Brazil.
[3]Departamento de Ciências Exatas, Universidade de São Paulo, Av. Pádua Dias 11, 13418-900 Piracicaba, SP, Brazil.

References

Adamidis, K, Loukas, S: A Lifetime distribution with decreasing failure rate. Stat. Probab. Lett. **39**, 35–42 (1998)
Barreto-Souza, W, de Morais, AL, Cordeiro, GM: The Weibull-geometric distribution. J. Stat. Comput. Simul. **81**, 645–657 (2011)
Bebbington, M, Lai, CD, Zitikis, R: A flexible Weibull extension. Reliab. Eng. Syst. Saf. **92**, 719–726 (2007)
Carrasco, JMF, Ortega, EMM, Cordeiro, GM: A generalized modified Weibull distribution for lifetime modeling. Comput. Stat. Data Anal. **53**, 450–462 (2008)
Chen, G, Balakrishnan, N: A general purpose approximate goodness-of-fit test. J. Qual. Technol. **27**, 154–161 (1995)
Chen, ZA: A new two-parameter lifetime distribution with bathtub shape or increasing failure rate function. Stat. Probab. Lett. **49**, 155–61 (2000)
Cordeiro, GM, Ortega, EMM, Silva, GO: The Beta Weibull geometric distribution. Statistics. **47**, 817–834 (2013)
de Santana, TVF, Ortega, EMM, Cordeiro, GM, Silva, GO: The Kumaraswamy log-logistic distribution. J. Stat. Theory Appl. **11**, 265–291 (2012)
Doornik, JA: An Object-Oriented Matrix Language Ox 5. Timberlake Consultants Press, London (2007)
Gupta, RD, Kundu, D: Generalized exponential distribution. Australian and New Zealand J. Stat. **41**, 173–188 (1999)
Gupta, RD, Kundu, D: Exponetiated exponential family: an alternative to gamma and Weibull distributions. Biom. J. **43**, 117–130 (2001a)
Gupta, RD, Kundu, D: Generalized exponential distribution: Different method of estimations. J. Stat. Comput. Simul. **69**, 315–337 (2001b)
Gurvich, MR, Dibenedetto, AT, Ranade, SV: A new statistical distribution for characterizing the random strength of brittle materials. J. Mater. Sci. **32**, 2559–2564 (1997)
Kies, JA: Naval Research Laboratory (U.S.) and Naval Postgraduate School (U.S.): The Strength of Glass. NRL report. Naval Research Laboratory (1958). https://books.google.com.br/books?id=wVnYSAAACAAJ
Lanjoni, EM, Ortega, EMM, Cordeiro, GM: Extended Burr XII regression models: theory and applications. J. Agric. Biol. Environ. Stat (2015). doi:10.1007/s13253-015-0236-z
Lai, CD, Xie, M, Murthy, DNP: A modified Weibull distribution. Trans. Reliab. **52**, 33–37 (2003)
Lawless, JF: Statistical Models and Methods for Lifetime Data (2003)
Mudholkar, GS, Srivastava, DK, Friemer, M: The exponentiated Weibull family: A reanalysis of the bus-motor-failure data. Technometrics. **37**, 436–445 (1995)
Mudholkar, GS, Srivastava, DK, Kollia, GD: A generalization of the Weibull distribution with application to the analysis of survival data. J. Am. Stat. Assoc. **91**, 1575–1583 (1996)
Nadarajah, S, Kotz, S: On some recent modifications of Weibull distribution. IEEE Trans. Reliab. **54**, 561–562 (2005)
Nadarajah, S, Kotz, S: The beta exponential distribution. Reliab. Eng. Syst. Saf. **91**, 689–697 (2006)
Nikulin, M, Haghighi, F: A Chi-squared test for the generalized power Weibull family for the head-and-neck cancer censored data. J. Math. Sci. **133**, 1333–1341 (2006)
Pham, H: A bathtub-shaped hazard rate function with applications to system safety. Int. J. Reliabil. Appl. **3**, 1–16 (2002)
Pham, H, Lai, CD: On recent generalizations of the Weibull distribution. IEEE Trans. Reliab. **56**, 454–458 (2007)
Phani, KK: A new modified Weibull distribution. Commun. Am. Ceramic Soc. **70**, 182–184 (1987)
Prudnikov, AP, Brychkov, YA, Marichev, OI: Integrals and Series, volume 1 (1986)
Silva, GO, Ortega, EMM, Cordeiro, GM: The Beta modified Weibull distribution. Lifetime Data Anal. **16**, 409–430 (2010a)
Silva, RB, Barreto-Souza, W, Cordeiro, GM: A new distribution with decreasing, increasing and upside-down bathtub failure rate. Comput. Stat. Data Anal. **54**, 935–944 (2010b)
Slymen, DJ, Lachenbruch, PA: Survival distributions arising from two families and generated by transformations. Commun. Stat. Theory Methods. **13**, 1179–1201 (1984)
Smith, RM, Bain, LJ: An exponential power life testing distributions. Commun. Stat. **4**, 469–481 (1975)
Xie, M, Lai, CD: Reliability analysis using an additive Weibull model with bathtub-shaped failure rate function. Reliab. Eng. Syst. Saf. **52**, 87–93 (1995)
Xie, M, Tang, Y, Goh, TN: A modified Weibull extension with bathtub failure rate function. Reliab. Eng. Syst. Saf. **76**, 279–285 (2002)
Wang, M, Elbatal, I: The modified Weibull geometric distribution. Metron. **73**, 303–315 (2015)
White, JS: The moments of log-Weibull order statistics. Technometrics. **11**, 373–386 (1969)

10

Generalized log-logistic proportional hazard model with applications in survival analysis

Shahedul A. Khan* and Saima K. Khosa

*Correspondence: s.khan@usask.ca
Department of Mathematics and Statistics, University of Saskatchewan, Wiggins Road, Saskatoon, SK S7N 5E6, Canada

Abstract

Proportional hazard (PH) models can be formulated with or without assuming a probability distribution for survival times. The former assumption leads to parametric models, whereas the latter leads to the semi-parametric Cox model which is by far the most popular in survival analysis. However, a parametric model may lead to more efficient estimates than the Cox model under certain conditions. Only a few parametric models are closed under the PH assumption, the most common of which is the Weibull that accommodates only monotone hazard functions. We propose a generalization of the log-logistic distribution that belongs to the PH family. It has properties similar to those of log-logistic, and approaches the Weibull in the limit. These features enable it to handle both monotone and nonmonotone hazard functions. Application to four data sets and a simulation study revealed that the model could potentially be very useful in adequately describing different types of time-to-event data.

Keywords: Cox PH, Log-logistic distribution, Parametric model, Proportional hazard, Semi-parametric model, Time-to-event data, Weibull distribution

AMS Subject Classification: Primary 62N01; Secondary 62P10

Introduction

Proportional hazard (PH) models play a vital role in analyzing time-to-event data. A key assumption in the PH model is that the hazard ratio comparing any two specifications of covariates is constant over time (commonly known as PH assumption). Although the PH assumption may not hold for one or more covariates over the entire study period, it may hold in shorter time intervals. Therefore, violation of the PH assumption may be handled using time-dependent covariates (Kleinbaum and Klein 2012). One of the appealing features of PH models is that the regression coefficients have relative risk interpretation, which is preferred by many clinicians.

The Cox PH model (Cox 1972) is the most popular in survival analysis mainly because of two reasons: (a) no assumption is required about the probability distribution of survival times (i.e., a semi-parametric model), and (b) it usually fits the data well no matter which parametric model is appropriate. In contrast, distributional assumption is required for a fully parametric PH model (Kalbfleisch and Prentice 2002; Lawless 2002). This also leads to the added requirement of checking the appropriateness of the chosen distribution. Nevertheless, as demonstrated by Efron (1977) and Oakes (1977), parametric models lead to more efficient estimates than Cox's model under certain conditions. More specifically, if the distributional assumption is valid, a parametric model leads to smaller standard errors of the estimates than would be in the absence of a distributional assumption (Collett 2003). Moreover, the use of Cox PH in joint modeling of time-to-event and longitudinal data (Wulfsohn and Tsiatis 1997) usually leads to an underestimation of the standard errors of the parameter estimates (Hsieh et al. 2006; Rizopoulos 2012), and therefore most methods for joint modeling are based on parametric response distributions (Hwang and Pennell 2014). Regarding the choice between a parametric and Cox's PH model, Nardi and Schemper (2003) suggested to use a richer parametric model or simply the Cox's model in case of an unsatisfactory fit of the chosen probability distribution.

The most commonly used parametric time-to-event models are the Weibull, log-logistic and log-normal distributions. The log-logistic and log-normal distributions belong to the accelerated failure time (AFT) family, and are useful in modeling nonmonotone hazard rates (Lawless 2002). Note that the log-logistic also accommodates decreasing hazard functions. Only a few parametric models are closed under PH assumption, the most common of which is the Weibull that accommodates only monotone hazard functions. In fact, Weibull is the only distribution that is closed under both AFT and PH families (Kalbfleisch and Prentice 2002). Mudholkar et al. (1996) proposed a generalization of the Weibull distribution which permits parametric PH regression modeling. It is a three-parameter distribution and is capable of modeling both monotone and nonmonotone hazard functions. One difficulty with this model is that it is nonregular (the support depends on some parameters) in the case of increasing hazard functions, and therefore the standard maximum likelihood asymptotics do not hold. In this paper, we propose a simple extension of the log-logistic model which is closed under the PH relationship. The proposed generalized log-logistic model is a three-parameter distribution, and has characteristics similar to those of the log-logistic model. Moreover, it approaches the Weibull in the limit. These features enable it to satisfactorily handle both monotone (increasing and decreasing) and nonmonotone (unimodal) hazard functions. In Section 1, we introduce the generalized log-logistic model and discuss estimation and testing of the parameters using the maximum likelihood method. The proposed method is then illustrated with applications to four data sets, one of which involves joint modeling of time-to-event and longitudinal data (Section Examples). In Section Simulations, a simulation study is presented to evaluate the performance of generalized log-logistic in comparison with other commonly used PH models to describe different types of time-to-event data. We conclude in Section Conclusion by summarizing our findings.

The generalized log-logistic model

The generalized log-logistic distribution for a nonnegative random variable T can be conveniently specified in terms of the hazard function as follows:

$$h(t;\boldsymbol{\alpha}) = \frac{\kappa\rho(\rho t)^{\kappa-1}}{1+(\gamma t)^{\kappa}}, t > 0, \tag{1}$$

where $\rho > 0$, $\kappa > 0$ and $\gamma > 0$ are parameters and $\boldsymbol{\alpha} = (\kappa, \gamma, \rho)'$. If γ depends on ρ via $\gamma = \rho$ and $\gamma = \rho\eta^{-1/\kappa}$ with $\eta > 0$, then (1) reduces to the hazard function of the log-logistic (Lawless 2002) and Burr XII (Wang et al. 2008) distributions, respectively. Taking γ not dependent on ρ, it is easy to verify that (1) is closed under PH relationship (see below). The hazard function is monotone decreasing when $\kappa \leq 1$, and unimodal when $\kappa > 1$ (i.e., $h(t; \boldsymbol{\alpha}) = 0$ at $t = 0$, increases to a maximum at $t = [(\kappa - 1)/\gamma^{\kappa}]^{1/\kappa}$, and then approaches zero monotonically as $t \to \infty$). Note that (1) approaches the Weibull hazard function as $\gamma^{\kappa} \to 0$. This particular feature of the generalized log-logistic model enables it to handle monotone increasing hazard satisfactorily via $\kappa > 1$ and γ small (close to zero).

The survivor function, probability density function and cumulative hazard function of the generalized log-logistic distribution are, respectively,

$$S(t; \boldsymbol{\alpha}) = [1 + (\gamma t)^{\kappa}]^{-\frac{\rho^{\kappa}}{\gamma^{\kappa}}}, \tag{2}$$

$$f(t; \boldsymbol{\alpha}) = \frac{\kappa\rho(\rho t)^{\kappa-1}}{[1 + (\gamma t)^{\kappa}]^{\frac{\rho^{\kappa}}{\gamma^{\kappa}}+1}}, \tag{3}$$

$$H(t; \boldsymbol{\alpha}) = \frac{\rho^{\kappa}}{\gamma^{\kappa}} \log[1 + (\gamma t)^{\kappa}]. \tag{4}$$

The median of the distribution is $\dfrac{\left(2^{\frac{\gamma^{\kappa}}{\rho^{\kappa}}} - 1\right)^{\frac{1}{\kappa}}}{\gamma}$, and the rth moment is

$$E(T^r) = \frac{\rho^{\kappa}}{\gamma^{\kappa+r}} \frac{\Gamma\left(\frac{\rho^{\kappa}}{\gamma^{\kappa}} - \frac{r}{\kappa}\right)\Gamma\left(\frac{r}{\kappa} + 1\right)}{\Gamma\left(\frac{\rho^{\kappa}}{\gamma^{\kappa}} + 1\right)} \quad \text{provided} \quad \frac{\kappa\rho^{\kappa}}{\gamma^{\kappa}} > r.$$

In particular, the mean is $E(T) = \frac{\rho^{\kappa}}{\gamma^{\kappa}} \frac{\Gamma\left(\frac{\rho^{\kappa}}{\gamma^{\kappa}} - \frac{1}{\kappa}\right)\Gamma\left(\frac{1}{\kappa}+1\right)}{\Gamma\left(\frac{\rho^{\kappa}}{\gamma^{\kappa}}+1\right)}$ provided $\frac{\kappa\rho^{\kappa}}{\gamma^{\kappa}} > 1$.

For the family of PH models with covariates $\mathbf{z} = (z_1, z_2, \ldots, z_p)'$, the hazard function for T can be expressed as

$$h(t; \mathbf{z}) = h_0(t; \boldsymbol{\alpha})\, e^{\mathbf{z}'\boldsymbol{\beta}}, \tag{5}$$

where $h_0(t; \boldsymbol{\alpha})$ is the baseline hazard function (i.e., the hazard function when $\mathbf{z} = \mathbf{0}$) characterized by the vector of parameters $\boldsymbol{\alpha}$, and $\boldsymbol{\beta} = (\beta_1, \beta_2, \ldots, \beta_p)'$ is the vector of regression coefficients. A fully parametric PH model can be formulated by specifying $h_0(t; \boldsymbol{\alpha})$ parametrically. If $h_0(t; \boldsymbol{\alpha})$ is specified by the generalized log-logistic hazard function (1), then (5) takes the form

$$h(t; \mathbf{z}) = \frac{\kappa\rho^*(\rho^* t)^{\kappa-1}}{1 + (\gamma t)^{\kappa}}, \tag{6}$$

where $\rho^* = e^{\mathbf{z}'\boldsymbol{\beta}/\kappa}$. Thus the generalized log-logistic is closed under proportionality of hazards. Another widely used parametric PH family is the Weibull, for which $h_0(t; \boldsymbol{\alpha}) = \kappa\rho(\rho t)^{\kappa}$. Note that the Cox PH model is semiparametric, for which the baseline hazard function in (5) is left arbitrary and is denoted by $h_0(t)$.

Estimation

Suppose that a censored random sample consisting of data $(t_i, \delta_i, \mathbf{z}_i)$, $i = 1, 2, \ldots, n$, is available, where t_i is a lifetime or censoring time according to whether $\delta_i = 1$ or 0, respectively, and $\mathbf{z}_i = (z_{i1}, z_{i2}, \ldots, z_{ip})'$ is the vector of covariates for the ith individual. Letting $m = \sum_{i=1}^{n} \delta_i$, $a_i = \exp(\mathbf{z}_i' \boldsymbol{\beta})$ and $b_i = (\gamma t_i)^{\kappa}$, the log-likelihood function for the generalized log-logistic PH can be written as

$$\ell(\boldsymbol{\theta}) = m \log \kappa + m\kappa \log \rho + (\kappa - 1) \sum_{i=1}^{n} \delta_i \log t_i - \sum_{i=1}^{n} \delta_i \log (1 + b_i) + \sum_{i=1}^{n} \delta_i \log a_i - \left(\frac{\rho}{\gamma}\right)^{\kappa} \sum_{i=1}^{n} a_i \log (1 + b_i), \tag{7}$$

where $\boldsymbol{\theta} = (\boldsymbol{\alpha}', \boldsymbol{\beta}')'$. The first derivatives of the log-likelihood function are

$$\frac{\partial \ell(\boldsymbol{\theta})}{\partial \kappa} = \frac{m}{\kappa} + m \log \rho + \sum_{i=1}^{n} \delta_i \log t_i - \frac{1}{\kappa} \sum_{i=1}^{n} \delta_i b_i c_i - \left(\frac{\rho}{\gamma}\right)^{\kappa} \left(\frac{1}{\kappa}\right) \sum_{i=1}^{n} a_i b_i c_i - \left(\frac{\rho}{\gamma}\right)^{\kappa} \log \left(\frac{\rho}{\gamma}\right) \sum_{i=1}^{n} a_i \log (1 + b_i), \tag{8}$$

$$\frac{\partial \ell(\boldsymbol{\theta})}{\partial \gamma} = -\left(\frac{\kappa}{\gamma}\right) \sum_{i=1}^{n} \delta_i d_i - \left(\frac{\kappa}{\gamma}\right) \left(\frac{\rho}{\gamma}\right)^{\kappa} \sum_{i=1}^{n} a_i d_i - \left(\frac{\kappa}{\gamma}\right) \left(\frac{\rho}{\gamma}\right)^{\kappa} \sum_{i=1}^{n} a_i \log (1 - d_i), \tag{9}$$

$$\frac{\partial \ell(\boldsymbol{\theta})}{\partial \rho} = \frac{m\kappa}{\rho} - \left(\frac{\kappa}{\rho}\right) \left(\frac{\rho}{\gamma}\right)^{\kappa} \sum_{i=1}^{n} a_i \log (1 + b_i), \tag{10}$$

$$\frac{\partial \ell(\boldsymbol{\theta})}{\partial \beta_j} = \sum_{i=1}^{n} \delta_i z_{ij} - \left(\frac{\rho}{\gamma}\right)^{\kappa} \sum_{i=1}^{n} a_i \log (1 + b_i) z_{ij} \text{ for } j = 1, 2, \ldots, p, \tag{11}$$

where $c_i = \log b_i/(1 + b_i)$ and $d_i = b_i/(1 + b_i)$ (see Appendix). To improve the convergence of iterative procedures for maximum likelihood estimation and the accuracy of large-sample methods, we remove range restrictions on parameters through the parameterizations $\boldsymbol{\alpha}^* = (\kappa^*, \gamma^*, \rho^*)'$, where $\kappa^* = \log \kappa$, $\gamma^* = \log \gamma$ and $\rho^* = \log \rho$. The maximum likelihood estimate of $\boldsymbol{\theta}^* = (\boldsymbol{\alpha}^{*\prime}, \boldsymbol{\beta}')'$ can then be obtained by solving the equations $\partial \ell(\boldsymbol{\theta}^*)/\partial \kappa^* = 0$, $\partial \ell(\boldsymbol{\theta}^*)/\partial \gamma^* = 0$, $\partial \ell(\boldsymbol{\theta}^*)/\partial \rho^* = 0$ and $\partial \ell(\boldsymbol{\theta}^*)/\partial \beta_j = 0$ iteratively, where (see Appendix)

$$\frac{\partial \ell(\boldsymbol{\theta}^*)}{\partial \kappa^*} = \left[\kappa \left(\frac{\partial \ell(\boldsymbol{\theta})}{\partial \kappa}\right)\right]_{\boldsymbol{\alpha} = \exp(\boldsymbol{\alpha}^*)}, \quad \frac{\partial \ell(\boldsymbol{\theta}^*)}{\partial \gamma^*} = \left[\gamma \left(\frac{\partial \ell(\boldsymbol{\theta})}{\partial \gamma}\right)\right]_{\boldsymbol{\alpha} = \exp(\boldsymbol{\alpha}^*)},$$

$$\frac{\partial \ell(\boldsymbol{\theta}^*)}{\partial \rho^*} = \left[\rho \left(\frac{\partial \ell(\boldsymbol{\theta})}{\partial \rho}\right)\right]_{\boldsymbol{\alpha} = \exp(\boldsymbol{\alpha}^*)}, \quad \frac{\partial \ell(\boldsymbol{\theta}^*)}{\partial \beta_j} = \left[\frac{\partial \ell(\boldsymbol{\theta})}{\partial \beta_j}\right]_{\boldsymbol{\alpha} = \exp(\boldsymbol{\alpha}^*)}.$$

Many software packages have reliable optimization procedures to maximize log-likelihood functions. We wrote our computer code in R (R Core Team 2016), and used the function *nlminb* for optimization (see the Additional file 1).

Initial values

We may use Weibull, log-logistic and Cox PH fits to generate initial values in solving the equations $\partial \ell(\boldsymbol{\theta}^*)/\partial \kappa^* = 0$, $\partial \ell(\boldsymbol{\theta}^*)/\partial \gamma^* = 0$, $\partial \ell(\boldsymbol{\theta}^*)/\partial \rho^* = 0$ and $\partial \ell(\boldsymbol{\theta}^*)/\partial \beta_j = 0$. Let $\hat{\kappa}_1$ and $\hat{\rho}_1$ be the maximum likelihood estimates of the Weibull shape and scale parameters, respectively, $\hat{\kappa}_2$ and $\hat{\rho}_2$ the maximum likelihood estimates of the log-logistic shape and scale parameters, respectively, and $\hat{\boldsymbol{\beta}}^*$ the estimates of the regression coefficients for the

Cox PH model. Note that maximum likelihood methods for the Weibull, log-logistic and Cox PH models are available in many statistical softwares, including R (R Core Team 2016). We propose to use $\log \hat{\kappa}_1$, $\log |\hat{\kappa}_1 - \hat{\kappa}_2|$, $\log \hat{\rho}_1$ and $\hat{\boldsymbol{\beta}}^*$ as initial values for κ^*, γ^*, ρ^* and $\boldsymbol{\beta}$, respectively. If convergence is not achieved with these initial values, we propose to replace $\log \hat{\kappa}_1$ and $\log \hat{\rho}_1$ by $\log \hat{\kappa}_2$ and $\log \hat{\rho}_2$, respectively. In fitting the generalized log-logistic model to many data sets, we have not experienced any difficulty in obtaining convergence with this technique.

Tests and confidence intervals

Tests and interval estimates for the model parameters are based on the approximate normality of the maximum likelihood estimators. The asymptotic distribution of $\hat{\boldsymbol{\theta}}^*$ is approximately a $(p+3)$-variate normal distribution with mean $\boldsymbol{\theta}^*$ and covariance matrix $\Sigma = I(\hat{\boldsymbol{\theta}}^*)^{-1}$, where

$$I(\hat{\boldsymbol{\theta}}^*) = -\begin{bmatrix} \frac{\partial^2 \ell(\boldsymbol{\theta}^*)}{\partial \kappa^{*2}} & \frac{\partial^2 \ell(\boldsymbol{\theta}^*)}{\partial \kappa^* \partial \gamma^*} & \cdots & \frac{\partial^2 \ell(\boldsymbol{\theta}^*)}{\partial \kappa^* \partial \beta_p} \\ \frac{\partial^2 \ell(\boldsymbol{\theta}^*)}{\partial \gamma^* \partial \kappa^*} & \frac{\partial^2 \ell(\boldsymbol{\theta}^*)}{\partial \gamma^{*2}} & \cdots & \frac{\partial^2 \ell(\boldsymbol{\theta}^*)}{\partial \gamma^* \partial \beta_p} \\ \vdots & \vdots & \vdots & \vdots \\ \frac{\partial^2 \ell(\boldsymbol{\theta}^*)}{\partial \beta_p \partial \kappa^*} & \frac{\partial^2 \ell(\boldsymbol{\theta}^*)}{\partial \beta_p \partial \gamma^*} & \cdots & \frac{\partial^2 \ell(\boldsymbol{\theta}^*)}{\partial \beta_p^2} \end{bmatrix}_{\boldsymbol{\theta}^* = \hat{\boldsymbol{\theta}}^*}.$$

is the $(p+3) \times (p+3)$ observed information matrix (second derivatives of $\ell(\boldsymbol{\theta}^*)$ are given in Appendix). By the multivariate delta method, the asymptotic distribution of $\hat{\boldsymbol{\theta}}$ is also approximately normal with mean $\boldsymbol{\theta}$ and covariance matrix $D\Sigma D'$, where D is the $(p+3) \times (p+3)$ diagonal matrix $diag(\hat{\boldsymbol{\alpha}}, 1, 1, \ldots, 1)$ and $\hat{\boldsymbol{\alpha}} = \exp(\hat{\boldsymbol{\alpha}}^*)$.

Generalized log-logistic distribution in joint modeling

Joint models are used to quantify association between an internal time-dependent covariate and time until an event of interest occurs (Wulfsohn and Tsiatis 1997). It involves two separate models: a model that takes into account measurement error in the time-dependent covariate to estimate its true values (longitudinal model), and another model that uses these estimated values to quantify the association between this covariate and the time to the occurrence of the event (time-to-event model). The idea behind the joint modeling technique is to couple the time-to-event model with the longitudinal model. The general framework of the maximum likelihood method and large sample theory can be found in Rizopoulos (2012). Maximization of the log-likelihood function for joint modeling is computationally challenging, as it involves evaluating multiple integrals that do not have an analytical solution, except in very special cases. The R package **JM** has been developed by Rizopoulos (2010) to fit joint models using Weibull baseline hazard, piecewise-constant baseline hazard, spline approximation of the baseline hazard and unspecified baseline hazard functions. We have modified the source codes for Weibull to fit joint models using the generalized log-logistic baseline hazard function. The application of the generalized log-logistic distribution in joint modeling is illustrated with an example in Section 1.

Goodness of fit

The nonparametric estimates are useful for assessing the quality of fit of a particular parametric time-to-event model (Lawless 2002). For a model without covariate, we use the approach to simultaneously examine plots of parametric and nonparametric estimates of the survival function, superimposed on the same graph. Let $S(t;\hat{\boldsymbol{\theta}})$ and $\hat{S}(t)$ be the estimates of the survivor functions based on the parametric model of interest and the Kaplan-Meier method (Kaplan and Meier 1958), respectively. The estimates $S(t;\hat{\boldsymbol{\theta}})$ as a function of t should be close to $\hat{S}(t)$ if the parametric model is adequate. For a model with covariates, we consider residual diagnostic plots, where the residuals are defined based on the cumulative hazard function $H(t;\boldsymbol{\theta})$. If $\hat{S}(H(t;\hat{\boldsymbol{\theta}}))$ is the Kaplan-Meier estimate of $H(t;\hat{\boldsymbol{\theta}})$, then a plot of $-\log\hat{S}(H(t;\hat{\boldsymbol{\theta}}))$ versus $H(t;\hat{\boldsymbol{\theta}})$ should be roughly a straight line with unit slope when the model is adequate (Lawless 2002).

We also use the Akaike's information criterion (AIC) (Akaike 1974) to compare the fits of different models. The AIC is defined by

$$\text{AIC} = -2\log(\text{maximized likelihood}) + 2(p+k),$$

where p is the number of covariates and k is the number of parameters of the assumed probability distribution ($k = 3$ for the generalized log-logistic model). In general, when comparing two or more models, we prefer the one with the lowest AIC value. A rule of thumb is that if $\Delta_{\text{M}} = \text{AIC}_{\text{M}} - \text{AIC}_{\text{min}} > 2$, then there is considerably less support for Model M compared to the model with minimum AIC (Burnham and Anderson 2002).

Examples

Three data sets are taken from the literature to demonstrate the ability of the generalized log-logistic distribution in modeling time-to-event data. The application of the generalized log-logistic PH in joint modeling is illustrated using another data set of AIDS patients. We first use the scaled TTT transform of failure times to detect the shape of the hazard function (Mudholkar et al. 1996). The scaled TTT transform is given by $\phi(\nu/n) = \left[\sum_{i=1}^{\nu} T_{(i)} + (n-\nu)T_{(\nu)}\right] / \left(\sum_{i=1}^{n} T_{(i)}\right)$, where $T_{(i)}$ represent the order statistics of the sample, and $\nu = 1, 2, \ldots, n$. The hazard function is increasing, decreasing and unimodal if the plot of $(\nu/n, \phi(\nu/n))$ is concave, convex, and concave followed by convex, respectively. For the first three examples (Sections Example 1: Head and neck cancer data-Example 3: Vaginal cancer mortality in rats), we first fit the generalized log-logistic, Weibull and log-logistic models (without covariate) and check the appropriateness of the distributional assumption using diagnostic plots. Then, we analyze the data using regression models, and compare the fits via residual plots. Note that the regression model based on the log-logistic distribution is given by $\log T = \beta_0 + \beta_1 z_1 + \ldots + \beta_p z_p + \tau W$ where $\tau = 1/\kappa$, $\beta_0 = -\log\rho$ and W has the logistic distribution with density $f(w) = e^w/(1+e^w)^2$, $-\infty < w < \infty$. This model has an accelerated life interpretation (Lawless 2002), whereas the generalized log-logistic and Weibull PH models have relative risk interpretation. In the fourth example (Section Example 4: AIDS data), we consider joint models based on the generalized log-logistic, Weibull and piecewise-constant baseline hazard functions.

Example 1: Head and neck cancer data

Data description, hazard shape and distributional assumption

Efron (1988) described a randomized clinical trial to compare radiation therapy alone (arm A) versus radiation plus chemotherapy (arm B) in treating head and neck cancer patients. Survival times (in days) for 51 patients in arm A (9 observations were censored) and 45 in arm B (14 were censored) were reported. The TTT plot in Fig. 1(a) suggests a

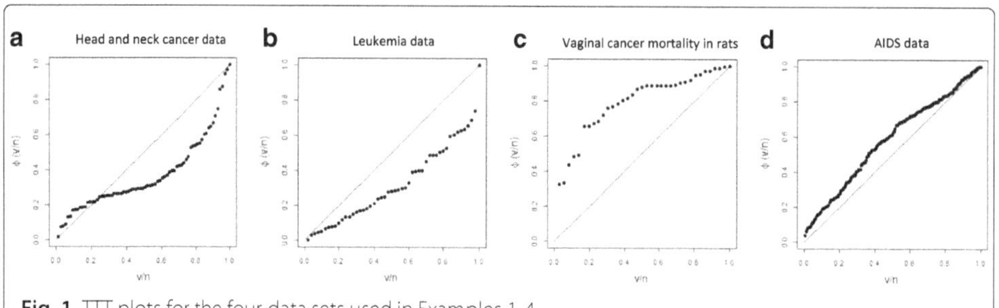

Fig. 1 TTT plots for the four data sets used in Examples 1-4

unimodal hazard shape of the survival times. Plots of $S(t;\hat{\theta})$ and $\hat{S}(t)$ (Fig. 2(a–c)) indicate more support for the generalized log-logistic distribution in comparison with the Weibull and log-logistic distributions in describing the head and neck cancer data.

Regression analysis

Letting $z_i = \text{I(treatment = radiation therapy)}$ that equals 1 if the treatment involves radiation therapy alone and 0 otherwise, we fit the generalized log-logistic PH, Weibull PH and log-logistic AFT models to the head and neck cancer data (numerical results are summarized in Table 1). The standard error of $\hat{\beta}$ for the generalized log-logistic model is smaller that those for the Weibull and log-logistic models, and therefore the generalized log-logistic would be preferred on grounds of efficiency. We also see that the generalized log-logistic has the lowest AIC value, which is supported by the residual plots (Fig. 2(d–f)): residuals lying closely to the unit-slope line for generalized log-logistic indicate its superiority over the Weibull and log-logistic models. In summary, the generalized log-logistic fits the data adequately and is the best among the three models under consideration.

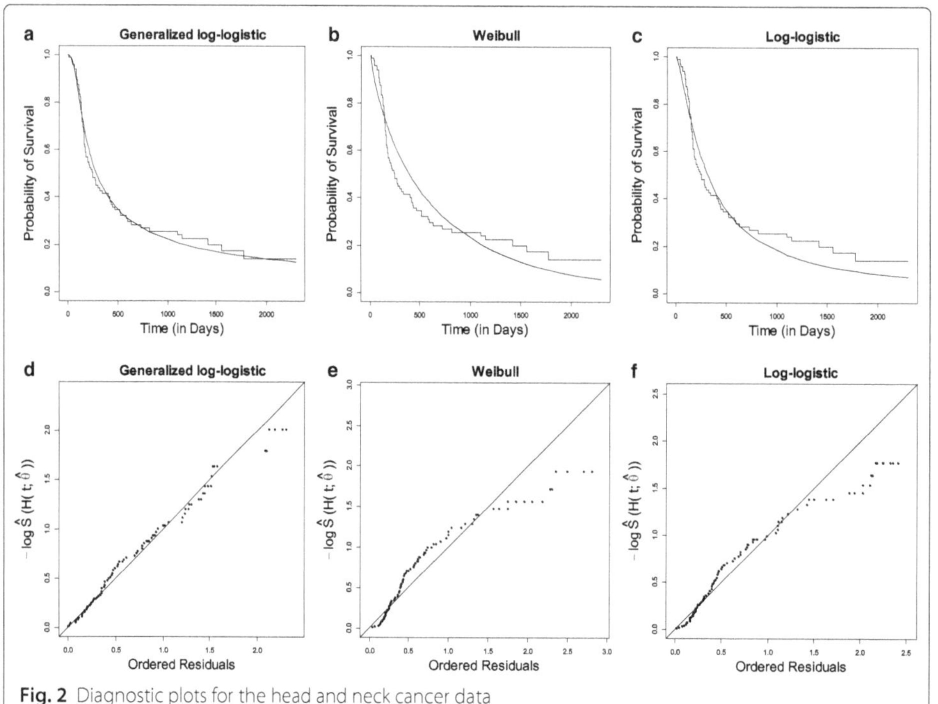

Fig. 2 Diagnostic plots for the head and neck cancer data

Table 1 Generalized log-logistic, Weibull and log-logistic fits for the head and neck cancer data

	Generalized log-logistic PH (AIC = 1053.39)		Weibull PH (AIC = 1082.52)		Log-logistic AFT (AIC = 1067.25)	
Parameter	Estimate	SE	Estimate	SE	Estimate	SE
β	0.5459	0.2382	0.6686	0.2415	−0.5549	0.2779
$\log \kappa$	0.9790	0.1986	−0.1619	0.0921	0.2764	0.0971
$\log \rho$	−5.2692	0.1844	−6.8248	0.2112	−6.0492	0.2128
$\log \gamma$	−4.6497	0.1755				

Example 2: Autologous and allogeneic bone marrow transplants

Data description, hazard shape and distributional assumption

Klein and Moeschberger (2003) described a study involving a sample of 101 patients with advanced acute myelogenous leukemia. Fifty-one of these patients had received an autologous (auto) bone marrow transplant, whereas 50 an allogeneic (allo) transplant. Survival times (in months) for 28 auto transplant and 22 allo transplant patients were censored. Careful inspection of the TTT plot in Fig. 1(b) reveals an indication of the unimodality of the hazard function. A comparison of the diagnostic plots (without covariate) in Fig. 3(a–c) suggests that the assumption of generalized log-logistic is more appropriate than the assumption of Weibull or log-logistic in describing these data.

Regression analysis

For regression analysis, we consider the covariate z_i = I(transplant = allo). The fits via the generalized log-logistic PH, Weibull PH and log-logistic AFT are summarized in Table 2. The generalized log-logistic has the lowest AIC value, suggesting it produced the best-fitting model. The residual plots (Fig. 3(d–f)) also support this fact. It is interesting

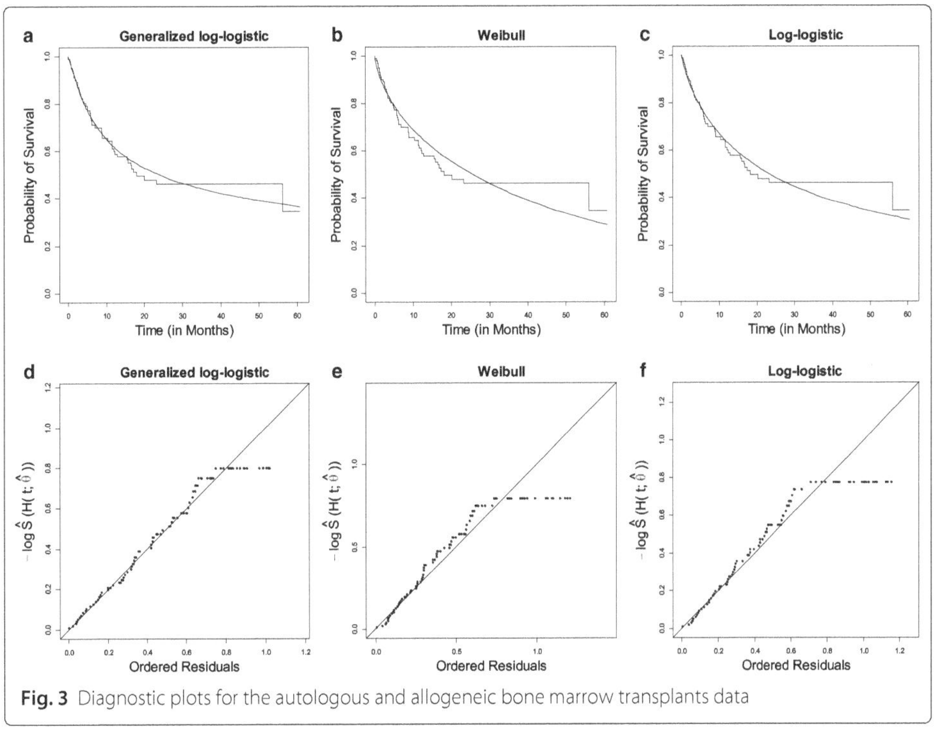

Fig. 3 Diagnostic plots for the autologous and allogeneic bone marrow transplants data

Table 2 Generalized log-logistic, Weibull and log-logistic fits for the bone marrow transplants data

	Generalized log-logistic PH (AIC = 444.64)		Weibull PH (AIC = 450.08)		Log-logistic AFT (AIC = 446.46)	
Parameter	Estimate	SE	Estimate	SE	Estimate	SE
β	0.1981	0.2854	0.2535	0.2854	−0.0808	0.4481
$\log \kappa$	0.2148	0.2376	−0.3878	0.1229	−0.1694	0.1213
$\log \rho$	−2.4055	0.4917	−3.9683	0.3300	−3.1847	0.3474
$\log \gamma$	−1.3188	0.6253				

to note here that both the Weibull and log-logistic suggest a decreasing hazard function (estimate of the shape parameter is less than 1), whereas the generalized log-logistic captures the unimodal shape of the hazard function ($\hat{\kappa} = e^{0.2148} = 1.24 > 1$).

Example 3: Vaginal cancer mortality in rats

Data description, hazard shape and distributional assumption

Pike (1966) described a laboratory experiment involving the development of vaginal cancer in rats insulted with the carcinogen DMBA. There were 19 rats in group 1 and 21 in group 2. Seventeen rats in group 1 and 19 in group 2 had developed tumours at the time the data were collected (i.e., two observations in each group were censored). There were reasonable scientific grounds for believing that there might be a threshold value before which no tumour could be detected. For this reason, Lawless (2002) considered the values $t' = t - 100$ to analyze these data. We also consider here the transformed version of the original observations. The TTT plot in Fig. 1(c) suggests an increasing hazard function for T'. Figure 4(a–c) shows diagnostic plots for the generalized log-logistic, Weibull and log-logistic fits (without covariate). We see that the generalized log-logistic and Weibull fits are similar, and provide slightly better description of the data compared to the log-logistic model.

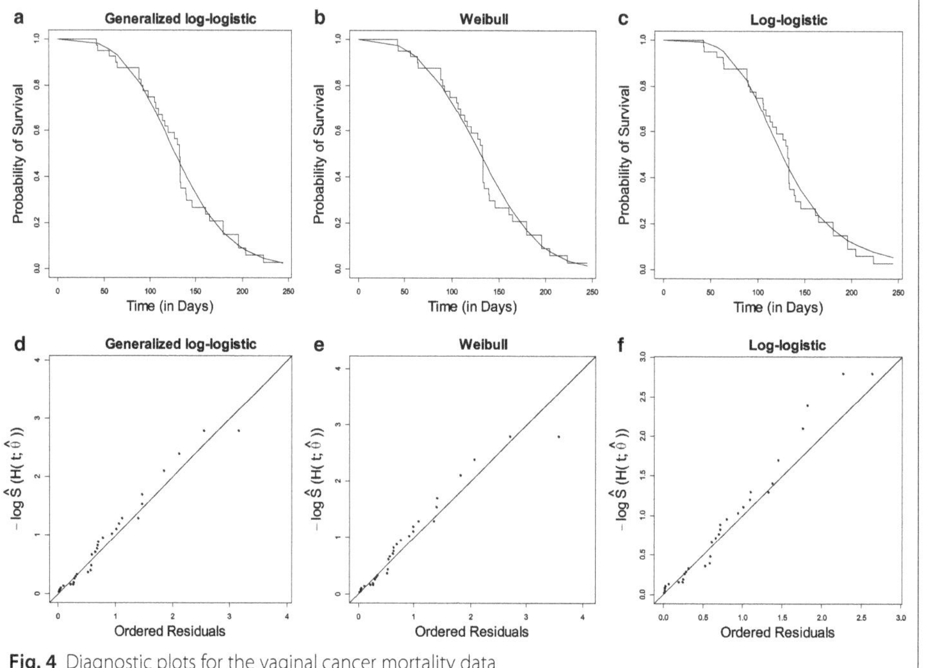

Fig. 4 Diagnostic plots for the vaginal cancer mortality data

Regression analysis

For regression analysis, we consider the covariate z_i = I(group = group 1). Table 3 gives the estimates of the parameters and associated standard errors from the generalized log-logistic, Weibull and log-logistic fits. As demonstrated by the TTT plot, the Weibull PH suggests an increasing hazard rate ($\hat{\kappa} = e^{1.1308} = 3.098$). Note that a small value of $\hat{\gamma}$ ($\hat{\gamma} = e^{-5.0190} = 0.007$) and $\hat{\kappa} = e^{1.2568} = 3.514 > 1$ for generalized log-logistic also support this fact. Although the AIC values (Table 3) suggest no obvious preference of one model over the other, the residual plots (Fig. 4(d–f)) clearly indicate more support for the generalized log-logistic and Weibull models. This example demonstrates that the generalized log-logistic has the ability to satisfactorily fit data which exhibit increasing hazard rates.

Example 4: AIDS data

Data description and hazard shape

This example illustrates the use of the generalized log-logistic distribution in joint modeling. Rizopoulos (2012) described a study involving 467 human immunodeficiency virus (HIV) infected patients who had failed or were intolerant to zidovudine therapy (ZT). The main objective was to compare two antiretroviral drugs to prevent the progression of HIV infections: didanosine (ddI) and zalcitabine (ddC). Patients were randomly assigned to receive either ddI or ddC and followed until death or the end of the study, resulting in 188 complete and 279 censored observations. It was also of interest to quantify the association between CD4 cell counts (internal time-dependent covariate) measured at $t = 0, 2,$ 6, 12 and 18 months, and time to death. The TTT plot in Fig. 1(d) indicates an increasing hazard shape.

Regression analysis

For regression analysis, Rizopoulos (2012) considered joint models of the form

$$h_i(t; \mathbf{z}_i) = h_0(t; \boldsymbol{\alpha}) \exp\{\beta_0 + \beta_1 \text{drug}_i + \beta_2 \text{sex}_i + \beta_3 \text{ZT}_i + \beta_4 \text{CD4}_i(t)\}, \tag{12}$$

$$\text{CD4}_i(t) = b_0 + b_1 t + b_2(t \times \text{drug}_i) + b_{0i} + b_{1i} t + \epsilon_i(t), \tag{13}$$

where (12) is the time-to-event model with drug_i = I(drug = ddI), sex_i = I(sex = male) and ZT_i = I(ZT = failure); and (13) is the longitudinal model with b_0, b_1 and b_2 being the fixed-effects parameters, b_{0i} and b_{1i} the random-effects parameters, and $\epsilon_i(t)$ the random error component. We have reanalyzed the data here using generalized log-logistic, Weibull and piecewise-constant (six knots placed at equally spaced percentiles of the observed event times (Rizopoulos 2012)) baseline hazard functions in (12). Note that

Table 3 Generalized log-logistic, Weibull and log-logistic fits for the vaginal cancer mortality data

	Generalized log-logistic PH (AIC = 391.35)		Weibull PH (AIC = 389.87)		Log-logistic AFT (AIC = 391.89)	
Parameter	Estimate	SE	Estimate	SE	Estimate	SE
β	0.6254	0.3485	0.6599	0.3474	−0.1861	0.1203
$\log \kappa$	1.2568	0.2168	1.1308	0.1300	1.5077	0.1429
$\log \rho$	−5.3516	0.5889	−5.0864	0.0754	−4.9301	0.0846
$\log \gamma$	−5.0190	0.1154				

$h_0(t;\boldsymbol{\alpha}) = \kappa t^{\kappa-1}/[1+(\gamma t)^{\kappa}]$ and $\kappa t^{\kappa-1}$ for generalized log-logistic and Weibull, respectively, and so $\beta_0 = \kappa \log \rho$ for both these models. For piecewise-constant baseline hazard, $h_0(t;\boldsymbol{\alpha}) = \sum_{q=1}^{7} \xi_q \mathrm{I}(\nu_{q-1} < t \leq \nu_q)$ and $\beta_0 = 0$ in (12), where $0 = \nu_0 < \nu_1 < \ldots < \nu_7$ is the split of the time scale and ξ_q is the value of the hazard in the interval $(\nu_{q-1}, \nu_q]$. The estimates of the parameters and standard errors for the time-to-event process are presented in Table 4. We see that the estimates of the coefficients (i.e., $\hat{\beta}_1$, $\hat{\beta}_2$, $\hat{\beta}_3$ and $\hat{\beta}_4$) and their standard errors are broadly similar under the three competing models. The AIC values and residual plots (Fig. 5) also suggest no obvious preference of one model over the other. Although we see no obvious preference of the generalized log-logistic model for this example for which the hazard function is monotone increasing, generalized log-logistic could be useful in joint modeling where the shape of the hazard function is unimodal.

Simulations

Four covariates in a PH regression framework were considered in all simulations: two continuous covariates (z_1 and z_2), each generated from the standard normal distribution; and two binary covariates (z_3 and z_4), each generated from the *Bernoulli*(0.5) distribution. Regression parameter values were chosen to be $\boldsymbol{\beta} = (0.50, -0.50, 0.75, -0.75)'$ corresponding to the covariate vector $\mathbf{z} = (z_1, z_2, z_3, z_4)'$. To evaluate the performance of the generalized log-logistic model, we considered three simulation scenarios based on the shape of the hazard function. For each scenario (see below), lifetime data were generated from the generalized Weibull distribution with probability density function

$$f(t;\boldsymbol{\alpha},\boldsymbol{\beta}) = \kappa\rho(\rho t)^{\kappa-1} \exp(\mathbf{z}'\boldsymbol{\beta}) \left[1-\gamma(\rho t)^{\kappa}\right]^{\frac{\exp(\mathbf{z}'\boldsymbol{\beta})}{\gamma}-1}, \tag{14}$$

where $\rho > 0$, $\kappa > 0$ and $-\infty < \gamma < \infty$ are distributional parameters and $\boldsymbol{\alpha} = (\kappa, \gamma, \rho)'$; the support of the distribution is $t > 0$ for $\gamma \leq 0$ and $0 < t < \frac{1}{\rho\gamma^{\kappa}}$ for $\gamma > 0$. Note that the hazard function of the generalized Weibull distribution is (a) monotone increasing for $\kappa \geq 1$ and $\gamma \geq 1$, (b) monotone decreasing for $0 < \kappa \leq 1$ and $\gamma \leq 1$, and (c) unimodal for $\kappa > 1$ and $\gamma < 0$. The simulation scenarios are then specified as follows.

1. **Scenario 1: Decreasing hazard.** Lifetimes were generated from generalized Weibull with $\kappa = 0.5$, $\gamma = -0.1$ and $\rho = 0.1$, and censoring times were generated from the exponential distribution with rate parameter $\lambda = 0.045$.

Table 4 AIDS data: estimates and standard errors for the time-to-event process of joint models

	Generalized log-logistic (AIC = 8699.61)		Weibull (AIC = 8699.26)		Piecewise-constant (AIC = 8711.61)	
Parameter	Estimate	SE	Estimate	SE	Estimate	SE
β_0	−3.1615	0.4411	−2.9477	0.3898	–	–
β_1	0.3690	0.1575	0.3727	0.1576	0.3647	0.1573
β_2	−0.3647	0.2583	−0.3619	0.2591	−0.3364	0.2585
β_3	0.3372	0.1556	0.3455	0.1555	0.3329	0.1555
β_4	−0.2824	0.0382	−0.2784	0.0378	−0.2860	0.0382
$\log\kappa$	0.3838	0.7709	0.2377	0.0732		
$\log\gamma$	−2.8874	0.1333				

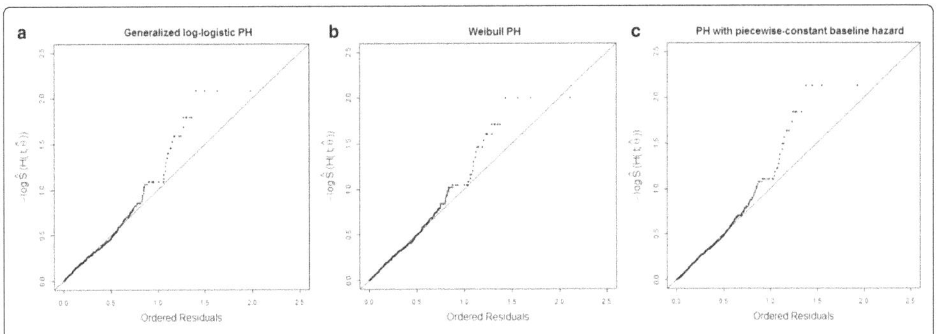

Fig. 5 Residual plots for the AIDS data

2. **Scenario 2: Increasing hazard.** Lifetimes were generated from generalized Weibull with $\kappa = 2$, $\gamma = 0.1$ and $\rho = 0.1$, and censoring times were generated from the exponential distribution with rate parameter $\lambda = 0.060$.
3. **Scenario 3: Unimodal hazard.** Lifetimes were generated from generalized Weibull with $\kappa = 2$, $\gamma = -0.1$ and $\rho = 0.1$, and censoring times were generated from the exponential distribution with rate parameter $\lambda = 0.060$.

Our choice of the parameter values led to, on average, 39.99, 40.69 and 42.99% censored observations for Scenarios 1-3, respectively. Given the covariates and censoring indicator, we then fit the generalized log-logistic, Weibull and Cox PH models to the simulated lifetimes. Note that since the Cox model is robust (usually fits the data well no matter which parametric model is appropriate), we consider this in our simulation study to compare model performance. For each scenario, 500 data sets (each of size $n = 100$) were generated, and the average of each of the estimated model parameters across these data sets was calculated. Absolute bias (AB) and mean square error (MSE) were then computed for model comparison (numerical results are summarized in Table 5).

Results for scenario 1. For the continuous covariates (z_1 and z_2), all three models produced estimates with similar MSE, whereas for the binary covariates (z_3 and z_4), the generalized log-logistic demonstrated the smallest MSE. In terms of bias, generalized log-logistic and Cox PH were roughly equivalent, and both were superior to Weibull.

Results for scenario 2. For the regression coefficients, the generalized log-logistic produced estimates with the smallest bias. We also see that the generalized log-logistic and Weibull produced estimates with similar MSE, and both were superior to the Cox PH model. Note that the generalized log-logistic estimates for κ and γ were 2.269 and 0.006, respectively (i.e., the estimate of γ^κ is close to zero), supporting the fact that the hazard function is monotone increasing.

Results for scenario 3. In terms of bias, the generalized log-logistic and Cox PH produced comparable estimates of the regression coefficients. However, the generalized log-logistic produced the most accurate estimates in terms of MSE, mostly as a consequence of smaller standard deviations of the estimates. As expected, the Weibull produced the least accurate estimates in terms of both bias and MSE for Scenario 3 (i.e., unimodal hazard).

A simulation study with about 20% censored observations per data set also led to similar conclusions (data not shown). In summary, our simulation study has demonstrated

Table 5 Model performance and comparison using simulation study ($n = 100$) with about 40% censored observations

Scenarios	Parameter	True	Generalized log-logistic PH			Weibull PH			Cox PH		
			Mean	AB	MSE	Mean	AB	MSE	Mean	AB	MSE
Scenario 1	β_1	0.50	0.524	0.024	0.024	0.531	0.031	0.024	0.524	0.024	0.025
(True model:	β_2	−0.50	−0.538	0.038	0.032	−0.545	0.045	0.033	−0.536	0.036	0.032
generalized	β_3	0.75	0.798	0.048	0.088	0.808	0.058	0.090	0.795	0.045	0.091
Weibull)	β_4	−0.75	−0.780	0.030	0.081	−0.792	0.042	0.083	−0.782	0.032	0.083
	ρ	0.10	0.148			0.103					
	κ	0.50	0.550			0.508					
	γ	−0.10	0.073								
Scenario 2	β_1	0.50	0.516	0.016	0.027	0.518	0.018	0.027	0.532	0.032	0.031
(True model:	β_2	−0.50	−0.522	0.022	0.035	−0.523	0.023	0.035	−0.533	0.033	0.039
generalized	β_3	0.75	0.785	0.035	0.099	0.788	0.038	0.099	0.813	0.063	0.112
Weibull)	β_4	−0.75	−0.765	0.015	0.082	−0.768	0.018	0.082	−0.796	0.046	0.092
	ρ	0.10	0.107			0.106					
	κ	2.00	2.269			2.249					
	γ	0.10	0.006								
Scenario 3	β_1	0.50	0.519	0.019	0.025	0.530	0.030	0.026	0.516	0.016	0.026
(True model:	β_2	−0.50	−0.548	0.048	0.038	−0.557	0.057	0.039	−0.547	0.047	0.039
generalized	β_3	0.75	0.791	0.041	0.099	0.811	0.061	0.103	0.791	0.041	0.102
Weibull)	β_4	−0.75	−0.790	0.040	0.089	−0.811	0.061	0.093	−0.792	0.042	0.095
	ρ	0.10	0.103			0.098					
	κ	2.00	2.130			2.016					
	γ	−0.10	0.024								

that the generalized log-logistic could potentially be a very useful parametric model to adequately describe different types of time-to-event data.

Conclusion

In this paper, we proposed a simple extension of the log-logistic distribution to a PH model by appending an additional parameter. As described in Section 1, the proposed model naturally accommodates decreasing and unimodal hazard functions. The log-logistic distribution is known to be useful to describe unimodal hazard functions (Lawless 2002). As demonstrated in Examples 1 and 2, it turns out that the generalized log-logistic may provide better fits in describing unimodal hazard functions compared to the log-logistic distribution. Moreover, our simulation study revealed that the generalized log-logistic could produce more accurate results compared to the Weibull and Cox PH models in describing monotone decreasing and unimodal hazard functions. In summary, the flexibility provided by the generalized log-logistic model could be very useful in adequately describing different types of time-to-event data.

Appendix: Derivatives of the log-likelihood function

Let $m = \sum_{i=1}^{n} \delta_i$, $a_i = \exp(\mathbf{z}_i'\boldsymbol{\beta})$, $b_i = (\gamma t_i)^\kappa$, $c_i = \log b_i/(1+b_i)$ and $d_i = b_i/(1+b_i)$. We have

- $$\log(\gamma t_i) = \frac{\log b_i}{\kappa}, \tag{15}$$

- $$(\gamma t_i)^\kappa \log(\gamma t_i) = \frac{b_i \log b_i}{\kappa}, \tag{16}$$

- $$\frac{\partial b_i}{\partial \kappa} = \frac{\partial}{\partial \kappa}(\gamma t_i)^\kappa = (\gamma t_i)^\kappa \log(\gamma t_i) = \frac{b_i \log b_i}{\kappa}, \tag{17}$$

- $$\frac{\partial \log b_i}{\partial \kappa} = \frac{\log b_i}{\kappa}, \tag{18}$$

- $$\frac{\partial \log(1+b_i)}{\partial \kappa} = \frac{b_i \log b_i}{\kappa(1+b_i)} = \frac{b_i c_i}{\kappa}, \tag{19}$$

- $$\frac{\partial b_i \log b_i}{\partial \kappa} = \frac{b_i \log b_i}{\kappa} \log b_i + b_i \frac{\log b_i}{\kappa} = \frac{b_i (\log b_i)(1+\log b_i)}{\kappa}, \tag{20}$$

- $$\frac{\partial c_i}{\partial \kappa} = \frac{\partial}{\partial \kappa} \frac{\log b_i}{1+b_i} = \frac{\log b_i}{\kappa(1+b_i)} \left(1 - \frac{b_i \log b_i}{1+b_i}\right) = \frac{c_i(1-b_i c_i)}{\kappa}, \tag{21}$$

- $$\frac{\partial b_i c_i}{\partial \kappa} = \frac{b_i c_i (1 - b_i c_i + \log b_i)}{\kappa} = \frac{b_i c_i (1+c_i)}{\kappa}, \tag{22}$$

- $$\frac{\partial d_i}{\partial \kappa} = \frac{\partial}{\partial \kappa} \frac{b_i}{1+b_i} = \frac{b_i \log b_i}{\kappa(1+b_i)} \left(1 - \frac{b_i}{1+b_i}\right) = \frac{c_i d_i}{\kappa}, \tag{23}$$

- $$\frac{\partial \log(1-d_i)}{\partial \kappa} = \frac{\partial}{\partial \kappa} \log(1+b_i)^{-1} = -\frac{\partial}{\partial \kappa} \log(1+b_i) = -\frac{b_i c_i}{\kappa}, \tag{24}$$

- $$\frac{\partial b_i}{\partial \gamma} = \frac{\partial}{\partial \gamma}(\gamma t_i)^\kappa = \kappa \gamma^{\kappa-1} t_i^\kappa = \frac{\kappa}{\gamma} b_i, \tag{25}$$
- $$\frac{\partial d_i}{\partial \gamma} = \frac{\partial}{\partial \gamma} \frac{b_i}{1+b_i} = \frac{\kappa}{\gamma} \frac{b_i}{1+b_i}\left(1 - \frac{b_i}{1+b_i}\right) = \frac{\kappa}{\gamma} d_i(1-d_i), \tag{26}$$
- $$\frac{\partial}{\partial \gamma} \log(1-d_i) = -\frac{\partial}{\partial \gamma} \log(1+b_i) = -\frac{\kappa}{\gamma} \frac{b_i}{1+b_i} = -\frac{\kappa}{\gamma} d_i. \tag{27}$$

Using (7) and (15)-(27), we can derive the first and second derivatives of the log-likelihood function as follows.

$$\frac{\partial \ell(\boldsymbol{\theta})}{\partial \kappa} = \frac{m}{\kappa} + m \log \rho + \sum_{i=1}^n \delta_i \log t_i - \frac{1}{\kappa} \sum_{i=1}^n \delta_i b_i c_i - \left(\frac{\rho}{\gamma}\right)^\kappa \left(\frac{1}{\kappa}\right) \sum_{i=1}^n a_i b_i c_i$$
$$- \left(\frac{\rho}{\gamma}\right)^\kappa \log\left(\frac{\rho}{\gamma}\right) \sum_{i=1}^n a_i \log(1+b_i).$$

$$\frac{\partial \ell(\boldsymbol{\theta})}{\partial \gamma} = -\left(\frac{\kappa}{\gamma}\right) \sum_{i=1}^n \delta_i d_i - \left(\frac{\kappa}{\gamma}\right)\left(\frac{\rho}{\gamma}\right)^\kappa \sum_{i=1}^n a_i d_- \left(\frac{\kappa}{\gamma}\right)\left(\frac{\rho}{\gamma}\right)^\kappa \sum_{i=1}^n a_i \log(1-d_i).$$

$$\frac{\partial \ell(\boldsymbol{\theta})}{\partial \rho} = \frac{m\kappa}{\rho} - \left(\frac{\kappa}{\rho}\right)\left(\frac{\rho}{\gamma}\right)^\kappa \sum_{i=1}^n a_i \log(1+b_i).$$

$$\frac{\partial \ell(\boldsymbol{\theta})}{\partial \beta_j} = \sum_{i=1}^n \delta_i z_{ij} - \left(\frac{\rho}{\gamma}\right)^\kappa \sum_{i=1}^n a_i \log(1+b_i) z_{ij} \text{ for } j = 1, 2, \ldots, p.$$

$$\frac{\partial^2 \ell(\boldsymbol{\theta})}{\partial \kappa^2} = \frac{\partial}{\partial \kappa} \frac{\partial \ell(\boldsymbol{\theta})}{\partial \kappa}$$
$$= -\frac{m}{\kappa^2} + \frac{1}{\kappa^2} \sum_{i=1}^n \delta_i b_i c_i - \frac{1}{\kappa^2} \sum_{i=1}^n \delta_i b_i c_i (1+c_i) - \left(\frac{\rho}{\gamma}\right)^\kappa \log\left(\frac{\rho}{\gamma}\right)\left(\frac{1}{\kappa}\right) \sum_{i=1}^n a_i b_i c_i$$
$$+ \left(\frac{\rho}{\gamma}\right)^\kappa \left(\frac{1}{\kappa^2}\right) \sum_{i=1}^n a_i b_i c_i - \left(\frac{\rho}{\gamma}\right)^\kappa \left(\frac{1}{\kappa^2}\right) \sum_{i=1}^n a_i b_i c_i (1+c_i)$$
$$- \left(\frac{\rho}{\gamma}\right)^\kappa \log\left(\frac{\rho}{\gamma}\right)\left(\frac{1}{\kappa}\right) \sum_{i=1}^n a_i b_i c_i - \left(\frac{\rho}{\gamma}\right)^\kappa \left\{\log\left(\frac{\rho}{\gamma}\right)\right\}^2 \sum_{i=1}^n a_i \log(1+b_i)$$
$$= -\frac{m}{\kappa^2} - \frac{1}{\kappa^2} \sum_{i=1}^n \delta_i b_i c_i^2 - \left(\frac{\rho}{\gamma}\right)^\kappa \left(\frac{1}{\kappa^2}\right) \sum_{i=1}^n a_i b_i c_i^2 - \left(\frac{2}{\kappa}\right)\left(\frac{\rho}{\gamma}\right)^\kappa \log\left(\frac{\rho}{\gamma}\right) \sum_{i=1}^n a_i b_i c_i$$
$$- \left(\frac{\rho}{\gamma}\right)^\kappa \left\{\log\left(\frac{\rho}{\gamma}\right)\right\}^2 \sum_{i=1}^n a_i \log(1+b_i).$$

$$\frac{\partial^2 \ell(\boldsymbol{\theta})}{\partial \gamma^2} = \frac{\partial}{\partial \gamma} \frac{\partial \ell(\boldsymbol{\theta})}{\partial \gamma}$$
$$= \left(\frac{\kappa}{\gamma^2}\right) \sum_{i=1}^n \delta_i d_i - \left(\frac{\kappa}{\gamma}\right)^2 \sum_{i=1}^n \delta_i d_i (1-d_i) + \kappa \rho^\kappa \left(\frac{\kappa+1}{\gamma^{\kappa+2}}\right) \sum_{i=1}^n a_i d_i$$
$$- \left(\frac{\kappa}{\gamma}\right)^2 \left(\frac{\rho}{\gamma}\right)^\kappa \sum_{i=1}^n a_i d_i (1-d_i) + \kappa \rho^\kappa \left(\frac{\kappa+1}{\gamma^{\kappa+2}}\right) \sum_{i=1}^n a_i \log(1-d_i)$$
$$+ \left(\frac{\kappa}{\gamma}\right)^2 \left(\frac{\rho}{\gamma}\right)^\kappa \sum_{i=1}^n a_i d_i$$
$$= \left(\frac{\kappa}{\gamma^2}\right) \sum_{i=1}^n \delta_i d_i - \left(\frac{\kappa}{\gamma}\right)^2 \sum_{i=1}^n \delta_i d_i (1-d_i)$$
$$+ \frac{\kappa(\kappa+1)}{\gamma^2} \left(\frac{\rho}{\gamma}\right)^\kappa \sum_{i=1}^n a_i [d_i + \log(1-d_i)] + \left(\frac{\kappa}{\gamma}\right)^2 \left(\frac{\rho}{\gamma}\right)^\kappa \sum_{i=1}^n a_i d_i^2.$$

$$\frac{\partial^2 \ell(\boldsymbol{\theta})}{\partial \rho^2} = \frac{\partial}{\partial \rho}\frac{\partial \ell(\boldsymbol{\theta})}{\partial \rho} = -\frac{m\kappa}{\rho^2} - \frac{\kappa(\kappa-1)}{\rho^2}\left(\frac{\rho}{\gamma}\right)^{\kappa}\sum_{i=1}^{n} a_i \log(1+b_i).$$

$$\frac{\partial^2 \ell(\boldsymbol{\theta})}{\partial \beta_j \partial \beta_{j'}} = \frac{\partial}{\partial \beta_j}\frac{\partial \ell(\boldsymbol{\theta})}{\partial \beta_{j'}} = -\left(\frac{\rho}{\gamma}\right)^{\kappa}\sum_{i=1}^{n} a_i \log(1+b_i) z_{ij} z_{ij'} \text{ for } j, j' = 1, 2, \ldots, p.$$

$$\begin{aligned}
\frac{\partial^2 \ell(\boldsymbol{\theta})}{\partial \kappa \partial \gamma} &= \frac{\partial}{\partial \kappa}\frac{\partial \ell(\boldsymbol{\theta})}{\partial \gamma} \\
&= -\left(\frac{1}{\gamma}\right)\sum_{i=1}^{n}\delta_i d_i - \left(\frac{\kappa}{\gamma}\right)\left(\frac{1}{\kappa}\right)\sum_{i=1}^{n}\delta_i c_i d_i - \left(\frac{\kappa}{\gamma}\right)\left(\frac{\rho}{\gamma}\right)^{\kappa}\left(\frac{1}{\kappa}\right)\sum_{i=1}^{n} a_i c_i d_i \\
&\quad -\left(\frac{1}{\gamma}\right)\left(\frac{\rho}{\gamma}\right)^{\kappa}\sum_{i=1}^{n} a_i d_i - \left(\frac{\kappa}{\gamma}\right)\left(\frac{\rho}{\gamma}\right)^{\kappa}\log\left(\frac{\rho}{\gamma}\right)\sum_{i=1}^{n} a_i d_i + \left(\frac{\kappa}{\gamma}\right)\left(\frac{\rho}{\gamma}\right)^{\kappa}\left(\frac{1}{\kappa}\right)\sum_{i=1}^{n} a_i b_i c_i \\
&\quad -\left(\frac{1}{\gamma}\right)\left(\frac{\rho}{\gamma}\right)^{\kappa}\sum_{i=1}^{n} a_i \log(1-d_i) - \left(\frac{\kappa}{\gamma}\right)\left(\frac{\rho}{\gamma}\right)^{\kappa}\log\left(\frac{\rho}{\gamma}\right)\sum_{i=1}^{n} a_i \log(1-d_i)
\end{aligned}$$

$$\begin{aligned}
&= -\left(\frac{1}{\gamma}\right)\sum_{i=1}^{n}\delta_i d_i(1+c_i) - \left(\frac{1}{\gamma}\right)\left(\frac{\rho}{\gamma}\right)^{\kappa}\sum_{i=1}^{n} a_i[d_i + \log(1-d_i) + c_i(d_i - b_i)] \\
&\quad -\left(\frac{\kappa}{\gamma}\right)\left(\frac{\rho}{\gamma}\right)^{\kappa}\log\left(\frac{\rho}{\gamma}\right)\sum_{i=1}^{n} a_i[d_i + \log(1-d_i)].
\end{aligned}$$

$$\begin{aligned}
\frac{\partial^2 \ell(\boldsymbol{\theta})}{\partial \kappa \partial \rho} &= \frac{\partial}{\partial \kappa}\frac{\partial \ell(\boldsymbol{\theta})}{\partial \rho} \\
&= \frac{m}{\rho} - \left(\frac{1}{\rho}\right)\left(\frac{\rho}{\gamma}\right)^{\kappa}\sum_{i=1}^{n} a_i \log(1+b_i) - \left(\frac{\kappa}{\rho}\right)\left(\frac{\rho}{\gamma}\right)^{\kappa}\log\left(\frac{\rho}{\gamma}\right)\sum_{i=1}^{n} a_i \log(1+b_i) \\
&\quad -\left(\frac{\kappa}{\rho}\right)\left(\frac{\rho}{\gamma}\right)^{\kappa}\left(\frac{1}{\kappa}\right)\sum_{i=1}^{n} a_i b_i c_i \\
&= \frac{m}{\rho} - \left(\frac{1}{\rho}\right)\left(\frac{\rho}{\gamma}\right)^{\kappa}\sum_{i=1}^{n} a_i\left[b_i c_i + \log(1+b_i)\right] \\
&\quad -\left(\frac{\kappa}{\rho}\right)\left(\frac{\rho}{\gamma}\right)^{\kappa}\log\left(\frac{\rho}{\gamma}\right)\sum_{i=1}^{n} a_i \log(1+b_i).
\end{aligned}$$

$$\frac{\partial^2 \ell(\boldsymbol{\theta})}{\partial \kappa \partial \beta_j} = \frac{\partial}{\partial \beta_j}\frac{\partial \ell(\boldsymbol{\theta})}{\partial \kappa} = -\left(\frac{1}{\kappa}\right)\left(\frac{\rho}{\gamma}\right)^{\kappa}\sum_{i=1}^{n} a_i b_i c_i z_{ij} - \left(\frac{\rho}{\gamma}\right)^{\kappa}\log\left(\frac{\rho}{\gamma}\right)\sum_{i=1}^{n} a_i \log(1+b_i) z_{ij}.$$

$$\begin{aligned}
\frac{\partial^2 \ell(\boldsymbol{\theta})}{\partial \gamma \partial \rho} &= \frac{\partial}{\partial \gamma}\frac{\partial \ell(\boldsymbol{\theta})}{\partial \rho} \\
&= \left(\frac{\kappa}{\rho}\right)\left(\frac{\kappa}{\gamma}\right)\left(\frac{\rho}{\gamma}\right)^{\kappa}\sum_{i=1}^{n} a_i \log(1+b_i) - \left(\frac{\kappa}{\rho}\right)\left(\frac{\kappa}{\gamma}\right)\left(\frac{\rho}{\gamma}\right)^{\kappa}\sum_{i=1}^{n} a_i d_i \\
&= -\left(\frac{\kappa}{\rho}\right)\left(\frac{\kappa}{\gamma}\right)\left(\frac{\rho}{\gamma}\right)^{\kappa}\sum_{i=1}^{n} a_i \log(1-d_i) - \left(\frac{\kappa}{\rho}\right)\left(\frac{\kappa}{\gamma}\right)\left(\frac{\rho}{\gamma}\right)^{\kappa}\sum_{i=1}^{n} a_i d_i \\
&= -\left(\frac{\kappa}{\rho}\right)\left(\frac{\kappa}{\gamma}\right)\left(\frac{\rho}{\gamma}\right)^{\kappa}\sum_{i=1}^{n} a_i[d_i + \log(1-d_i)].
\end{aligned}$$

$$\frac{\partial^2 \ell(\boldsymbol{\theta})}{\partial \gamma \partial \beta_j} = \frac{\partial}{\partial \beta_j} \frac{\partial \ell(\boldsymbol{\theta})}{\partial \gamma}$$

$$= -\left(\frac{\kappa}{\gamma}\right)\left(\frac{\rho}{\gamma}\right)^{\kappa} \sum_{i=1}^{n} a_i d_i z_{ij} - \left(\frac{\kappa}{\gamma}\right)\left(\frac{\rho}{\gamma}\right)^{\kappa} \sum_{i=1}^{n} a_i \log(1-d_i) z_{ij}$$

$$= -\left(\frac{\kappa}{\gamma}\right)\left(\frac{\rho}{\gamma}\right)^{\kappa} \sum_{i=1}^{n} a_i [\, d_i + \log(1-d_i)]\, z_{ij}.$$

$$\frac{\partial^2 \ell(\boldsymbol{\theta})}{\partial \rho \partial \beta_j} = \frac{\partial}{\partial \beta_j} \frac{\partial \ell(\boldsymbol{\theta})}{\partial \rho}$$

$$= -\left(\frac{\kappa}{\rho}\right)\left(\frac{\rho}{\gamma}\right)^{\kappa} \sum_{i=1}^{n} a_i \log(1+b_i) z_{ij}.$$

The maximum likelihood estimate of $\boldsymbol{\theta}^* = (\boldsymbol{\alpha}^{*\prime}, \boldsymbol{\beta}')'$ is obtained by solving the equations $\partial \ell(\boldsymbol{\theta}^*)/\partial \kappa^* = 0$, $\partial \ell(\boldsymbol{\theta}^*)/\partial \gamma^* = 0$, $\partial \ell(\boldsymbol{\theta}^*)/\partial \rho^* = 0$ and $\partial \ell(\boldsymbol{\theta}^*)/\partial \beta_j = 0$ iteratively. The first and second derivatives of $\ell(\boldsymbol{\theta}^*)$ can be derived by noting that

$$\frac{\partial \ell}{\partial \log u} = u\left(\frac{\partial \ell}{\partial u}\right), \tag{28}$$

$$\frac{\partial^2 \ell}{\partial \log u\, \partial \log v} = u\left(\frac{\partial v}{\partial u}\right)\left(\frac{\partial \ell}{\partial v}\right) + uv\left(\frac{\partial^2 \ell}{\partial u \partial v}\right), \tag{29}$$

$$\frac{\partial^2 \ell}{\partial \log u\, \partial v} = u\left(\frac{\partial^2 \ell}{\partial u \partial v}\right). \tag{30}$$

Using (29)-(30), the first and second derivatives of $\ell(\boldsymbol{\theta}^*)$ can be expressed as

$$\frac{\partial \ell(\boldsymbol{\theta}^*)}{\partial \kappa^*} = \left[\kappa\left(\frac{\partial \ell(\boldsymbol{\theta})}{\partial \kappa}\right)\right]_{\alpha=e^{\alpha^*}}, \quad \frac{\partial \ell(\boldsymbol{\theta}^*)}{\partial \gamma^*} = \left[\gamma\left(\frac{\partial \ell(\boldsymbol{\theta})}{\partial \gamma}\right)\right]_{\alpha=e^{\alpha^*}},$$

$$\frac{\partial \ell(\boldsymbol{\theta}^*)}{\partial \rho^*} = \left[\rho\left(\frac{\partial \ell(\boldsymbol{\theta})}{\partial \rho}\right)\right]_{\alpha=e^{\alpha^*}}, \quad \frac{\partial \ell(\boldsymbol{\theta}^*)}{\partial \beta_j} = \left[\frac{\partial \ell(\boldsymbol{\theta})}{\partial \beta_j}\right]_{\alpha=e^{\alpha^*}},$$

$$\frac{\partial^2 \ell(\boldsymbol{\theta}^*)}{\partial \kappa^{*2}} = \left[\kappa\left(\frac{\partial \ell(\boldsymbol{\theta})}{\partial \kappa}\right) + \kappa^2\left(\frac{\partial^2 \ell(\boldsymbol{\theta})}{\partial \kappa^2}\right)\right]_{\alpha=e^{\alpha^*}},$$

$$\frac{\partial^2 \ell(\boldsymbol{\theta}^*)}{\partial \gamma^{*2}} = \left[\gamma\left(\frac{\partial \ell(\boldsymbol{\theta})}{\partial \gamma}\right) + \gamma^2\left(\frac{\partial^2 \ell(\boldsymbol{\theta})}{\partial \gamma^2}\right)\right]_{\alpha=e^{\alpha^*}},$$

$$\frac{\partial^2 \ell(\boldsymbol{\theta}^*)}{\partial \rho^{*2}} = \left[\rho\left(\frac{\partial \ell(\boldsymbol{\theta})}{\partial \rho}\right) + \rho^2\left(\frac{\partial^2 \ell(\boldsymbol{\theta})}{\partial \rho^2}\right)\right]_{\alpha=e^{\alpha^*}},$$

$$\frac{\partial^2 \ell(\boldsymbol{\theta}^*)}{\partial \beta_j \partial \beta_{j'}} = \left[\frac{\partial^2 \ell(\boldsymbol{\theta})}{\partial \beta_j \partial \beta_{j'}}\right]_{\alpha=e^{\alpha^*}}, \quad \frac{\partial^2 \ell(\boldsymbol{\theta}^*)}{\partial \kappa^* \partial \gamma^*} = \left[\kappa\gamma\left(\frac{\partial^2 \ell(\boldsymbol{\theta})}{\partial \kappa \partial \gamma}\right)\right]_{\alpha=e^{\alpha^*}},$$

$$\frac{\partial^2 \ell(\boldsymbol{\theta}^*)}{\partial \kappa^* \partial \rho^*} = \left[\kappa\rho\left(\frac{\partial^2 \ell(\boldsymbol{\theta})}{\partial \kappa \partial \rho}\right)\right]_{\alpha=e^{\alpha^*}}, \quad \frac{\partial^2 \ell(\boldsymbol{\theta}^*)}{\partial \kappa^* \partial \beta_j} = \left[\kappa\left(\frac{\partial^2 \ell(\boldsymbol{\theta})}{\partial \kappa \partial \beta_j}\right)\right]_{\alpha=e^{\alpha^*}},$$

$$\frac{\partial^2 \ell(\boldsymbol{\theta}^*)}{\partial \gamma^* \partial \rho^*} = \left[\gamma\rho\left(\frac{\partial^2 \ell(\boldsymbol{\theta})}{\partial \gamma \partial \rho}\right)\right]_{\alpha=e^{\alpha^*}}, \quad \frac{\partial^2 \ell(\boldsymbol{\theta}^*)}{\partial \gamma^* \partial \beta_j} = \left[\gamma\left(\frac{\partial^2 \ell(\boldsymbol{\theta})}{\partial \gamma \partial \beta_j}\right)\right]_{\alpha=e^{\alpha^*}},$$

$$\frac{\partial^2 \ell(\boldsymbol{\theta}^*)}{\partial \rho^* \partial \beta_j} = \left[\rho\left(\frac{\partial^2 \ell(\boldsymbol{\theta})}{\partial \rho \partial \beta_j}\right)\right]_{\alpha=e^{\alpha^*}}.$$

Acknowledgements
The authors acknowledge the comments and suggestions of the editor and the reviewers. This work was partially supported by NSERC through Discovery Grant (#368532) to SA Khan, and the University of Saskatchewan through New Faculty Start-up Operating Fund to SA Khan.

Authors' contributions
SAK, the principal investigator, conceptually developed the proposed distribution with related mathematical results and R code for computation, and drafted the manuscript. SKK contributed in developing mathematical results and writing Sections 1-3. Both authors read and approved the final manuscript.

Competing interests
The authors declare that they have no competing interests.

References
Akaike, H: A new look at the statistical model identification. IEEE Trans. Autom. Control. **19**, 716–723 (1974)
Burnham, KP, Anderson, DR: Model Selection and Multimodel Inference: A Practical Information-Theoretic Approach. Springer, New York (2002)
Collett, D: Modelling Survival Data in Medical Research. Chapman and Hall/CRC, Florida (2003)
Cox, DR: Regression models and life-tables. J. R. Stat. Soc. Ser. B. **34**, 187–220 (1972)
Efron, B: The efficiency of Cox's likelihood function for censored data. J. Am. Stat. Assoc. **72**, 557–565 (1977)
Efron, B: Logistic regression, survival analysis, and the kaplan-meier curve. J. Am. Stat. Assoc. **83**, 414–425 (1988)
Hsieh, F, Tseng, YK, Wang, JL: Joint modeling of survival and longitudinal data: Likelihood approach revisited. Biometrics. **62**, 1037–1043 (2006)
Hwang, BS, Pennell, ML: Semiparametric bayesian joint modeling of a binary and continuous outcome with applications in toxicological risk assessment. Stat. Med. **33**, 1162–1175 (2014)
Kalbfleisch, JD, Prentice, RL: The Statistical Analysis of Failure Time Data. John Wiley & Sons, New Jersey (2002)
Kaplan, EL, Meier, P: Nonparametric estimation from incomplete observations. J. Am. Stat. Assoc. **53**, 457–481 (1958)
Klein, JP, Moeschberger, ML: Survival Analysis: Techniques for Censored and Truncated Data. Springer, New York (2003)
Kleinbaum, DG, Klein, M: Survival Analysis: A Self-Learning Text. Springer, New York (2012)
Lawless, JF: Statistical Models and Methods for Lifetime Data. John Wiley & Sons, New Jersey (2002)
Mudholkar, GS, Srivastava, DK, Kollia, GD: A generalization of the weibull distribution with application to the analysis of survival data. J. Am. Stat. Assoc. **91**, 1575–1583 (1996)
Nardi, A, Schemper, M: Comparing Cox and parametric models in clinical studies. Stat. Med. **22**, 3597–3610 (2003)
Oakes, D: The asymptotic information in censored survival data. Biometrika. **64**, 441–448 (1977)
Pike, MC: A method of analysis of certain classes of experiments in carcinogenesis. Biometrics. **22**, 142–161 (1966)
R Core Team: R: A Language and Environment for Statistical Computing. Foundation for Statistical Computing, R. R Core Team, Vienna (2016)
Rizopoulos, D: Jm: An r package for the joint modelling of longitudinal and time-to-event data. J. Stat. Softw. **35**, 1–33 (2010)
Rizopoulos, D: Joint Models for Longitudinal and Time-to-Event Data With Applications in R. Chapman and Hall/CRC, Florida (2012)
Wang, Y, Hossain, AM, Zimmer, WJ: Useful properties of the three-parameter Burr XII distribution. In: Ahsanullah, M (ed.) Applied Statistics Research Progress, pp. 11–20. Nova Science Publishers, New York, (2008)
Wulfsohn, MS, Tsiatis, AA: A joint model for survival and longitudinal data measured with errror. Biometrics. **53**, 330–339 (1997)

11

Geometric disintegration and star-shaped distributions

Wolf-Dieter Richter

Correspondence:
wolf-dieter.richter@uni-rostock.de
Institute of Mathematics, University of Rostock, Ulmenstraße 69, Haus 3, 18057 Rostock, Germany

Abstract

Geometric and stochastic representations are derived for the big class of p-generalized elliptically contoured distributions, and (generalizing Cavalieris and Torricellis method of indivisibles in a non-Euclidean sense) a geometric disintegration method is established for deriving even more general star-shaped distributions. Applications to constructing non-concentric elliptically contoured and generalized von Mises distributions are presented.

Keywords: p-generalized elliptically contoured distributions; Non-concentric elliptically contoured distributions; Star-generalized von Mises distributions; p-generalized ellipsoids and ellipsoidal coordinates; Stochastic random vector representation; Geometric measure representation; Star-generalized surface measure; Star-generalized uniform distribution; Intersection-proportion function; Ball number function

AMS subject classification: Primary 60E05; 60D05; secondary 28A50; 28A75; 51F99

1 Introduction

The needs of statistical practice and challenging probabilistic questions in the interplay of measure theory and several other mathematical disciplines stimulate the development of statistical distribution theory. In part, well established mathematical strategies are followed to enlarge known families of distributions, and partly new types of distributions are derived by new methods.

Numerous studies on multivariate probability distributions with a view towards their statistical applications are closely connected with notions like decomposition or disintegration of probability laws, invariant measures on groups and related manifolds, and cross sections. Among the basic references in this field, we refer to (Barndorff-Nielsen et al. 1989; Eaton 1983, 1989; Farrell 1976, 1985; Koehn 1970; Muirhead 1982; Nachbin 1976; Wijsman 1967, 1986, 1990). In the spirit of those works, generalizations of elliptically contoured distributions where the contours are described by a positive function that is positive homogeneous or are arbitrary cross sections are discussed in (Balkema and Nolde 2010; Fernandez et al. 1995) and in (Kamiya et al. 2008; Takemura and Kuriki 1996) respectively.

Zonoid trimming for multivariate distributions is considered in (Koshevoy and Mosler 1997; Mosler 2002). In (Balkema et al. 2010; Balkema and Nolde 2010; Joenssen and Vogel

2012; Kinoshita and Resnick 1991; Mosler 2013; Nolde 2014) the authors describe the way in which star-shaped sets and, correspondingly, star-shaped distributions occur in limit set theory, in meta density analysis, power studies for goodness-of-fit tests, depth analysis and analysis of residual dependence between extreme values in the presence of asymptotic independence, respectively. Norm-contoured sets and related estimation problems of identifying structures in high-dimensional data sets are dealt with in (Scholz 2002).

The authors in (Fang et al. 1990; Kallenberg 2005; Schindler 2003) present studies of symmetric laws from different points of view. The class of $l_{n,p}$-symmetric distributions, $n \in \mathbb{N} = \{1, 2, \ldots\}$, $p \geq 1$, was introduced in (Osiewalski and Steel 1993) and studied in (Gupta and Song 1997; Song and Gupta 1997; Schechtman and Zinn 1990; Rachev and Rüschendorf 1991; Szablowski 1998). Applications of these distributions are discussed in (Nardon and Pianca 2009; Pogány and Nadarajah 2010).

Geometric measure representations for $l_{n,p}$-symmetric distributions, $n \in \mathbb{N}$, $p > 0$, and for heteroscedastic Gaussian distributions were derived in (Richter 2009, 2013), respectively.

An extension of the class of $l_{n,p}$-symmetric distributions to the class of skewed $l_{n,p}$-symmetric distributions has been derived in (Arellano-Valle and Richter 2012). A general approach to geometric representations of skewed $l_{n,2}$-symmetric distributions can be found for $n = 2$ in (Günzel et al. 2012) and for arbitrary n in (Richter and Venz 2014). Another definition of power exponential distributions than the one used here was given in (Gómez et al. 1998), where a special case of elliptically contoured distributions is dealt with. Densities of p-generalized elliptically contoured distributions and of more general star-shaped distributions have been considered in (Balkema and Nolde 2010; Fernandez et al. 1995).

In the present paper, geometric and stochastic representations are derived for the big class of p-generalized elliptically contoured distributions. Generalizing Cavalieri's and Torricelli's method of indivisibles in a non-Euclidean sense, a geometric disintegration method is established for deriving even more general star-shaped distributions. Basic properties of these distributions are studied, applications of the new representations to constructing non-concentric elliptically contoured distributions and to generalizing the von Mises distribution are discussed, and the necessary background from non-Euclidean metric geometry is developed.

Many authors use iterated integration in distribution theory by first integrating with respect to (w.r.t.) a radius variable and then w.r.t. certain directional coordinates. In the present paper, we shall use basically the inverse order of integration. This way, we shall make use of the star-sphere intersection-proportion function (ipf) of a given set. The ipf is essentially based upon a suitably defined non-Euclidean surface content on a star sphere. The latter notion needs therefore the most effort in the present work. Areas from probability theory and mathematical statistics where the ipf successfully applies are surveyed in (Richter 2012). Applying this function allows to study the contours of mass concentration of a probability distribution independently from the tail behavior of the distribution, and often leads to a numerical stabilization of the evaluation of probability integrals. Further, the ipf allows a non-Euclidean surface measure interpretation of certain sector measures considered in the literature.

The paper is organized as follows. After quoting some preliminary facts in Section 2, we deal with the notion of a star-generalized surface content in Section 3. This notion will be studied both based upon a local definition and in terms of an integral in Section 3.1. The latter definition makes use of a preliminary system of coordinates which moreover enables a generalization of the method of indivisibles. Then Section 3.2. deals exclusively with the new surface measure on p-generalized ellipsoids. After much technical work, Theorem 5 finally proves that the local approach to the star generalized surface content results in the same quantity as a suitably defined non-Euclidean surface content in terms of an integral defined using a modified standard approach of differential geometry. To this end, some more coordinate systems are introduced and exploited. This includes consideration of star-generalized trigonometric functions and several Jacobians. In Section 4, star-shaped distributions are introduced in several steps and some of their basic properties are studied. The most explicit results are presented for the class of p-generalized elliptically contoured distributions in Section 4.7. For this specific class, all of the more general results of the preceding parts of Section 4, including the main results in Theorems 7 and 8, allow an additional interpretation which in each case is based upon a suitable non-Euclidean geometry. Moreover, the ball number function will be extended in Section 4.5 and characteristic functions are discussed in Section 4.6. In the two-dimensional case, some consequences from the preceding sections concerning the new class of non-concentric elliptically contoured distributions and a star generalization of the von Mises distribution are drawn in Sections 5.1 and 5.2, respectively. The paper ends with some concluding remarks in Section 6, basically indicating some possible future work.

2 Preliminaries

The main considerations of this paper are most easily understood by making use of a relatively easy coordinate transformation. For showing the deeper meaning of several results derived this way, we shall make use, however, of different rather technical systems of coordinates which will be introduced in later sections. Here, we begin with some preliminary notions, including the preliminary coordinate system mentioned in the Introduction.

Throughout this paper, $K \subset \mathbb{R}^n$ denotes a star body, i.e. a nonempty star-shaped set that is compact and is equal to the closure of its interior, having the origin 0_n in its interior. Its topological boundary will be denoted by S. The functional $h_K : \mathbb{R}^n \to [0, \infty)$ defined by $h_K(x) = \inf\{\lambda > 0 : x \in \lambda K\}$, $x \in \mathbb{R}^n$ where $\lambda K = \{(\lambda x_1, \ldots, \lambda x_n)^T : (x_1, \ldots, x_n)^T \in K\}$ is known as the Minkowski functional of the star body K. We assume that h_K is positive-homogeneous of degree one, i.e. $h_K(\lambda x) = \lambda h_K(x), \lambda > 0$, which is the case if, e.g., h_K is a norm or an antinorm. For the latter notion, we refer to (Moszyńska and Richter 2012), and for the role which homogeneous functionals generally play in stochastics, we refer to (Hoffmann-Jorgensen 1994).

Let us consider $K(r) = rK = \{x \in \mathbb{R}^n : h_K(x) \leq r\}$ and its boundary $S(r) = rS$ as the star ball and star sphere of Minkowski radius or star radius $r > 0$, respectively. A countable collection $\mathfrak{F} = \{C_1, C_2, ...\}$ of pairwise disjoint cones C_j with vertex being the origin 0_n and $\mathbb{R}^n = \bigcup_j C_j$ will be called a fan. By $\mathfrak{B}_n$ we denote the Borel-σ-field in $\mathbb{R}^n$. We put $S_j = S \cap C_j$, $S_j \cap \mathfrak{B}_n = \mathfrak{B}_{S,j}$ and $\mathfrak{B}_S = \sigma\{\mathfrak{B}_{S,1}, \mathfrak{B}_{S,2}, ...\}$. We shall consider only star bodies K and sets $A \in \mathfrak{B}_S$ satisfying the following condition.

Assumption 1. *The star body K and the set $A \in \mathfrak{B}_S$ are chosen such that for every j the set*

$$G(A \cap S_j) = \left\{ \vartheta \in \mathbb{R}^{n-1} : \exists \eta \text{ with } \theta = \left(\vartheta^T, \eta\right)^T \in A \cap S_j \right\}$$

is well defined and such that for every $\vartheta = (\vartheta_1, \ldots, \vartheta_{n-1})^T \in G(A \cap S_j)$ there is a uniquely determined $\eta > 0$ satisfying $h_K\left((\vartheta_1, \ldots, \vartheta_{n-1}, \eta)^T\right) = 1$.

The latter quantity will be denoted by $\eta_j(\vartheta)$, $j = 1, 2, \ldots$.

For every $x \in \mathbb{R}^n, x \neq 0$ there are uniquely determined $r > 0$ and $\theta \in S$ such that $x = r\theta$. For $x \in rS_j$, we have $x = r\left(\vartheta^T, \eta_j(\vartheta)\right)^T$, and we will write $r\eta_j(\vartheta) = y_j(\vartheta)$. Consequently, $h_K(x) = h_K(r\theta) = rh_K\left(\left(\vartheta^T, \eta_j(\vartheta)\right)^T\right) = r$.

For $j = 1, 2, \ldots$ the star spherical coordinate transformation $StSph_j : [0, \infty) \times G(S_j) \to C_j$ is defined by $x_i = r\vartheta_i, i = 1, \ldots, n-1, x_n = y_j(\vartheta)$. The equations $r = h_K(x), \vartheta_i = x_i/r$, $i = 1, \ldots, n-1$ define a.e. uniquely the inverse map of $StSph_j$.

Note that if K is convex or an axes aligned p-generalized ellipsoid, $p > 0$, see Section 3.2.1, one may assume the sets $S \cap C_{1(2)}$ to be the upper and lower hemi-spheres, $S^{+(-)} = \left\{\theta = (\theta_1, \ldots, \theta_n)^T \text{with } \theta_n > (<) 0\right\}$, respectively.

Lemma 1. *The absolute value of the Jacobian of the star-spherical coordinate transformation is $J(r, \vartheta) = r^{n-1} J_j^*(\vartheta)$, with $J_j^*(\vartheta) = |\eta_j(\vartheta) - \sum_{i=1}^{n-1} \vartheta_i \frac{\partial}{\partial \vartheta_i} \eta_j(\vartheta)|$ for every $r > 0, \vartheta \in G(S_j), j = 1, 2, \ldots$*

Proof. The formula for $J(r, \vartheta) = |\frac{d(x_1, \ldots, x_n)}{d(r, \vartheta)}|$ given in the lemma can be checked immediately by determining all partial derivatives and evaluating the resulting determinant. □

The coordinate system introduced here will be the basis of our considerations in Section 3.1 dealing with a general local notion of surface content. A specific integral notion of surface content dealt with in Section 3.2.3 will make use of another system of coordinates which will be introduced in Section 3.2.2. For the comparison study of the two seemingly rather different two approaches to measuring surfaces in Section 3.2.4, we will consider again suitable coordinates.

An essential part of the message of Lemma 1 is that the Jacobian allows a factorization into a term not depending on the radius coordinate and one that is independent of the directional coordinates.

Later in this paper, the restriction of the star spherical coordinate transformation to the case $r = 1$ will be denoted by $StSph^*$.

3 The star-generalized surface measure

3.1 Basics

The results in (Richter 2009, 2013) reflect the basic role which a suitable notion of non-Euclidean surface content plays for the study of non-spherical distributions. Here, we give first formally a local definition of a generalized surface measure which allows us to derive geometric and stochastic representations of star-shaped distributions and correspondingly distributed random vectors, respectively. For a more advanced understanding of the notions and results, we refer to Remark 6 below.

For $A \in \mathfrak{B}_S$, we introduce the central projection cone $CPC(A) = \{x \in \mathbb{R}^n : x/h_K(x) \in A\}$ and the star sector of star radius ϱ, $sector(A,\varrho) = CPC(A) \cap K(\varrho)$. We are now ready to introduce the first basic notion of this paper. To this end, let μ be the Lebesgue measure in $\mathbb{R}^n$.

Definition 1. *The star-generalized surface measure is defined on* $\varrho \cdot \mathfrak{B}_S$ *by* $\mathfrak{O}_S(A) = f'(\varrho)$ *where* $f(\varrho) = \mu(sector(A,\varrho))$.

If K is the Euclidean unit ball, and thus S is the Euclidean unit sphere, then $\mathfrak{O}_S$ equals the usual Euclidean surface content measure. The equation in Definition 1 should be well known for this case, but, astonishing enough, numerous authors do not make this very clear to their readers.

In contrast to the usual differential geometric definition of the notion of surface content in terms of an integral, the approach in Definition 1 is based upon a derivative. The equation

$$\mu(sector(A,R)) = \int_0^R \mathfrak{O}_S(rA)dr, A \in \mathfrak{B}_S \tag{1}$$

is an immediate consequence of the fundamental theorem of calculus and might seem therefore to be of no special interest, here. If, however, non-trivial explanations for $\mathfrak{O}_S$ are available as in (Richter 2009, 2013) where K is an $l_{n,p}$–ball or an ellipsoidal ball, respectively, then things change noticably. In both cases, a particular non-Euclidean geometry was identified such that the correspondingly modified integral notion of surface content based upon this non-Euclidean geometry coincides with the locally defined surface measure $\mathfrak{O}_S$. This allows a non-Euclidean interpretable extension of Cavalieri's and Torricelli's method of indivisibles, see (Richter 1985, 2009). Later in this paper, we shall observe this for a bigger class of star bodies. Moreover, we remark that $\mathfrak{O}_S(A) = n\mu(sector(A,1)), \forall A \in \mathfrak{B}_S$, meaning much more than just $\mathfrak{O}_S(S) = n\mu(K)$.

Theorem 1. *For sets* $A \in \mathfrak{B}_S$ *satisfying Assumption 1 in Section 2, the star-generalized surface measure allows the representation*

$$\mathfrak{O}_S(A) = \sum_j \int_{G(A\cap S_j)} J_j^*(\vartheta)d\vartheta.$$

Proof. Using star-spherical coordinates, and that $G(A \cap S_j) = StSph^{*-1}(A \cap S_j)$, we get according to Lemma 1

$$\mu(sector(A,\varrho)) = \int_{sector(A,\varrho)} dx = \int_0^\varrho \sum_j \int_{StSph^{*-1}(A\cap S_j)} r^{n-1}J_j^*(\vartheta)d(\vartheta,r).$$

Definition 1 applies. □

Remark 1. *(a) With the notations* $\mathfrak{O}_S(A) = \int_A \mathfrak{O}_S(d\theta)$, *and*

$$\sum_j \int_{G(A\cap S_j)} J_j^*(\vartheta)d\vartheta = \int_{G(A)} J^*(\vartheta)d\vartheta,$$

an alternative expression of Theorem 1 is

$$\int_A \mathfrak{O}_S(d\theta) = \int_{G(A)} J^*(\vartheta)d\vartheta.$$

If f is integrable then we write $\int_A f(\theta)\mathfrak{O}_S(d\theta) = \int_{G(A)} f\left(\left(\vartheta^T, \eta(\vartheta)\right)^T\right) J^*(\vartheta)d\vartheta.$

(b) The sector measure on $\mathfrak{B}_S$, *i.e. the measure* $sm_K(A) = \frac{\mu(sector(A,1))}{\mu(K)}$, *satisfies the representation* $sm_K(A) = \frac{\mathfrak{O}_S(A)}{\mathfrak{O}_S(S)}, A \in \mathfrak{B}_S$.

(c) A class of examples where Theorem 1 applies is given by all star bodies K corresponding to norms or antinorms for which there exist countably many pairwise disjoint sets A_j *satisfying Assumption 1 and* $S = \bigcup_j A_j$.

The following consequence of Theorem 1 follows using Fubini's theorem and can be read in the special case $f = 1$ as a disintegration formula for the Lebesgue measure. For a certain survey of such formulas, see (Richter 2012). These representations may also be considered as closely connected with a generalized method of indivisibles with the latter being defined as the intersections of a Borel set B with the star spheres $S(r), r > 0$. Constructions of such type are called cross sections by several authors, see (Eaton 1983; Farrell 1976, 1985; Koehn 1970; Wijsman 1967, 1986, 1990) and (Takemura and Kuriki 1996).

Corollary 1. *Let the star body K satisfy Assumption 1. Then*

(a) For $B \in \mathfrak{B}_n$ *and integrable f,* $\int_B f(x)dx = \int_0^\infty \left[r^{n-1} \int_{[\frac{1}{r}B]\cap S} f(r\theta)\mathfrak{O}_S(d\theta) \right] dr.$

(b) For bounded measurable B, $\int_B dx = \int_0^\infty \mathfrak{O}_S(B \cap S(r))dr.$

Proof. Changing from Cartesian to star spherical coordinates yields

$$\int_B f(x)dx = \int_0^\infty \left[r^{n-1} \sum_j \int_{G([\frac{1}{r}B]\cap S_j)} f(StSph_j(r,\vartheta))J_j^*(\vartheta)d\vartheta \right] dr$$

$$= \int_0^\infty \left[r^{n-1} \sum_j \int_{[\frac{1}{r}B]\cap S_j} f(r\theta)\mathfrak{O}_S(d\theta) \right] dr.$$

The rest follows with $f = 1$ and the notation in Remark 1 □

Corollary 1 may be rewritten using the following second basic notion of this paper.

Definition 2. *The star sphere intersection-proportion function (ipf) of the set* $B \in \mathfrak{B}_n$ *is defined as* $\mathfrak{F}_S(B, r) = \mathfrak{O}_S\left(\left[\frac{1}{r}B\right] \cap S\right) / \mathfrak{O}_S(S)$, $r > 0$.

The ipf was first introduced in (Richter 1985, 1987, 1991) for Gaussian and spherical distributions, respectively, i.e. for cases where S is the Euclidean unit sphere, and generalized later in (Richter 2007) to the case that S is an $l_{n,p}-$sphere. Moreover, the ipf corresponding to an asymmetric sphere S was considered for the case that K is the shifted positive part of an $l_{n,1}-$ ball, i.e. a simplex, and for the case that K is a, possibly asymmetric, polygon or Platonic body, respectively.

Corollary 2. *If the conditions of Corollary 1(b) are satisfied,*

$$\int\limits_B dx = \mathfrak{O}_S(S)\int\limits_0^\infty r^{n-1}\mathfrak{F}_S(B,r)dr.$$

Proof. It follows from Corollary 1 that

$$\int\limits_B dx = \int\limits_0^\infty r^{n-1}\mathfrak{O}_S\left(\left[\frac{1}{r}B\right]\cap S\right)dr.$$

The rest follows by Definition 2. □

Remark 2. *According to Remark 1(b) and Definition 2, the ipf allows the sector measure interpretation* $\mathfrak{F}_S(B,r) = sm_K\left(\left[\frac{1}{r}B\right]\cap S\right),\ r>0.$

Whether one prefers the interpretation of the ipf according to the definition of the sector measure sm_K in terms of volumes or according to Definition 2 in terms of star-generalized surface contents may depend on several aspects. The authors in (Barthe et al. 2003; Naor 2007; Schechtman and Zinn 1990) use the notion of cone measure in similar situations.

As already mentioned in the first part of the present section, one is naturally interested in a fully differential geometric explanation of the star-generalized surface measure $\mathfrak{O}_S$ in terms of an integral. Such an explanation will be given in Section 3.2.3 when K is an element of a class of generalized ellipsoids which are star-shaped but not necessarily convex.

3.2 The star-generalized surface content of *p*-generalized ellipsoids

3.2.1 Volumes of p-generalized ellipsoids

Because the notion of the star-generalized surface content is derived from that of volumes, we first study volumes of p-generalized ellipsoids in this section. To this end, let $\mathfrak{b} = \{\mathfrak{b}_1,\ldots,\mathfrak{b}_n\}$ be any orthonormal basis (onb) in $\mathbb{R}^n$ and put $x = \sum\limits_{i=1}^n \xi_i\mathfrak{b}_i$ for $x\in\mathbb{R}^n$. Moreover, let $a=(a_1,\ldots,a_n)^T$ be an arbitrary vector having positive components, p a positive real number, $|.|_{a,p}:\mathbb{R}^n\to[0,\infty)$ the function defined by $|x|_{a,p} = \left(\sum\limits_1^n|\frac{\xi_i}{a_i}|^p\right)^{1/p}, x\in\mathbb{R}^n$ and $B_{a,p}=\left\{x\in\mathbb{R}^n : |x|_{a,p}\le 1\right\}$ the corresponding unit ball w.r.t. $\mathfrak{b}$. Its topological boundary $E_{a,p}$ is a generalized ellipsoid having form parameter p and main axes being aligned with the coordinate axes and having lengths $2a_i, i=1,\ldots,n$. One may consider $E_{a,p}$ also as a sphere w.r.t. the function $|.|_{a,p}$ which is a norm if $p\ge 1$ and an antinorm if $0<p<1$.

The $\mathfrak{b}_i$-axis may be interpreted in the sense of main axis from principal component analysis. For a discussion of these notions in connection with that of correlation, we refer to (Dietrich et al. 2013). The set $B_{a,p}(R) = RB_{a,p}$ will be called a p-generalized ellipsoidal ball, or simply p-generalized ellipsoid, of $|.|_{a,p}$-radius R, $R > 0$, and w.r.t. the basis $\mathfrak{b}$.

The evaluation of the volume of $B_{a,p}(R)$ may be immediately reduced to that of an $l_{n,p}$-ball having a suitable p-radius. To this end, we denote the $l_{n,p}$-ball of p-radius R by $K_{n,p}(R) = RK_{n,p}$ where $K_{n,p} = B_{\mathbb{1},p}$, $\mathbb{1} = (1, \ldots, 1)^T \in \mathbb{R}^n$, and its topological boundary, the $l_{n,p}$–sphere of p-radius R, by $S_{n,p}(R) = RS_{n,p}$. Moreover, we put $a_i^* = \prod\limits_{j=1, j\neq i}^{n} a_j$, $i = 1, \ldots, n$ and let $diag\left(a_1^*, \ldots, a_n^*\right)$ denote a diagonal $n \times n$-matrix whose diagonal entries are $a_2 \cdot \ldots \cdot a_n, \ \ldots, a_1 \cdot \ldots \cdot a_{n-1}$, respectively. If $\mathfrak{b}$ is the standard onb in $\mathbb{R}^n$, then $diag\left(a_1^*, \ldots, a_n^*\right) B_{a,p}(R) = K_{n,p}(a_1 ... a_n R)$. Changing variables $u = diag\left(a_1^*, \ldots, a_n^*\right) x$ in the integral $\mu(B_{a,p}(R)) = \int\limits_{\{x \in R^n: |x|_{a,p} \le R\}} dx$ gives

$$\mu(B_{a,p}(R)) = \int\limits_{K_{n,p}(Ra_1 \ldots a_n)} \frac{du}{(a_1 \ldots a_n)^{n-1}}.$$

Hence, $\mu(B_{a,p}(R)) = \frac{\mu(K_{n,p}(Ra_1 ... a_n))}{(a_1 ... a_n)^{n-1}} = a_1 \ldots a_n \frac{\omega_{n,p}}{n} R^n$ where, in accordance with (Richter 2009), $\omega_{n,p} = \frac{2^n\left(\Gamma(\frac{1}{p})\right)^n}{p^{n-1}\Gamma(\frac{n}{p})} = O_{p,q}(S_{n,p})$ with $\frac{1}{p} + \frac{1}{q} = 1$ is the $l_{n,q}$-surface content of the $l_{n,p}$–unit sphere $S_{n,p}$. The following theorem has thus been proved.

Theorem 2. *The p-generalized ellipsoid of* $|.|_{a,p}$*-radius R has the volume*

$$\mu(B_{a,p}(R)) = a_1 \ldots a_n \frac{\omega_{n,p}}{n} R^n.$$

Corollary 3. *The star-generalized surface content of a p-generalized ellipsoid with axes of lengths* $2a_i, i = 1, \ldots, n$ *is* $\mathfrak{O}_S(E_{a,p}) = a_1 ... a_n \omega_{n,p}$.

This corollary is an immediate consequence of Definition 1.

Notice that this formula for the star-generalized surface content of $E_{a,p}$ proves that the parameters p and a have separate influence on the result. Moreover, it makes no use of elliptic integrals, whereas the Euclidean surface content of $E_{a,p}$ does.

Similarly, as Equation (1), the equation

$$\mu(B_{a,p}(R)) = \int\limits_0^R \mathfrak{O}_S(E_{a,p}(r))dr \tag{2}$$

where $E_{a,p}(r) = rE_{a,p}$, might seem to be of no special interest, at this stage of our study. We shall show, however, later in this paper that $\mathfrak{O}_S$ allows a non-trivial interpretation as the surface measure w.r.t. a well defined, non-Euclidean, metric geometry. This allows us to re-define $\mathfrak{O}_S$ in a well established differential geometric approach. This will be done in the next but one section. We shall make use of a specific coordinate system which will be defined in the next section.

Following the notation in (Richter 2013), we will call the star-generalized surface measure alternatively the $E_{a,p}$-generalized surface measure if K is a p-generalized ellipsoid with axes of lengths $2a_i, i = 1, \ldots, n$.

3.2.2 The p-generalized ellipsoidal coordinates

We recall that $l_{n,p}$-generalized and ellipsoidal generalized trigonometric functions and coordinates have been shown in (Richter 2007, 2009) and (Richter 2011b, 2013) to be powerful tools for studying $l_{n,p}$-symmetric and elliptically contoured distributions, respectively. The coordinates which we define in this section are in some sense combinations and generalizations of the aforementioned ones. They will be used in Section 3.2.4 for showing the equivalence of two approaches to the star-generalized surface measure $\mathfrak{O}_S$: the local one presented already in Definition 1, and an integral one which will be given later.

Let us assume for a moment that $n = 2$ and $(x, y)^T = x\mathfrak{b}_1 + y\mathfrak{b}_2$ where $\{\mathfrak{b}_1, \mathfrak{b}_2\}$ is an onb in $\mathbb{R}^2$. In the following definition, ϕ can be interpreted as the angle between $pos(x, 0)$ and $pos(x, y)$ where $pos(\xi, \eta) = \{(\gamma\xi, \gamma\eta)^T : \gamma > 0\}$.

Definition 3. *The $E_{a,b;p}$-generalized trigonometric functions are defined as*

$$\cos_{(a,b;p)}(\phi) = \frac{(\cos\phi)/a}{N_{a,b;p}(\phi)} \text{ and } \sin_{(a,b;p)}(\phi) = \frac{(\sin\phi)/b}{N_{a,b;p}(\phi)},\ \phi \in [0, 2\pi)$$

for positive a, b, p and where $N_{a,b;p}(\phi) = (|(\cos\phi)/a|^p + |(\sin\phi)/b|^p)^{1/p}$.

Remark 3. *These generalized trigonometric functions may be extended to functions on the whole real line with period 2π. Basic analytical and geometric interpretations of these functions follow from the representations*

$$\cos_{(a,b;p)}(\phi) = \frac{x/a}{(|x/a|^p + |y/b|^p)^{1/p}},\ |\cos_{(a,b;p)}(\phi)| = \frac{|(\cos_{(a,b)}(\phi), 0)|_p}{|(\cos_{(a,b)}(\phi), \sin_{(a,b)}(\phi))|_p},$$

and

$$\sin_{(a,b;p)}(\phi) = \frac{y/b}{(|x/a|^p + |y/b|^p)^{1/p}},\ |\sin_{(a,b;p)}(\phi)| = \frac{|(0, \sin_{(a,b)}(\phi))|_p}{|(\cos_{(a,b)}(\phi), \sin_{(a,b)}(\phi))|_p},$$

where $|.|_p = |.|_{\mathbb{1},p}$ and $\sin_{(a,b)}$ and $\cos_{(a,b)}$ are defined in (Richter 2011b).

Remark 4. *Euler's formula is generalized by* $|\cos_{(a,b;p)}(\phi)|^p + |\sin_{(a,b;p)}(\phi)|^p = 1$.

Remark 5. *For every ϕ,*

$$\cos'_{(a,b;p)}(\phi) = -\frac{\sin_{a,b;p}(\phi)|\sin_{a,b;p}(\phi)|^{p-2}}{abN^2_{(a,b;p)}(\phi)},\ \sin'_{(a,b;p)}(\phi) = \frac{\cos_{a,b;p}(\phi)|\cos_{a,b;p}(\phi)|^{p-2}}{abN^2_{(a,b;p)}(\phi)}.$$

We assume again that $x = x_1\mathfrak{b}_1 + ... + x_n\mathfrak{b}_n$, $x \in \mathbb{R}^n$.

Definition 4. *The p-generalized ellipsoidal coordinate transformation $T^E_{a,p} = T^E_{a,p}(n)$, $T^E_{a,p} : M_n \to \mathbb{R}^n$, with $M_n = [0, \infty) \times M^*_n, M^*_n = [0, \pi)^{\times(n-2)} \times [0, 2\pi)$ is defined by*

$$x_1 = a_1 r \cos_{(a_1,a_2;p)}(\phi_1), x_2 = a_2 r \sin_{(a_1,a_2;p)}(\phi_1)\cos_{(a_2,a_3;p)}(\phi_2), \ldots,$$

$$x_{n-1} = a_{n-1} r \sin_{(a_1,a_2;p)}(\phi_1) ... \sin_{(a_{n-2},a_{n-1};p)}(\phi_{n-2})\cos_{(a_{n-1},a_n;p)}(\phi_{n-1}),$$

$$x_n = a_n r \sin_{(a_1,a_2;p)}(\phi_1) ... \sin_{(a_{n-2},a_{n-1};p)}(\phi_{n-2})\sin_{(a_{n-1},a_n;p)}(\phi_{n-1}).$$

Theorem 3. *The map $T^E_{a,p}$ is almost one-to-one, its inverse is a.e. given by*

$$r = \left(\sum_1^n \left(\left|\frac{x_i}{a_i}\right|\right)^p\right)^{1/p}, \phi_j = \arccos_{(a_j,a_{j+1};p)}\left(\frac{x_j/a_j}{\left(\sum_{i=j}^n |x_i/a_i|^p\right)^{1/p}}\right), \textit{and, if } x_{n-1} \neq 0,$$

$\arctan|\frac{x_n}{x_{n-1}}| = \phi_{n-1}$ *if* $(x_{n-1}, x_n) \in Q_1$, $= \pi - \phi_{n-1}$ *in* Q_2, $= -\pi + \phi_{n-1}$ *in* Q_3, *and* $= 2\pi - \phi_{n-1}$ *in* Q_4.

Here, $\arccos_{(a_j,a_{j+1};p)}$ *denotes the function inverse to* $\cos_{(a_j,a_{j+1};p)}$ *and* Q_1 *up to* Q_4 *denote anti-clockwise enumerated quadrants from* $\mathbb{R}^2$.

Proof. The proof of this theorem is quite similar to that of Theorem 1 in (Richter 2007) and is therefore omitted. □

Theorem 4. *The Jacobian of the coordinate transformation $T^E_{a,p}$ is*

$$J\left(T^E_{a,p}\right)(r, \phi_1, \ldots, \phi_{n-1}) = r^{n-1} J^*\left(T^E_{a,p}\right)(\phi_1, \ldots, \phi_{n-1}),$$

$$J^*\left(T^E_{a,p}\right)(\phi_1, \ldots, \phi_{n-1}) = \frac{1}{a_2 \cdot \ldots \cdot a_{n-1}} \prod_{i=1}^{n-1} \frac{\left(\sin_{(a_i,a_{i+1};p)}(\phi_i)\right)^{n-1-i}}{N^2_{(a_i,a_{i+1};p)}(\phi_i)}.$$

Proof. The proof will be given in four steps. First, we change variables $\frac{x_i}{a_i} = y_i$, $i = 1, \ldots, n$. The Jacobian of this transformation is $\left|\frac{D(x_1,\ldots,x_n)}{D(y_1,\ldots,y_n)}\right| = a_1 \cdot \ldots \cdot a_n$.

Next, we change variables $y_1 = \tilde{r}\mu_1, y_2 = \tilde{r}\left(1 - |\mu_1|^p\right)^{1/p} \mu_2, \ldots,$

$$y_{n-1} = \tilde{r}\left(1 - |\mu_1|^p\right)^{1/p} \cdot \ldots \cdot \left(1 - |\mu_{n-2}|^p\right)^{1/p} \mu_{n-1},$$

$$y_n = +(-)\tilde{r}\left(1 - |\mu_1|^p\right)^{1/p} \cdot \ldots \cdot \left(1 - |\mu_{n-2}|^p\right)^{1/p}\left(1 - |\mu_{n-1}|^p\right)^{1/p}.$$

As it was shown in the proof of Theorem 2 in the afore mentioned paper, the Jacobian of this transformation is $\left|\frac{D(y_1,\ldots,y_n)}{D(\tilde{r},\mu_1,\ldots,\mu_{n-1})}\right| = \tilde{r}^{n-1} \prod_{i=1}^{n-1} \left(1 - |\mu_i|^p\right)^{(n-p-i)/p}$.

Third, we change variables $\tilde{r} = r, \mu_i = \cos_{(a_i,a_{i+1};p)}(\phi_i), i = 1, \ldots, n-1$. The Jacobian of this transformation is

$$\left|\frac{D(\tilde{r}, \mu_1, \ldots, \mu_{n-1})}{D(r, \phi_1, \ldots, \phi_{n-1})}\right| = \left|\det diag\left(1, \frac{d}{d\phi_1}\cos_{(a_1,a_2;p)}(\phi_1), \ldots, \frac{d}{d\phi_{n-1}}\cos_{(a_{n-1},a_n;p)}(\phi_{n-1})\right)\right|.$$

It follows from the properties of the $E_{a,b;p}$-generalized trigonometric functions that

$$\left|\frac{D(\tilde{r}, \mu_1, \ldots, \mu_{n-1})}{D(r, \phi_1, \ldots, \phi_{n-1})}\right| = \prod_{i=1}^{n-1} \frac{\sin_{(a_i,a_{i+1};p)}(\phi_i)|\sin_{(a_i,a_{i+1};p)}(\phi_i)|^{p-2}}{a_i a_{i+1} N^2_{(a_i,a_{i+1};p)}(\phi_i)}.$$

On combining all three transformations, we get finally

$$J\left(T^E_{a,p}\right)(r, \phi_1, \ldots, \phi_{n-1}) = a_1 \ldots a_n \cdot r^{n-1} \prod_{i=1}^{n-1} \frac{\left(\sin_{(a_i,a_{i+1};p)}(\phi_i)\right)^{d-i-1}}{a_i a_{i+1} N^2_{(a_i,a_{i+1};p)}(\phi_i)}.$$

□

Corollary 4. *If* $n = 2$ *then* $J\left(T^E_{a,p}\right)(r, \phi) = \frac{r}{N^2_{(a_1,a_2;p)}(\phi)}$, *and if* $n = 3$ *then* $J\left(T^E_{a,p}\right)(r, \phi_1, \phi_2) = \frac{r^2 \sin \phi_1}{a_2^2 N^2_{(a_2,a_3;p)}(\phi_2) N^3_{(a_1,a_2;p)}(\phi_1)}$.

For the corresponding results in the case $p = 2$, we refer to (Richter 2011b, 2013).

3.2.3 Integral approach to the star-generalized surface measure on p-generalized ellipsoids

Let us recall that measuring the Euclidean surface content of $E_{a,p}(R)$ necessarily involves certain elliptic integrals. In this paper, however, we make use of a non-Euclidean definition of surface content which avoids such integrals. To this end, we shall consider the ellipsoid $E_{a,p}(R)$ as a subset of the generalized Minkowski space $\left(\mathbb{R}^n, |.|_{\frac{1}{a},q}\right)$ where $\frac{1}{a} = \left(\frac{1}{a_1}, \ldots, \frac{1}{a_n}\right)^T$ and p and q are connected with each other by the equation $\frac{1}{p} + \frac{1}{q} = 1$. We will introduce now the notion of the $|.|_{\frac{1}{a},q}$-surface content of $E_{a,p}(R)$ in a similar way as the notion of the $l_{n,q}$-surface content was introduced in (Richter 2009) for $l_{n,p}$-spheres. Notice that effects coming from scaling axes with the help of the parameter vector a and effects being due to the form parameter p are dealt with here in a separate way when introducing the function $|.|_{\frac{1}{a},q}$.

Let $\mathfrak{b}$ be the standard onb, and let y be defined as the positive solution of $\sum_{i=1}^{n-1} |x_i/a_i|^p + |y/a_n|^p = R^p$ where $\sum_{i=1}^{n-1} |x_i/a_i|^p < R^p$. At $(x_1, \ldots, x_{n-1}, y)^T$, $y > 0$, the vector normal to the upper half $E^+_{a,p}(R)$ of the ellipsoid $E_{a,p}(R)$ is $N(x_1, \ldots, x_{n-1}) = (-1)^n \left(\sum_{i=1}^{n-1} \frac{\partial y}{\partial x_i} e_i - e_n\right)$. Since it always will be clear how to deal with the case $y < 0$, we will not further mention this case.

Definition 5. *Let $A \subset E^+_{a,p}(R) \cap \mathfrak{B}_n$. The integral (or $|.|_{\frac{1}{a},q}$-) surface content of the set A is defined by $O_{a,p,q}(A) = \int_{G(A)} |N(x_1, \ldots, x_{n-1})|_{\frac{1}{a},q}\, dx_1 \ldots dx_{n-1}$ where $G(A) = \left\{ (x_1, \ldots, x_{n-1})^T : (x_1, \ldots, x_n)^T \in A \right\}$.*

Later in this paper, this definition will be called the integral approach to the notion of star-generalized surface content. This will be justified in the next section. Let us mention that if $a = \mathbb{1}$ then the surface measure $O_{a,p,q}$ which is based upon the geometry of the ellipsoid $E_{\frac{1}{a},q}$ equals the surface measure $O_{p,q}$ in (Richter 2009) being based upon the geometry of the $l_{n,q}$-ball $K_{n,q}$ which is dual to $K_{n,p}$. Two-dimensional special cases of Definition 5 were dealt with, e.g., in (Richter 2011a, 2011b) for arbitrary star discs and ellipses, respectively.

Lemma 2. *The $|.|_{\frac{1}{a},q}$-surface content of the whole generalized ellipsoid $E_{a,p}(R)$ of $|.|_{a,p}$-radius R is $O_{a,p,q}\left(E_{a,p}(R)\right) = a_1 \ldots a_n \omega_{n,p} R^{n-1}$.*

Proof. It follows from $\frac{\partial y}{\partial x_j} = -\frac{a_n |x_j|^{p-1}}{a_j^p \left(R^p - \sum_{i=1}^{n-1} (|x_i/a_i|)^p\right)^{(p-1)/p}}$, $j = 1, \ldots, n-1$

and with $q = p/(p-1)$ that

$$|N(x_1, \ldots, x_{n-1})|^q_{\frac{1}{a},q} = \sum_{j=1}^{n-1} \frac{a_n^q |x_j/a_j|^p}{R^p - \sum_{i=1}^{n-1} |x_i/a_i|^p)} + a_n^q = \frac{a_n^q R^p}{R^p - \sum_{i=1}^{n-1} |x_i/a_i|^p}.$$

Hence, because of symmetry,

$$O_{a,p,q}(E_{a,p}(R)) = 2a_n R^{p-1} \int\limits_{G(E^+_{a,p}(R))} \frac{d(x_1...x_{n-1})}{\left(R^p - \sum\limits_{i=1}^{n-1} |x_i/a_i|^p\right)^{(p-1)/p}}.$$

For suitably transforming this integral, we shall introduce now another system of coordinates. Let the p-generalized $(n-1)$-dimensional standard elliptical coordinate transformation

$$T_{a,p} : M_{n-1} \to \mathbb{R}^{n-1}, M_{n-1} = [0,\infty) \times M^*_{n-1}, M^*_{n-1} = [0,\pi)^{\times(n-3)} \times [0,2\pi)$$

be defined by

$$x_1 = a_1 r \cos_p(\phi_1), x_2 = a_2 r \sin_p(\phi_1)\cos_p(\phi_2), \ldots,$$
$$x_{n-2} = a_{n-2} r \sin_p(\phi_1)\ldots\sin_p(\phi_{n-3})\cos_p(\phi_{n-2}),$$
$$x_{n-1} = a_{n-1} r \sin_p(\phi_1)\ldots\sin_p(\phi_{n-3})\sin_p(\phi_{n-2})$$

where the p-generalized trigonometric functions $\sin_p$ and $\cos_p$ are defined in (Richter 2007). If $a = 1\!\!1 \in \mathbb{R}^{n-1}$ then this transformation coincides with the $l_{n-1,p}$-spherical coordinate transformation $SPH_p^{(n-1)}$, the Jacobian of which is given in (Richter 2007). If we write $J(T)$ for the Jacobian of a coordinate transformation T then $J(T_{a,p})(r,\phi) = \left|\frac{d(x_1,\ldots,x_{n-1})}{d(r,\phi_1,\ldots,\phi_{n-2})}\right| = a_1 \cdot \ldots \cdot a_{n-1} J\left(SPH_p^{(n-1)}\right)(r,\phi)$.

Moreover, let $J^*\left(SPH_p^{(n-1)}\right)(\phi) = J\left(SPH_p^{(n-1)}\right)(1,\phi)$ be the restriction of the Jacobian of $SPH_p^{(n-1)}$ to the sphere defined by $r = 1$.

Changing from Cartesian to p-generalized standard elliptical coordinates gives

$$O_{a,p,q}(E_{a,p}(R)) = 2a_1 \ldots a_n R^{p-1} \int\limits_0^R \frac{r^{n-2}}{(R^p - r^p)^{(p-1)/p}} dr$$

$$\times \int\limits_0^\pi \ldots \int\limits_0^\pi \int\limits_0^{2\pi} J^*\left(SPH_p^{(n-1)}\right)(\phi_1,\ldots,\phi_{n-2}) d\phi_{n-2} \ldots d\phi_1.$$

Because of

$$\int\limits_0^R \frac{r^{n-2} dr}{(R^p - r^p)^{(p-1)/p}} = R^{n-p} \int\limits_0^1 \frac{t^{n-2} dt}{(1-t^p)^{(p-1)/p}} = \frac{1}{p} R^{n-p} B\left(\frac{1}{p}, \frac{n-1}{p}\right)$$

and $\frac{1}{p}B\left(\frac{1}{p},\frac{n-1}{p}\right)\omega_{n-1,p} = \frac{1}{2}\omega_{n,p}$, it follows that $O_{a,p,q}(E_{a,p}(R)) = a_1 \ldots a_n \omega_{n,p} R^{n-1}$ □

Hence, for the specific sets $E_{a,b}(R)$, the local and the integral approaches to the star-generalized surface content lead to the same result. In the next section, we will generalize this result. When doing this, we will again make use of a modified coordinate system.

3.2.4 Comparing the local and integral approaches to generalized surface measures on p-generalized ellipsoids

In Section 3.2.3, the surface measure $O_{a,p,q}$ was used for measuring the whole p-generalized ellipsoid $E_{a,p}$ following a differential geometric, or integral or global approach. In the present section, however, we compare it for arbitrary $A \in \mathfrak{B}^E_{a,p} = \mathfrak{B}^n \cap E_{a,p}$ with the alternative local approach which makes use of derivatives and which was introduced in Definition 1. In this sense, we continue to follow the general method of analyzing the non-Euclidean geometry underlying a multivariate probability distribution which was developed in (Richter 2009, 2013). The following theorem says that the star-generalized surface measure coincides with the integral surface measure. For a comparison of these surface measures, it is sufficient to consider them for sets $A \in \mathfrak{B}^E_{a,p}$.

Theorem 5. *With* $S = E_{a,p}$ *and* $1/p + 1/q = 1$, $\mathfrak{O}_S(A) = O_{a,p,q}(A), \forall A \in \mathfrak{B}^E_{a,p}$.

Proof. W.l.o.g., we restrict our consideration to sets $A \in E^+_{a,p} \cap \mathfrak{B}^n$ and start from a slight generalization of the first result in the proof of Lemma 2,

$$O_{a,p,q}(A) = a_n \int\limits_{G(A)} \frac{d(x_1, \ldots, x_{n-1})}{\left(1 - \sum\limits_{i=1}^{n-1} |x_i/a_i|^p\right)^{(p-1)/p}}.$$

We change from Cartesian to p-generalized ellipsoidal coordinates in $(n-1)$ dimensions, $T^E_{a,p}(n-1)$: $(x_1, \ldots, x_{n-1}) \longrightarrow (r, \phi_1, \ldots, \phi_{n-2})$. Because of $\sum\limits_{i=1}^{n-1} |x_i/a_i|^p = r^p$,

$$O_{a,p,q}(A) = a_n \int\limits_{\left(T^E_{a,p}(n-1)\right)^{-1}(G(A))} \frac{r^{n-2}}{(1-r^p)^{(p-1)/p}} J^* \, d(r, \phi_1, \ldots, \phi_{n-2}),$$

$$J^* = J^*\left(T^E_{a,p}(n-1)\right)(\phi_1, \ldots, \phi_{n-2}) = a_{n-1} \prod_{i=1}^{n-2} \frac{\left(\sin_{(a_i,a_{i+1};p)}(\phi_i)\right)^{n-1-i}}{a_{i+1} N^2_{(a_i,a_{i+1};p)}(\phi_i)}.$$

$$\text{If } A = A(r_1, r_2, M^*) = \left\{ \left(y_1, \ldots, y_{n-1}, \left(1 - \sum_{i=1}^{n-1} |x_i/a_i|^p\right)^{1/p}\right)^T : \right.$$

$$\left. (y_1, \ldots, y_{n-1})^T = T^E_{a,p}(n-1)([r_1, r_2) \times M^*) \right\},$$

with $M^* = \{(\phi_1, \ldots, \phi_{n-2}) : \phi_{il} \le \phi_i \le \phi_{iu}, i = 1, \ldots, n-2\}$

$$\subset [0, \pi)^{\times(n-3)} \times [0, 2\pi) = M^*_{n-1}$$

then $$O_{a,p,q}(A) = a_n a_{n-1} \int\limits_{r1}^{r_2} \frac{r^{n-2}}{(1-r^p)^{(p-1)/p}} dr$$

$$\times \int\limits_{M^*} J^*\left(T^E_{a,p}(n-1)\right)(\phi_1, \ldots, \phi_{n-2}) d(\phi_1, \ldots, \phi_{n-2}).$$

In what follows, we use the coordinate transformation $\tilde{T}_{a,p} : (R, r, \phi) \rightarrow z[R, r, \phi], \phi = (\phi_1, \ldots, \phi_{n-2})$ defined by

$z_1 = a_1 Rr \cos_{(a_1,a_2;p)}(\phi_1), z_2 = a_2 Rr \sin_{(a_1,a_2;p)}(\phi_1) \cos_{(a_2,a_3;p)}(\phi_2), \ldots,$

$z_{n-2} = a_{n-2} Rr \sin_{(a_1,a_2;p)}(\phi_1) \cdot \ldots \cdot \sin_{(a_{n-3},a_{n-2};p)}(\phi_{n-3}) \cos_{(a_{n-2},a_{n-1};p)}(\phi_{n-2})$

$z_{n-1} = a_{n-1} Rr \sin_{(a_1,a_2;p)}(\phi_1) \cdot \ldots \cdot \sin_{(a_{n-3},a_{n-2};p)}(\phi_{n-3}) \sin_{(a_{n-2},a_{n-1};p)}(\phi_{n-2}),$

$z_n = a_n R(1 - r^p)^{1/p}.$

This transformation allows the representations

$A(r_1, r_2, M^*) = \tilde{T}_{a,p}(1, [r_1, r_2), M^*) = \{z[R, r, \phi] : R = 1, r \in [r_1, r_2), \phi \in M^*\}$ and

$sector(A(r_1, r_2, M^*), \rho) = \tilde{T}_{a,p}([0, \rho) \times [r_1, r_2) \times M^*)$

$= \{z[R, r, \phi] : 0 \le R < \rho, r \in [r_1, r_2), \phi \in M^*\}.$

The volume

$$\mu(sector(A(r_1, r_2, M^*), \rho)) = \int\limits_{sector(A(r_1,r_2,M^*),\rho)} dz$$

may therefore be written as

$$\mu(sector(A(r_1, r_2, M^*), \rho)) = \int\limits_{R=0}^{\rho} \int\limits_{r=r_1}^{r_2} \int\limits_{\phi\in M^*} J(\tilde{T}_{a,p})(R, r, \phi) dR dr d\phi.$$

Here, $J(\tilde{T}_{a,p})(R, r, \phi) = \frac{D(z_1,\ldots,z_n)}{D(R,r,\phi_1,\ldots,\phi_{(n-2)})}$ can be evaluated as in (Richter 2013) where the case $p = 2$ was dealt with:

$$\begin{aligned} J(\tilde{T}_{a,p})(R, r, \phi) &= |z_{nr} a_{n-1} (rR)^{n-2} J^* r - z_{nR} a_{n-1} (rR)^{n-2} J^* R| \\ &= \left| \left(-\frac{a_n R r^{p-1}}{(1-r^p)^{(p-1)/p}} (rR)^{n-2} r - a_n (1-r^p)^{1/p} (rR)^{n-2} R \right) J^* a_{n-1} \right| \\ &= a_{n-1} a_n J^* (T^E_{a,p}(n-1))(\phi) R^{n-1} \frac{r^{n-2}}{(1-r^p)^{1-1/p}}. \end{aligned}$$

It follows that

$$\lambda\big(sector\big(A\left(r_1, r_2, M^*\right), \rho\big)\big) = a_{n-1} a_n \int\limits_0^{\rho} R^{n-1} dR \int\limits_{r_1}^{r_2} \frac{r^{n-2}}{\sqrt{1-r^2}} dr \int\limits_{M^*} J^*\left(T^E_{a,p}(n-1)\right)(\phi) d\phi.$$

We consider now the local approach to the non-Euclidean surface content,

$$\frac{d}{d\rho} \mu(sector(A(r_1, r_2, M^*), \rho))|_{\rho=1} = a_{n-1} a_n \int\limits_{r_1}^{r_2} \frac{r^{n-2}}{\sqrt{1-r^2}} dr \int\limits_{M^*} J^*\left(T^E_{a,p}(n-1)\right)(\phi) d\phi,$$

and observe that $\mathfrak{O}_S(A(r_1, r_2, M^*)) = O_{a,p,q}(A(r_1, r_2, M^*))$.

The measures $\mathfrak{O}_S$ and $O_{a,p,q}$ coincide on the semi-algebra which is generated by the sets of the type $A(r_1, r_2, M^*)$. It follows from the measure extension theorem that these measures coincide on the whole Borel-σ-field $\mathfrak{B}^E_{a,p}$ on $E_{a,p}$, too. □

Remark 6. ***Reformulating the results of Section 3.1***

In the special case that $K = B_{a,p}, S = E_{a,p}$, just considered here, all the statements of Equations (1) and (2), Theorem 1, Corollaries 1-3 and Remarks 1 and 2 remain valid if the integral surface measure $O_{a,p,q}$ is used instead of the star-generalized surface measure $\mathfrak{O}_S$. The same is true for all those statements quoted below which are using the local notions of Section 2.

4 Star-shaped distributions and geometric disintegration

4.1 Star-shaped uniform distributions

In this section, we extend the method of indivisibles which was used so far for the Lebesgue measure to a class of probability laws which contains the families of elliptically contoured and $l_{n,p}$-symmetric distributions as special cases. This method was originally developed in (Richter 1985, 1987) for proving large deviation limit theorems for the multivariate standard Gaussian law.

We continue to use the notations from Section 2. Note that the following general consideration always covers the very well interpretable specific case that $S = E_{a,p}$ and thus $\mathfrak{O}_S = O_{a,p,q}$.

Definition 6. *The star-generalized uniform probability distribution on the Borel σ-field $\mathfrak{B}_S$ is defined as $\omega_S(A) = \mathfrak{O}_S(A)/\mathfrak{O}_S(S)$.*

Remark 7. *(a) Let $A \in \mathfrak{B}_S$. Then $\omega_S(A) = sm_K(A)$.*
(b) Let $B \in \mathfrak{B}_n$. Then $\omega_S\left(\left[\frac{1}{r}B\right] \cap S\right) = \mathfrak{F}_S(B, r)$, $r > 0$.

Let $(\Omega, \mathfrak{A}, P)$ be a probability space and $Y : \Omega \to \mathbb{R}^n$ a random vector being uniformly distributed on K, i.e.

$$P(Y \in B) = \mu(B)/\mu(K),\ \forall B \in K \cap \mathfrak{B}_n.$$

The a.s. defined normalized random vector $U_S = Y/h_K(Y)$ takes its values in $S, P(h_K(U_S) = 1) = 1$. Let us further put $R = h_K(Y)$.

Theorem 6. *(a) U_S follows the star-generalized uniform distribution, $U_S \sim \omega_S$.*
(b) The pdf of R is $f_R(r) = I_{(0,1)}(r)n \cdot r^{n-1}$.
(c) The random elements U_S and R are stochastically independent.

Proof. (a) Let $A \in \mathfrak{B}_S$. Then $P(U_S \in A) = P(Y \in sector(A, 1))$, and
$P(U_S \in A) = \frac{\mu(sector(A,1))}{\mu(K)} = \frac{1}{\mu(K)}\int_0^1 r^{n-1}dr \int_{StSph^{*-1}(A)} J^*(\vartheta)d\vartheta = \frac{\mathfrak{O}_S(A)}{n\mu(K)}$.
Because of $\mathfrak{O}_S(S) = n\mu(K)$, we have $P(U_S \in A) = \omega_S(A)$.
(b) For $0 < r < 1$, we consider the cumulative distribution function (cdf) of R,

$$P(R < r) = P(Y \in K(r)) = \mu(K(r))/\mu(K) = r^n I_{(0,1)}(r).$$

(c) The independence of U_S and R follows from $P(R < \varrho, U_S \in A)$
$= P(Y \in sector(A, \varrho)) = \int_{sector(A,\varrho)} P^Y(dx) = \frac{1}{\mu(K)} \int_{sector(A,\varrho)} d\mu$

$= \frac{1}{\mu(K)} \int_0^{\varrho} \int_{G(A)} r^{n-1} J^*(\vartheta)d\vartheta dr = \frac{1}{\mu(K)} \frac{\varrho^n}{n} \mathfrak{O}_S(A) = P(R < \varrho)P(U_S \in A)$ □

Remark 8. *(a) The pdf of R^2 is $\frac{d}{dr}P(R^2 < r) = I_{(0,1)}(r)\frac{n}{2}r^{n/2-1}$, $r \in R$.*
(b) The probability distribution of the random vector Y allows the representation

$$P(Y \in B) = \int_0^{\infty} P\left(U_S \in \frac{1}{r}B | R = r\right) dP(R < r)$$

which may be considered as a reformulation of Corollary 2 with

$$\mathfrak{F}_S(B, r) = \omega_S\left(\left[\frac{1}{r}B\right] \cap S\right) = P\left(U_S \in \frac{1}{r}B | R = r\right) \quad a.s. \tag{3}$$

That is why the family of probability measures $\mathfrak{P} = \{P_r, r > 0\}$ where P_r is defined on the Borel σ-field $\mathfrak{B}_n$ by $P_r(B) = \omega_S\left(\left[\frac{1}{r}B\right] \cap S\right) = P\left(U_S \in \frac{1}{r}B | R = r\right)$, may be called a geometric disintegration of P^Y w.r.t. P^R. The family $\mathfrak{P}$ may also be considered as a regular conditional probability.

4.2 Continuous star-shaped distributions

There are different ways to introduce more general classes of star-shaped distributions than the uniform ones considered so far. One of the possibilities is to continue with star-shaped distributions having a density, to derive their most basic properties and finally to introduce the class of all star-shaped distributions having just the latter as their defining properties. This way may be considered as formally generalizing the notion of norm-contoured distributions in (Richter, W.-D.: Norm contoured distributions in R^2, submitted), as well as being statistically well motivated by comparing empirical density level sets with level sets of Minkowski functionals of suitably chosen star bodies. This way will be followed in the present and in the following two sections. An alternative possibility would be just to introduce here the general class of star-shaped distributions and to restrict consideration to special classes of distributions like continuous ones only later.

Definition 7. *Let $g : \mathbb{R}^+ \to \mathbb{R}^+$ satisfy the assumptions $0 < I(g) < \infty$ where $I(g) = \int_0^\infty r^{n-1} g(r) dr$. We call g a density generating function (dgf), $\varphi_{g,K}(x) = C(g, K) g(h_K(x)), x \in \mathbb{R}^n$ a star-shaped density and K its contour defining star body.*

A probability measure having the density $\varphi_{g,K}$ will be denoted by $\Phi_{g,K}$. Let us emphasize that according to this definition 0_n may be any point from the set of all points w.r.t. which K is star-shaped, hence K needs not to be symmetric w.r.t. 0_n. Densities of such type have been studied already in (Balkema and Nolde 2010; Fernandez et al. 1995). Our more general considerations in Sections 4.3-4.7, however, seem to be new. The following theorem deals with a geometric measure representation of continuous star-shaped distributions.

Theorem 7. *For every $B \in \mathfrak{B}^n$, $\Phi_{g,K}(B) = C(g, K) \mathfrak{O}_S(S) \int_0^\infty r^{n-1} g(r) \mathfrak{F}_S(B, r) dr$.*

Proof. Because of $\Phi_{g,K}(B) = C(g, K) \int_B g(h_K(x)) dx$ it follows from Corollary 1(a) that

$\Phi_{g,K}(B) = C(g, K) \int_0^\infty r^{n-1} \int_{[\frac{1}{r}B] \cap S} g(h_K(r\theta)) \mathfrak{O}_S(d\theta) dr$. Hence,

$$\Phi_{g,K}(B) = C(g, K) \int_0^\infty r^{n-1} g(r) \mathfrak{O}_S \left(\left[\tfrac{1}{r}B\right] \cap S \right) dr. \qquad \square$$

Classes of dgfs are surveyed, e.g., in (Fang et al. 1990) and (Richter 2013). Numerous types of applications of special cases of the geometric measure representation in Theorem 7 are surveyed in (Richter 2009, 2012). Later applications are to be found in (Arellano-Valle and Richter 2012; Batún-Cutz et al. 2013) and (Günzel et al. 2012).

4.3 Stochastic representations

In this section, we consider that property of continuous star-shaped distributions which will serve in the next section to define a general class of star-shaped distributions.

Theorem 8. *If $Y \sim \Phi_{g,K}$ then Y allows the stochastic representation $Y \stackrel{d}{=} R_g U_S$ where R_g and U_S are stochastically independent, $U_S \sim \omega_S$ and R_g follows the density $f(r) = \frac{1}{I(g)} r^{n-1} g(r), r > 0$.*

Proof. $P(R < \varrho) = P(Y \in K(\varrho)) = C(g, K) \int_{K(\varrho)} g(h_K(x)) dx$

$$= 2C(g,K)\int_0^{\rho}\int_{G(S^+)} g\left(h_K\left(r\left(\vartheta^T,\eta^+(\vartheta)\right)^T\right)\right) r^{n-1}J^*(\vartheta)d\vartheta\, dr$$

$$= 2C(g,K)\int_0^{\rho} r^{n-1}g(r)dr \int_{G(S^+)} J^*(\vartheta)d\vartheta$$

□

Remark 9. *The normalizing constant* $C(g,K)$ *in Definition 7 allows according to Theorem 7 the representation* $C(g,K) = \frac{1}{\mathfrak{O}_S(S)I(g)}$ *and the statement of Theorem 7 may according to Theorem 8 be written as*

$$\Phi_{g,K}(B) = \int_0^{\infty} \mathfrak{F}_S(B,r)P(R \in dr),\ B \in \mathfrak{B}_n \tag{4}$$

where $\mathfrak{F}_S(B,r)$ *may be interpreted as in (3). Hence, (4) may be read as a generalization of Remark 8(b). Moreover,*

$$\varphi_{g,K}(x) = \frac{g(h_K(x))}{\mathfrak{O}_S(S)I(g)},\ x \in \mathbb{R}^n. \tag{5}$$

4.4 General star-shaped distributions

The results of the previous section may serve as a starting point for defining general star-shaped distributions. We follow the way in (Fang et al. 1990) and (Richter 2009, 2013) when we use the stochastic representation from Section 4.3 for defining now a large family of star-shaped distributions.

Definition 8. *A random vector* $Y : \Omega \to \mathbb{R}^n$ *is said to follow a star-shaped distribution if there are a star body K having the origin as an interior point,* $0_n \in int K$*, a vector* $\nu \in \mathbb{R}^n$*, and a random variable* $R : \Omega \to [0,\infty)$ *such that* $Y - \nu$ *satisfies the stochastic representation* $Y - \nu \overset{d}{=} R \cdot U_S$ *where* U_S *follows the star-generalized uniform distribution on the boundary S of K,* $U_S \sim \omega_S$*, and R and* U_S *are stochastically independent. If Y has a density with dgf g then by* $\Phi_{g,K,\nu}$ *the distribution law of Y is denoted,* $Y \sim \Phi_{g,K,\nu}$ *, and K is called a density contour defining star body of the star-shaped distribution* $\Phi_{g,K,\nu}$.

The set of all star-shaped distributions on $\mathfrak{B}_n$ will be denoted $StSh^{(n)}$ and its subset of continuous distributions by

$$CStSh^{(n)} = \{\Phi_{g,K,\nu} : \nu \in \mathbb{R}^n,\ K \text{ is a star body with } 0 \in \text{ int } K, g \text{ is a dgf}\}.$$

We recall that star-shaped sets are associated with multivariate stable distributions in (Molchanov 2009) to describe characteristic functions, thus playing there another role than in Definition 8. To finish this section, we remark that both the set of all star bodies having the origin as an interior point and the set $StSh^{(n)}$ are invariant w.r.t. any orthogonal transformation.

4.5 Extension of the ball number function

In (Richter 2011), the ball number function was defined for $l_{n,p}$-balls and the problem of extending it to balls being as general as possible was stated. It follows from the results in Section 3 that both the ratios $\frac{\mu(B_{a,p}(r))}{r^n}$ and $\frac{O_{a,p,q}(E_{a,p}(r))}{nr^{n-1}}$ do not depend on the star radius r, and that their constant values are one and the same number. This common value will be called the ball number $\pi(B_{a,p})$ of $B_{a,p}(r), r > 0$. Here, $\pi(B_{a,p}) = \mu(B_{a,p}) = a_1 \cdot ... \cdot a_n \cdot \frac{\omega_{n,p}}{n}$, hence the region where the ball number function is defined is extended here to all $B_{a,p}$ balls.

4.6 Characteristic functions

Let $Y : \Omega \to \mathbb{R}^n$ be a star-shaped distributed random vector which satisfies the stochastic representation $Y \overset{d}{=} R{\cdot}U_S$ where the non-negative random variable R is independent of the star-generalized uniformly distributed random vector U_S, and let moreover ϕ_Y and ϕ_{U_S} denote the characteristic functions (ch.f.) of the vectors Y and U_S, respectively. Further, let F_R denote the cdf of R.

Theorem 9. *The ch.f. of the star-shaped distributed random vector Y allows the integral representation* $\phi_Y(t) = \int\limits_0^\infty \phi_{U_S}(rt)dF_R(r),\ t \in \mathbb{R}^n$.

Proof. Because of the independence of R and U_S, Theorem 1.1.6 in (Sasvári 2013) applies. □

This theorem was proved first for spherically distributed vectors in (Schoenberg 1938) and later for $l_{n,1}$-symmetrically distributed random vectors in (Ng and Tian 2001) and for continuous $l_{n,p}$-symmetrically distributed vectors in (Kalke 2013).

Remark 10. *(a) The ch.f.* $\phi_{U_S}(t) = Ee^{it^TU_S},\ t \in \mathbb{R}^n$, *of* U_S *allows the integral representation* $\phi_{U_S}(t) = \int\limits_S \cos\left(t^T\theta\right) \omega_S(d\theta) + i \int\limits_S \sin\left(t^T\theta\right) \omega_S(d\theta)$.

(b) The ch.f. ϕ_Y *of a star-shaped distributed random vector Y having a density with dgf g and contour defining star body K allows the representation*

$$\mathfrak{O}_S(S)I(g)\phi_Y(t) = \int\limits_0^\infty \left[\sum_j \int\limits_{G(S_j)} \cos(t, r\begin{pmatrix} \vartheta \\ y(\vartheta) \end{pmatrix})J_j^*(\vartheta)d\vartheta \right] r^{n-1}g(r)dr$$
$$+ i \int\limits_0^\infty \left[\sum_j \int\limits_{G(S_j)} \sin(t, r\begin{pmatrix} \vartheta \\ y(\vartheta) \end{pmatrix})J_j^*(\vartheta)d\vartheta \right] r^{n-1}g(r)dr$$

where $(.,.)$ *denotes the Euclidean scalar product in* $\mathbb{R}^n$.

(c) On combining the representations in (a) and (b), and taking into account Remark 1, we get an alternative direct proof of Theorem 9 if Y has density $\varphi_{g,K}$.

(d) If U_S *is symmetrically distributed w.r.t. the origin,* $U_S \overset{d}{=} -U_S$, *then the imaginary parts of the integral representations in (a) and (b) vanish, and both* ϕ_{U_K} *and* ϕ_Y *are symmetric w.r.t. the origin.*

4.7 The class of p-generalized elliptically contoured distributions

The general principle for deriving geometric and stochastic representations of star-shaped distributions developed so far will be proved in this section to successfully apply to a class of distributions considered in Section 3.5 of (Arellano-Valle and Richter 2012) and including both the $l_{n,p}$-symmetric ones, accordingly represented in (Richter 2009), and the elliptically contoured ones, analogously dealt with in (Richter 2013).

According to Definition 6, Theorem 5 and Corollary 3, the p-generalized elliptically contoured uniform distribution on $\mathfrak{B}_{E_{a,p}}$ is defined for arbitrary $p > 0$ by

$$\omega_{a,p,q}(A) = D(n,a,p)O_{a,p,q}(A) \text{ with } D(n,a,p) = \frac{p^{n-1}\Gamma\left(\frac{n}{p}\right)}{a_1 \cdot \ldots \cdot a_n 2^n \left(\Gamma\left(\frac{1}{p}\right)\right)^n}$$

and with q satisfying $1/p + 1/q = 1$.

If a random vector $Y : \Omega \to \mathbb{R}^n$ follows the uniform probability distribution on $B_{a,p}$ then $U \sim \omega_{a,p,q} = \omega_{E_{a,p}}$ where U is a.s. defined as $U = Y/R$ and is independent of $R = h_{B_{a,p}}(Y)$. The latter, non-negative, random variable has the density described in Theorem 6(b).

Let $\mathfrak{O}(n)$ denote the set of all orthogonal $n \times n$ matrices. A random vector $Y : \Omega \to \mathbb{R}^n$ is said to follow a p-generalized elliptically contoured distribution $EC_{a,p,\nu,O}$ with parameters $a = (a_1, \ldots, a_n)^T, a_i > 0, i = 1, \ldots, n,\ p > 0,\ \nu \in \mathbb{R}^n$ and $O \in \mathfrak{O}(n)$ if there exists a random variable $R : \Omega \to [0, \infty)$ such that Y satisfies the stochastic representation

$$O^T(Y - \nu) \stackrel{d}{=} R \cdot U$$

where $U \sim \omega_{E_{a,p}}$ and U and R are stochastically independent. Note that Y has a density f_Y iff R has a density. If $O^T(Y - \nu)$ has the dgf g, i.e. if

$$f_Y(y) = C(g,a,p)g\left(\left|O^T(y-\nu)\right|_{a,p}\right), y \in \mathbb{R}^n$$

with $C(g,a,p) = C(g, B_{a,p})$ then R has the density

$$f_R(r) = I(g)^{-1}r^{n-1}g(r)I_{[0,\infty)}(r).$$

In this case, we write $Y \sim \Phi_{g,a,p,\nu,O}$ and $f_Y = \varphi_{g,a,p,\nu,O}$. Note that $\Phi_{g,a,p,0_n,I_n} = \Phi_{g,B_{a,p},0_n}$ where I_n denotes the $n \times n$-unit matrix. The measure $\Phi_{g,a,p,\nu,O}$ allows the geometric representation

$$\Phi_{g,a,p,\nu,O}(B) = \int_0^\infty \mathfrak{F}_{a,p}\left(O^T(B - \nu, r)\right) dF_R(r) \qquad (6)$$

where $\mathfrak{F}_{a,p}(M,r) = \omega_{E_{a,p}}\left(\left[\frac{1}{r}M\right] \cap E_{a,p}\right), r > 0.$

Example 1. *In the case of dimension $n = 2$, Figure 1 shows the density $\varphi_{g,a,p,\nu,O}$ and contours of its superlevel sets where $g(r) = \exp\left\{-\frac{r^p}{p}\right\}, a = (3,1)^T, p$ takes several values, $\nu = (0,0)^T$ and*

$$O = \begin{pmatrix} \cos\alpha & \sin\alpha \\ -\sin\alpha & \cos\alpha \end{pmatrix}, \quad \alpha = 5\pi/3.$$

The matrix O causes an anticlockwise rotation through an angle of size $\pi/3$.

Remark 11 (On independent coordinates). *Let $(R, \Phi_1, \ldots, \Phi_{n-1}) = \left(T_{a,p}^E\right)^{-1}(Y)$ be the random p-generalized ellipsoidal coordinates of Y where $Y \sim \Phi_{g,a,p,0_n,O}$. According to Theorem 3, the generalized radius is $\left(\sum_{i=1}^n |Y_i/a_i|^p\right)^{1/p} = R$, and by the density transformation formula and Theorem 4,*

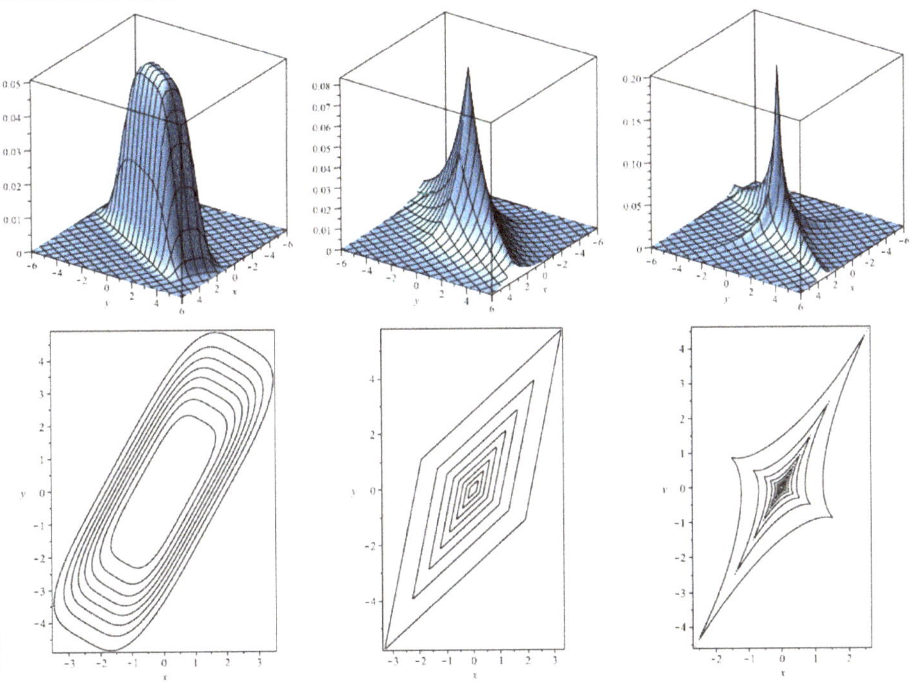

Figure 1 ***p*-generalized elliptically contoured densities for $p = 4$, 1 and 0.6, from left to right.**

$$f_{(R,\Phi_1,\ldots,\Phi_{n-1})}(r,\phi_1,\ldots,\phi_{n-1}) = \frac{C(g,a,p)}{a_2\cdot\ldots\cdot a_{n-1}}g(r)r^{n-1}\prod_{i=1}^{n-1}h_i(\phi_i)$$

with $h_i(\phi_i) = \frac{\left(\sin_{(a_i,a_{i+1};p)}(\phi_i)\right)^{n-i-1}}{N^2_{(a_i,a_{i+1};p)}(\phi_i)}$, $i = 1,\ldots,n-1$ *being integrable functions. Thus, suitably normalized, the functions* $h_0, h_1, \ldots, h_{n-1}$ *with* $h_0(r) = rg(r)I_{(0,\infty)}(r)$ *are the densities of* $R, \Phi_1, \ldots, \Phi_{n-1}$*, respectively.*

Hence, the random coordinates $R, \Phi_1, \ldots, \Phi_{n-1}$ *are stochastically independent.*

This result opens new perspectives for various applications which are not considered in the present paper.

5 Applications

5.1 The non-concentric elliptically contoured distribution class

The general distribution classes considered in the present paper include various interesting special cases to be studied only in detail in the future. Just to start with, let $0 < b < a$ and

$$K_{a,b} = \left\{(x,y)^T \in \mathbb{R}^2 : (x/a)^2 + (y/b)^2 \leq 1\right\}.$$

A point $(e,f)^T$ is from the inner part of $K_{a,b}$ iff it satisfies the inequality $(e/a)^2 + (f/b)^2 < 1$. For such points, we put $K_{a,b,e,f} = K_{a,b} - (e,f)^T$. As because $rK_{a,b,e,f} = K_{ar,br,er,fr}$, for arbitrary dgf g, the level sets of the density

$$\varphi^*_{g,a,b,e,f}(x,y) = C\left(g, K_{a,b,e,f}\right) g\left(h_{K_{a,b,e,f}}\left((x,y)^T\right)\right), (x,y)^T \in \mathbb{R}^2$$

are the boundaries of the sets $K_{ar,br,er,fr}$, $r > 0$. Note that if $(e,0)^T$ is a focal point of the elliptical disc $K_{a,b}$ then the origin $(0,0)^T$ is a focal point of $E_{ar,br,er,0}$, for all $r > 0$. In this case, we call the origin a position of the density $\varphi^*_{g,a,b,e,0}$. Similarly, the origin may also be called a position of the density $\varphi^*_{g,a,b,0,f}$ if $(0,f)^T$ is a focal point of $K_{a,b}$. The distribution class

$$NCEC = \left\{\varphi^*_{g,a,b,e,f}\ ,\ g \text{ is a dgf}, 0 < b < a, (e/a)^2 + (f/b)^2 < 1\right\}$$

will be called a non-concentric elliptically contoured distribution class. It is left to the reader to derive explicit expressions for the Minkowski functional $h_{K_{a,b,e,f}}$ and the normalizing constant $C(g, K_{a,b,e,f})$.

5.2 Circular distributions

In this section, we study directional distributions on further using the results of the present work. For a recent overview on circular distributions we refer to (Pewsey et al. 2013).

In the case of dimension two, the density of $\Phi_{g,K,\nu}$ is

$$\varphi_{g,K,\nu}(x,y) = C(g,K)g\left(h_K\left(\begin{pmatrix}x\\y\end{pmatrix} - \begin{pmatrix}\nu_1\\\nu_2\end{pmatrix}\right)\right),\ (x,y)^T \in \mathbb{R}^2.$$

We recall that star-generalized trigonometric functions and random polar coordinates are defined in (Richter 2011a) by $\cos_K(\phi) = \frac{\cos\phi}{h_K(\cos\phi,\sin\phi)}$, $\sin_K(\phi) = \frac{\sin\phi}{h_K(\cos\phi,\sin\phi)}$ and $X = R\cos_K(\Phi), Y = R\sin_K(\Phi)$, respectively. The K-generalized radius coordinate is $R = h_K(X,Y)$, and the angle Φ satisfies the representation of the usual polar angle,

$$\arctan(|Y/X|) = \Phi \text{ in } Q_1,\ = \pi - \Phi \text{ in } Q_2,\ = \Phi - \pi \text{ in } Q_3, = \ 2\pi - \Phi \text{ in } Q_4.$$

The Jacobian of this transformation is $rR_S^2(\phi)$ where the function $R_S(\phi) = 1/h_K(\cos\phi,\sin\phi)$ describes the boundary S of K:

$$S = \left\{R_S(\phi)\begin{pmatrix}\cos\phi\\\sin\phi\end{pmatrix}, 0 \le \phi < 2\pi\right\} = \left\{(x,y)^T : h_K(x,y) = 1\right\}.$$

With uniquely determined $\mu \in [0,2\pi)$ and $\lambda > 0$, the location vector $(\nu_1,\nu_2)^T$ can be represented as

$$\begin{pmatrix}\nu_1\\\nu_2\end{pmatrix} = \lambda\begin{pmatrix}\cos_K(\mu)\\\sin_K(\mu)\end{pmatrix}.$$

Thus, the density of $(R,\Phi)^T$ is

$$f_{(R,\Phi)}(r,\phi) = C(g,K)rR_S^2(\phi)g\left(h_K\left(\begin{pmatrix}r\cos_K(\phi) - \lambda\cos_K(\mu)\\r\sin_K(\phi) - \lambda\sin_K(\mu)\end{pmatrix}\right)\right)I_{[0,2\pi)}(\phi)I_{[0,\infty)}(r).$$

Integrating $f_{(R,\Phi)}$ w.r.t. ϕ, and dividing $f_{(R,\Phi)}$ by the latter result, gives $f_{\Phi|R}(\phi|r) = vMd_{g,K,r,\lambda,\mu}(\phi)$ where

$$vMd_{g,K,r,\lambda,\mu}(\phi) = \frac{R_S^2(\phi)g\left(h_K\left(\begin{pmatrix}r\cos_K(\phi) - \lambda\cos_K(\mu)\\r\sin_K(\phi) - \lambda\sin_K(\mu)\end{pmatrix}\right)\right)}{\int_0^{2\pi} R_S^2(\phi)g\left(h_K\left(\begin{pmatrix}r\cos_S(\phi) - \lambda\cos_S(\mu)\\r\sin_S(\phi) - \lambda\sin_S(\mu)\end{pmatrix}\right)\right)d\phi}. \tag{7}$$

This function will be called a star-shaped generalization of the von Mises density which itself appears as a special case for K being the Euclidean unit disc in $\mathbb{R}^2$ and $g(r) = \exp\left\{-r^2/2\right\}$, $r > 0$, c.f. (von Mises 1918).

Example 2. *For illustrating our principle of constructing generalized von Mises densities at the hand of a concrete example, let P_n denote the polygon having the n vertices $I_{n,i} = \left(\cos\left(\frac{2\pi}{n}(i-1)\right), \sin\left(\frac{2\pi}{n}(i-1)\right)\right)^T, i = 1, \ldots, n, n \geq 3$, and let K_n be the convex body circumscribed by P_n. The Minkowski functional of K_n has been dealt with in (Richter, W-D, Schicker, K: Circle numbers of centered regular convex polygons, submitted). Figure 2 shows, from the left to the right, in each row, the polygonally contoured density $\varphi_{g,K_n,\nu}(x,y), (x,y)^T \in \mathbb{R}^2$, the contours of this density (black shapes) together with the polygonally generalized circle consisting of all points $(x,y)^T \in \mathbb{R}^2$ satisfying the condition $h_{K_n}\left((x,y)^T\right) = r$ (blue shape), and the polygonally generalized von Mises density $vMd_{g,K_n,r,\lambda}$ where $g(\rho) = \exp\left\{-\frac{\rho^2}{2}\right\}, \rho > 0, r = 1, \mu = 3\pi/4$ and $(n,\lambda) = \left(3, \frac{1}{2}\right)$ in the first row, but $(n,\lambda) = \left(4, \frac{3}{2}\right)$ in the second row. Each of the arrows in the middle panel shows the way from the center of the blue drawn polygonally generalized circle to that of the black drawn ones.*

6 Concluding remarks

Exact distributions of numerous statistics like mean value statistic, Student statistic, Chi-squared statistic have been derived in a broad and well known literature under the assumption that the sample vector follows a multivariate normal law, or more generally, an elliptically contoured one. Similarly, exact statistical distributions have been derived when the sample distribution is exponential or one of its geometric generalizations. For

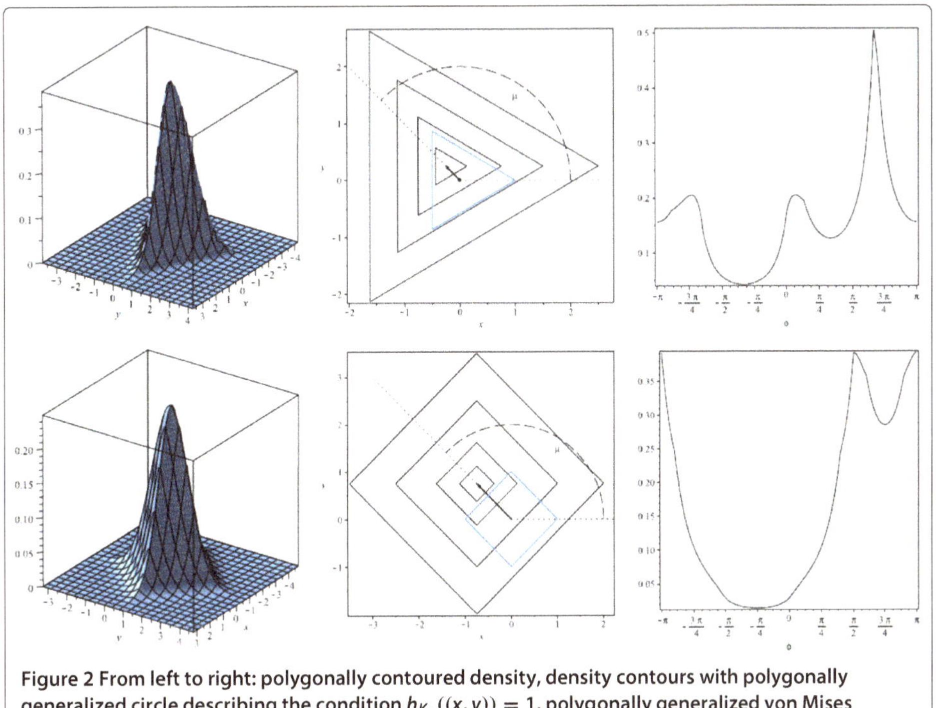

Figure 2 From left to right: polygonally contoured density, density contours with polygonally generalized circle describing the condition $h_{K_n}((x,y)) = 1$, polygonally generalized von Mises density, where $(n,\lambda) = (3, \frac{1}{2})$ in the first row and $= (4, \frac{3}{2})$ in the second row.

results of the latter type, we refer to (Henschel 2001, 2002; Henschel and Richter 2002). However, there seem, in general, not to be as many exact distributional results as in the Gaussian or elliptically contoured cases in the case of samples from any other multivariate distribution. Mathematical research without one of the above assumptions often deals with asymptotic considerations for large sample sizes. Many results of such work are again closely connected with properties of the Gaussian law which occurs often as a limit law.

The present paper provides new possibilities for deriving representations of exact statistical distributions if the sample vector follows a probability law coming from a rather big class of probability laws. It is not the place here to demonstrate in any detail all the possible applications of the present results. Instead, we refer to (Richter 2012) and several references given therein for getting a first overview. A recent example for deriving new results on distributions of functions of Gaussian vectors is given in (Richter 2014). This example might stimulate consideration of exact distributions of new classes of statistics.

Competing interests

The author declares that he has no competing interests.

Acknowledgements

The author is grateful to two Reviewers for their valuable hints which led to improvements of the manuscript, and to Thomas Dietrich for drawing the figures with *Maple 17*.

References

Arellano-Valle, RB, Richter, W-D: On skewed continuous $l_{n,p}$-symmetric distributions. Chilean J. Stat. **3**(2), 193–212 (2012)

Balkema, AA, Embrechts, P, Nolde, N: Meta densities and the shape of their sample clouds. J. Multivariate Anal. **101**, 1738–1754 (2010)

Balkema, AA, Nolde, N: Asymptotic independence for unimodal densities. Adv. Appl. Prob. **42**, 411–432 (2010)

Barndorff-Nielsen, OE, Blaesild, P, Eriksen, PS: Decomposition and Invariance of Measures, and Statistical Transformation Models, Lecture Notes in Statistics, Vol. 58. Springer, New York (1989)

Barthe, F, Csörnyei, M, Naor, A: A note on simulataneous polar and cartesian decomposition. In: Geometric Aspects of, Functional Analysis, Lecture Notes in Mathematics, vol. 1807, pp. 1–19. Springer, Berlin, (2003)

Batún-Cutz, J, Gonzáles-Farías, G, Richter, W-D: Maximum distributions for $l_{2,p}$-symmetric vectors are skewed $l_{1,p}$-symmetric distributions. Stat. Probab. Lett. **83**, 2260–2268 (2013)

Dietrich, T, Kalke, S, Richter, W-D: Stochastic representations and a geometric parametrization of the two-dimensional Gaussian law. Chilean J. Stat. **4**(2), 27–59 (2013)

Eaton, ML: Multivariate Statistics, a Vector Space Approach. Wiley, New York (1983)

Eaton, M L: Group invariance applications in statistics. In: Regional Conference, Series in Probability and Statistics, vol. 1, pp. 1–133. Institute of, Mathematical Statistics, Hayward, California, (1989)

Fang, K-T, Kotz, S, Ng, KW: Symmetric Multivariate and Related Distributions. Chapman & Hall, London (1990)

Farrell, RH: Techniques of, Multivariate Calculation. Lecture Notes in Mathematics 520. Springer, Berlin (1976)

Farrell, R H: Multivariate Calculation: Use of the Continuous Groups. Springer, Berlin (1985)

Fernandez, C, Osiewalski, J, Steel, MFJ: Modeling and inference with ν-spherical distributions. J. Amer. Stat. Assoc. **90**, 1331–1340 (1995)

Gómez, E, Gomez-Viilegas, MA, Marín, JM: A multivariate generalization of the power exponential family of distributions. Commun. Stat. Theory Methods. **27**(3), 589–600 (1998)

Günzel, T, Richter, W-D, Scheutzow, S, Schicker, K, Venz, J: Geometric approach to the skewed normal distribution. J. Stat. Plann. Inference. **142**, 3209–3224 (2012)

Gupta, A, Song, D: l_p-norm spherical distributions. J. Stat. Plann. Inference. **60**, 241–260 (1997)

Henschel, V: Exact distributions in the model of a regression line for the threshold parameter with exponential distribution of errors. Kybernetika. **37**(6), 703–723 (2001)

Henschel, V: Statistical inference in simplicially contoured sample distributions. Metrika. **56**, 215–228 (2002)

Henschel, V, Richter, W-D: Geometric generalization of the exponential law. J. Multivariate Anal. **81**, 189–204 (2002)

Hoffmann-Jorgensen, J: Probability with a View Towards Statistics II. Chapman & Hall, New York (1994)

Joenssen, DW, Vogel, J: A power study of goodness-of-fit tests for multivariate normality implemented in R. J. Stat. Comput. Simul. **75**, 93–107 (2012)

Kalke, S: Geometrische Strukturen $l_{n,p}$-symmetrischer charakteristischer Funktionen. Dissertation, Universität Rostock (2013)

Kallenberg, O: Probabilistic Symmetries and Invariance Principles. Springer, New York (2005)

Kamiya, H, Takemura, A, Kuriki, S: Star-shaped distributions and their generalizations. J. Stat. Plann. Inference. **138**, 3429–3447 (2008)

Kinoshita, K, Resnick, SI: Convergence of scaled random samples in R. Ann. Probab. **19**, 1640–1663 (1991)

Koehn, U: Global cross sections and the densities of maximal invariants. Ann. Math. Stat. **41**(6), 2045–2056 (1970)

Koshevoy, G, Mosler, K: Zonoid trimming for multivariate distributions. Ann. Stat. **9**, 1998–2017 (1997)

Molchanov, I: Convex and star-shaped sets associated with multivariate stable distributions I: Moments and densities. J. Multiv. Anal. **100**(10), 2195–2213 (2009)

Mosler, K: Multivariate Dispersion, Central, Regions and Depth: The Lift Zonoid Approach. Springer, New York (2002)
Mosler, K: Depth statistics (2013). http://arxiv.org/pdf/1207.4988.pdf
Moszyńska, M, Richter, W-D: Reverse triangle inequality. Antinorms and semi-antinorms. Studia Scientiarum Mathematicarum Hungarica. **49**(1), 120–138 (2012)
Muirhead, RJ: Aspects of Multivariate Statistical Theory. Wiley, New York (1982)
Nachbin, L: The Haar Integral. New York (1976)
Naor, A: The surface measure and cone measure on the sphere of l_p^n. Trans. Am. Math. Soc. **359**(3), 1045–1080 (2007)
Nardon, M, Pianca, P: Simulation techniques for generalized Gaussian densities. J. Stat. Comp. Simul. **11**, 1317–1329 (2009)
Ng, K-W, Tian, G: Characteristic functions of l_1-spherical and l_1-norm symmetric distributions and their applications. J. Multiv. Anal. **76**(2), 192–213 (2001)
Nolde, N: Geometric interpretation of the residual dependence coefficient. J. Multiv. Anal. **123**, 85–95 (2014)
Osiewalski, J, Steel, MFJ: Robust Basian inference in l_q-spherical models. Biometrika. **80**(2), 456–460 (1993)
Pewsey, A, Neuhäuser, M, Graeme, DR: Circular Statistics in R. Oxford (2013)
Pogány, TK, Nadarajah, S: On the characteristic function of the generalized normal distribution. C.R. Math., Acad. Sci. Paris. **203**(3), 203–206 (2010)
Rachev, ST, Rüschendorf, L: Approximate independence of distributions on spheres and their stability properties. Ann. Probab. **19**(3), 1311–1337 (1991)
Richter, W-D: Laplace Gauss integrals, Gaussian measure asymptotic behaviour and probabilities of moderate deviations. Z. Anal. Anwendungen. **4**(3), 257–267 (1985)
Richter, W-D: The Laplace integral and probabilities of moderate deviations in R^k. (In Russian). Translation: J. Math. Sci. **38**(6), 2391–2397 (1987)
Richter, W-D: Eine geometrische Methode in der Stochastik. Rostocker Mathematisches Kolloquium. **44**, 63–72 (1991)
Richter, W-D: Generalized spherical and simplicial coordinates. J. Math. Anal. Appl. **336**, 1187–1202 (2007)
Richter, W-D: Continuous $l_{n,p}$-symmetric distributions. Lithuanian Math. J. **49**, 93–108 (2009)
Richter, W-D: On the ball number function. Lithuanian Math. J. **51**(3), 440–449 (2011)
Richter, W-D: Circle numbers for star discs. ISRN Geometry. (16) (2011a). Article ID 479262
Richter, W-D: Ellipses numbers and geometric measure representations. J. Appl. Anal. **17**, 165–179 (2011b)
Richter, W-D: Exact distributions under non-standard model assumptions. AIP Conf. Proc. **1479**(442) (2012). doi:10.1063/1.4756160
Richter, W-D: Geometric and stochastic representations for elliptically contoured distributions. Commun. Stat.: Theory Methods. **42**(4), 579–602 (2013)
Richter, W-D: Classes of standard Gaussian random variables and their generalizations. In: Knif, J, Pape, B (eds.) Contributions to Mathematics, Statistics, Econometrics, and Finance Essays in Honour of Professor Seppo Pynnönen. Acta Wasaensia 296, pp. 53–69. University of Vaasa, Vaasa, (2014)
Richter, W-D, Venz, J: Geometric representations of multivariate skewed elliptically contoured distributions. Chilean J. Stat. **5**(2), 20 (2014)
Sasvári, Z: Multivariate Characteristic and Correlation Functions. De Gruyter, Berlin (2013)
Schechtman, G, Zinn, J: On the volume of the intersection of two l_p^n balls. Proc. Am. Math. Soc. **110**(1), 217–224 (1990)
Schindler, W: Measures with Symmetry Properties. Springer, Berlin-Heidelberg (2003)
Schoenberg, IJ: Metric spaces and completely monotone functions. Ann. Math. **39**, 811–841 (1938)
Scholz, SP: Robustheitskonzepte und -untersuchungen für Schätzer konvexer Körper. Dissertation, Universität Dortmund (2002)
Song, D, Gupta, AK: l_p-norm uniform distributions. Proc. Am.Math. Soc. **125**(2), 595–601 (1997)
Szablowski, PJ: Uniform distributions on spheres in finite dimensional l_α and their generalizations. J. Multivariate Anal. **64**, 103–107 (1998)
Takemura, A, Kuriki, S: Theory of cross sectionally contoured distributions and its applications. Discussion, Paper Series 96-F-15. Faculty of Economics, University of Tokyo (1996). http://www.stat.t.u-tokyo.ac.jp/~takemura/papers/dp96f15.pdf
von Mises, R: Über die Ganzzahligkeit der Atomgewichte und verwandte Fragen. Physikalische Zeitschrift. **19**, 490–500 (1918)
Wijsman, RA: Cross-sections of orbits and their application to densities of maximal invariants. In: Proc. Fifth Berkeley, Symp. on Math.Statist. and Prob., vol. 1, pp. 389–400. University of California Press, Berkeley and Los Angeles, (1967)
Wijsman, RA: Global cross sections as a tool for factorization of measures and distribution of maximal invariant. Sankhya. **48A**, 1–42 (1986)
Wijsman, R A: Invariant Measures on Groups and Their Use in Statistics, Hayward (1990)

12

Skewness-kurtosis adjusted confidence estimators and significance tests

Wolf-Dieter Richter

Correspondence:
wolf-dieter.richter@uni-rostock.de
Institute of Mathematics, University of Rostock, Ulmenstraße 69, Haus 3, 18057 Rostock, Germany

Abstract
First and second kind modifications of usual confidence intervals for estimating the expectation and of usual local alternative parameter choices are introduced in a way such that the asymptotic behavior of the true non-covering probabilities and the covering probabilities under the modified local non-true parameter assumption can be asymptotically exactly controlled. The orders of convergence to zero of both types of probabilities are assumed to be suitably bounded below according to an Osipov-type condition and the sample distribution is assumed to satisfy a corresponding tail condition due to Linnik. Analogous considerations are presented for the power function when testing a hypothesis concerning the expectation both under the assumption of a true hypothesis as well as under a modified local alternative. A limit theorem for large deviations by S.V. Nagajev/V.V. Petrov applies to prove the results. Applications are given for exponential families.

Keywords: Orders of confidence, Orders of modified local alternatives, True non-covering probabilities, Local non-true parameter choice, Covering probabilities, Linnik condition, Osipov-type condition, Skewness-kurtosis adjusted decisions, Order of significance, Error probabilities of first and second kind, Exponential family, Large deviations

Mathematics Subject Classification: 62E20; 62F05; 62F12; 60F10

1 Introduction

Asymptotic normality of the distribution of the suitably centered and normalized arithmetic mean of i.i.d. random variables is one of the best studied and most often exploited facts in asymptotic statistics. It is supplemented in local asymptotic normality theory by limit theorems for the corresponding distributions under the assumption that the mean is shifted of order $n^{-1/2}$. There are many successful simulations and real applications of both types of central limit theorems, and one may ask for a more detailed explanation of this success. The present note is aimed to present such additional theoretical explanation under certain circumstances. Moreover, the note is aimed to stimulate both analogous consideration in more general situations and checking the new results by simulation. Furthermore, based upon the results presented here, it might become attractive to search for additional explanation to various known simulation results in the area of asymptotic normality which is, however, behind the scope of the present note.

Based upon Nagajev's and Petrov's large deviation result in (Nagaev 1965; Petrov 1968), skewness-kurtosis modifications of usual confidence intervals for estimating the expectation and of usual local alternative parameter choices are introduced here in a way such that the asymptotic behavior of the true non-covering probabilities and the covering probabilities under the modified local non-true parameter assumption can be exactly controlled. The orders of convergence to zero of both types of probabilities are suitably bounded below by assuming an Osipov-type condition, see (Osipov 1975), and the sample distribution is assumed to satisfy a corresponding Linnik condition, see (Ibragimov and Linnik 1971; Linnik 1961).

Analogous considerations are presented for the power function when testing a hypothesis concerning the expectation both under the assumption of a true hypothesis and under a local alternative. Finally, applications are given for exponential families.

A concrete situation where the results of this paper apply is the case sensitive preparing of the machine settings of a machine tool. In this case, second and higher order moments of the manipulated variable do not change from one adjustment to another one and may be considered to be known over time.

It might be another aspect of stimulating further research if one asks for the derivation of limit theorems in the future being close to those in (Nagaev 1965; Petrov 1968) but where higher order moments are estimated.

Let $X_1, \ldots, X_n$ be i.i.d. random variables with the common distribution law from a shift family of distributions, $P_\mu(A) = P(A - \mu), A \in \mathcal{B}$, where $\mathcal{B}$ denotes the Borel σ-field on the real line, the expectation equals $\mu, \mu \in R$, and the variance is σ^2. It is well known that $T_n = \sqrt{n}(\bar{X}_n - \mu)/\sigma$ is asymptotically standard normally distributed, $T_n \sim AN(0,1)$. Hence, $P_\mu(T_n > z_{1-\alpha}) \to \alpha$, and under the local non-true parameter assumption, $\mu_{1,n} = \mu + \frac{\sigma}{\sqrt{n}}(z_{1-\alpha} - z_\beta)$, i.e. if one assumes that a sample is drawn with a shift of location (or with an error in the variable), then $P_{\mu_{1,n}}(T_n \le z_{1-\alpha}) = P_{\mu_{1,n}}\left(\sqrt{n}\frac{\bar{X}_n - \mu_{1,n}}{\sigma} \le z_\beta\right) \to \beta$ as $n \to \infty$, where z_q denotes the quantile of order q of the standard Gaussian distribution.

Let $ACI^u = \left[\bar{X}_n - \frac{\sigma}{\sqrt{n}} z_{1-\alpha}, \infty\right)$ denote the upper asymptotic confidence interval for μ where the true non-covering probabilities satisfy the asymptotic relation

$$P_\mu(ACI^u \textit{ does not cover } \mu) \to \alpha, \; n \to \infty.$$

Because $P_{\mu_{1,n}}\left(\bar{X}_n - \frac{\sigma}{\sqrt{n}} z_{1-\alpha} < \mu\right) = P_{\mu_{1,n}}\left(\sqrt{n}\frac{\bar{X}_n - \mu_{1,n}}{\sigma} \le z_\beta\right)$, the covering probabilities under $n^{-1/2}$-locally chosen non-true parameters satisfy

$$P_{\mu_{1,n}}(ACI^u \textit{ covers } \mu) \to \beta, \; n \to \infty.$$

The aim of this note is to prove refinements of the latter two asymptotic relations where $\alpha = \alpha(n) \to 0$ and $\beta = \beta(n) \to 0$ as $n \to \infty$, and to prove similar results for two-sided confidence intervals and for the power function when testing corresponding hypotheses.

2 Expectation estimation

2.1 First and second kind adjusted one-sided confidence intervals

According to (Ibragimov and Linnik 1971; Linnik 1961), it is said that a random variable X satisfies the Linnik condition of order $\gamma, 0 < \gamma < 1/2$, if

$$E_\mu \exp\left\{|X - \mu|^{\frac{4\gamma}{2\gamma+1}}\right\} < \infty. \tag{1}$$

Let us define the first kind (or first order) adjusted asymptotic Gaussian quantile by

$$z_{1-\alpha(n)}(1) = z_{1-\alpha(n)} + \frac{g_1}{6\sqrt{n}} z^2_{1-\alpha(n)}$$

where $g_1 = E(X - E(X))^3/\sigma^{3/2}$ is the skewness of X. Moreover, let the first kind (order) adjusted upper asymptotic confidence interval for μ be defined by

$$ACI^u(1) = \left[\bar{X}_n - \frac{\sigma}{\sqrt{n}} z_{1-\alpha(n)}(1), \infty\right)$$

and denote a first kind modified non-true local parameter choice by

$$\mu_{1,n}(1) = \mu_{1,n} + \frac{\sigma g_1}{6n}\left(z^2_{1-\alpha(n)} - z^2_{\beta(n)}\right).$$

Let us say that the probabilities $\alpha(n)$ and $\beta(n)$ satisfy an Osipov-type condition of order γ if

$$n^\gamma \exp\left\{\frac{n^{2\gamma}}{2}\right\} \cdot \min\{\alpha(n), \beta(n)\} \to \infty, \ n \to \infty. \quad (2)$$

This condition means that neither $\alpha(n)$ nor $\beta(n)$ tend to zero as fast as or even faster than $n^{-\gamma}\exp\{-n^{2\gamma}/2\}$, i.e. $\min\{\alpha(n), \beta(n)\} \gg n^{-\gamma}\exp\{-n^{2\gamma}/2\}$, and that $\max\{z_{1-\alpha(n)}, z_{1-\beta(n)}\} = o(n^\gamma), n \to \infty$. Here, $o(.)$ stands for the small Landau symbol.

If two functions f, g satisfy the relation $\lim_{n\to\infty} f(n)/g(n) = 1$ then this asymptotic equivalence will be expressed as $f(n) \sim g(n), n \to \infty$.

Theorem 1. *If $\alpha(n) \downarrow 0$, $\beta(n) \downarrow 0$ as $n \to \infty$ and conditions (1) and (2) are satisfied for $\gamma \in \left(\frac{1}{6}, \frac{1}{4}\right]$ then*

$$P_\mu(ACI^u(1) \text{ does not cover } \mu) \sim \alpha(n), \ n \to \infty$$

and

$$P_{\mu_{1,n}(1)}(ACI^u(1) \text{ covers } \mu) \sim \beta(n), \ n \to \infty.$$

Let us define the second kind adjusted asymptotic Gaussian quantile

$$z_{1-\alpha(n)}(2) = z_{1-\alpha(n)}(1) + \frac{3g_2 - 4g_1^2}{72n} z^3_{1-\alpha(n)}$$

where $g_2 = E(X - E(X))^4/\sigma^4 - 3$ is the kurtosis of X, the second kind adjusted upper asymptotic confidence interval for μ

$$ACI^u(2) = \left[\bar{X}_n - \frac{\sigma}{\sqrt{n}} z_{1-\alpha(n)}(2), \infty\right),$$

and a second kind modified non-true local parameter choice

$$\mu_{1,n}(2) = \mu_{1,n}(1) + \frac{\sigma\left(3g_2 - 4g_1^2\right)}{72n^{3/2}}\left(z^3_{1-\alpha(n)} - z^3_{\beta(n)}\right).$$

Theorem 2. *If $\alpha(n) \downarrow 0$, $\beta(n) \downarrow 0$ as $n \to \infty$ and conditions (1) and (2) are satisfied for $\gamma \in \left(\frac{1}{4}, \frac{3}{10}\right]$ then*

$$P_\mu(ACI^u(2) \text{ does not cover } \mu) \sim \alpha(n), \ n \to \infty$$

and

$$P_{\mu_{1,n}(2)}(ACI^u(2) \text{ covers } \mu) \sim \beta(n), \ n \to \infty.$$

Remark 1. Under the same assumptions, analogous results are true for lower asymptotic confidence intervals, i.e. for $ACI^l(s) = \left(-\infty, \bar{X}_n + \frac{\sigma}{\sqrt{n}} z^-_{1-\alpha}(s)\right), s = 1, 2:$

$$P_\mu(ACI^l(s) \text{ does not cover } \mu) \sim \alpha(n)$$

and

$$P_{\mu_{1,n}^{-}(s)}(ACI^{l}(s)\ covers\ \mu) \sim \beta(n),\ n \to \infty.$$

Here, $z_{1-\alpha}^{-}(s)$ means the quantity $z_{1-\alpha}(s)$ where g_1 is replaced by $-g_1, s = 1, 2$, and

$$\mu_{1,n}^{-}(s) = \mu - \frac{\sigma}{\sqrt{n}}(z_{1-\alpha} - z_\beta) + \frac{\sigma g_1}{6n}\left(z_{1-\alpha}^2 - z_\beta^2\right) - \frac{\sigma\left(3g_2 - 4g_1^2\right)}{72n^{3/2}}\left(z_{1-\alpha}^3 - z_\beta^3\right) I_{\{2\}}(s).$$

Remark 2. In many situations where limit theorems are considered as they were in Section 1, the additional assumptions (1) and (2) may, possibly unnoticed, be fulfilled. In such situations, Theorems 1 and 2, together with the following theorem, give more insight into the asymptotic relations stated in Section 1.

Theorem 3. *Large Gaussian quantiles satisfy the asymptotic representation*

$$z_{1-\alpha} = \sqrt{-2\ln\alpha - \ln|\ln\alpha| - \ln(4\pi)} \cdot \left(1 + O\left(\frac{\ln|\ln\alpha|}{(\ln\alpha)^2}\right)\right), \alpha \to +0.$$

Note that $O(.)$ means the big Landau symbol.

2.2 Two-sided confidence intervals

For $s \in \{1, 2\}, \alpha > 0$, put $L(s;\alpha) = \bar{X}_n - \frac{\sigma}{\sqrt{n}} z_{1-\alpha}(s)$ and $R(s;\alpha) = \bar{X}_n + \frac{\sigma}{\sqrt{n}} z_{1-\alpha}^{-}(s)$. Further, let $\alpha_i(n) > 0,\ i = 1, 2, \alpha_1(n) + \alpha_2(n) < 1$, and

$$ACI(s;\alpha_1(n), \alpha_2(n)) = [L(s;\alpha_1(n)), R(s;\alpha_2(n))].$$

If conditions (1) and (2) are fulfilled then $P_\mu((-\infty, L(s;\alpha_1(n)))\ covers\ \mu) \sim \alpha_1(n)$ and $P_\mu((R(s;\alpha_2(n)), \infty)\ covers\ \mu) \sim \alpha_2(n)$ as $n \to \infty$.

With more detailed notation $\mu_{1,n}(s) = \mu_{1,n}(s;\alpha,\beta)$ and $\mu_{1,n}^{-}(s) = \mu_{1,n}^{-}(s;\alpha,\beta)$,
$P_{\mu_{1,n}(s;\alpha_1(n),\beta_1(n))}((L(s;\alpha_1(n)), \infty)\ covers\ \mu) \sim \beta_1(n)$,
$P_{\mu_{1,n}^{-}(s;\alpha_2(n),\beta_2(n))}((-\infty, R(s;\alpha_2(n)))\ covers\ \mu) \sim \beta_2(n), n \to \infty$.
The following corollary has thus been proved.

Corollary 1. *If $\alpha_1(n) \downarrow 0, \alpha_2(n) \downarrow 0$ as $n \to \infty$ and conditions (1) and (2) are satisfied for $\gamma \in \left(\frac{1}{6}, \frac{1}{4}\right]$ if $s = 1$ and for $\gamma \in \left(\frac{1}{4}, \frac{3}{10}\right]$ if $s = 2$, and with $(\alpha(n), \beta(n)) = (\alpha_1(n), \alpha_2(n))$, then*

$$P_\mu(ACI(s;\alpha_1(n), \alpha_2(n))\ does\ not\ cover\ \mu) \sim (\alpha_1(n) + \alpha_2(n)),\ n \to \infty.$$

Moreover,

$$\max_{\nu \in \{\mu_{1,n}(s;\alpha_1(n),\beta_1(n)), \mu_{1,n}^{-}(s;\alpha_2(n),\beta_2(n))\}} P_\nu(ACI(s)\ covers\ \mu) \le \max\{\beta_1(n), \beta_2(n)\}.$$

3 Testing

3.1 Adjusted quantiles

Let us consider the problem of testing the hypothesis $H_0 : \mu \le \mu_0$ versus the alternative $H_A : \mu > \mu_0$. The first and second kind adjusted decision rules of the one-sided asymptotic Gauss test suggest to reject H_0 if $T_{n,0} > z_{1-\alpha(n)}(s)$ for $s = 1$ or $s = 2$, respectively, where $T_{n,0} = \sqrt{n}(\bar{X}_n - \mu_0)/\sigma$. Because

$$P_{\mu_0}(reject\ H_0) = P_{\mu_0}(ACI^u(s)\ does\ not\ cover\ \mu_0),$$

it follows from Theorems 1 and 2 that under the conditions given there the (sequence of) probabilities of an error of first kind satisfy the asymptotic relation

$$P_{\mu_0}(reject\ H_0) \sim \alpha(n), n \to \infty.$$

Concerning the power function of this test, because

$$P_{\mu_{1,n}(s)}(do\ not\ reject\ H_0) = P_{\mu_{1,n}(s)}(ACI^u(s)\ covers\ \mu_0),$$

it follows under the same assumptions that the probabilities of a second kind error in the case that the sequence of the modified local parameters is $(\mu_{1,n}(s))_{n=1,2,...}$, satisfy

$$P_{\mu_{1,n}(s)}(do\ not\ reject\ H_0) \sim \beta(n), n \to \infty.$$

Similar consequences for testing $H_1 : \mu > \mu_0$ or $H_2 : \mu \neq \mu_0$ are omitted, here.

3.2 Adjusted statistics

Let $T_n(1) = T_n - \frac{g_1}{6\sqrt{n}} T_n^2$ and $T_n(2) = T_n(1) - \frac{3g_2 - 8g_1^2}{72n} T_n^3$ be the first and second kind adjusted asymptotically Gaussian statistics, respectively, where $T_n = \frac{\sqrt{n}}{\sigma}(\bar{X}_n - \mu)$.

Theorem 4. *If the conditions* (1) *and* (2) *are satisfied for a certain* $\gamma \in \left(\frac{s}{2s+4}, \frac{s+1}{2s+6}\right]$ *where* $s \in \{1, 2\}$ *then*

$$P_{\mu_0}\left(T_n(s) > z_{1-\alpha(n)}\right) \sim \alpha(n),\ n \to \infty$$

and

$$P_{\mu_{1,n}(s)}\left(T_n(s) \leq z_{1-\alpha(n)}\right) \sim \beta(n),\ n \to \infty.$$

Clearly, the results of this theorem apply to both hypothesis testing and confidence estimation in a similar way as described in the preceding sections.

The material of the present paper is part of a talk presented by the author at the Conference of European Statistics Stakeholders, Rome 2014, see Abstracts of Communication, p.90, and arXiv:1504.02553. A more advanced 'testing-part' of this talk is presented in (Richter 2016) and deals with higher order comparisons of statistical tests.

4 Application to exponential families

Let ν denote a σ-finite measure and assume that the distribution P_ϑ has the Radon-Nikodym density $\frac{dP_\vartheta}{d\nu}(x) = \frac{e^{\vartheta x}}{\int e^{\vartheta x}\nu(dx)} = e^{\vartheta x - B(\vartheta)}$, say. For basics on exponential families we refer to Brown (1986). We assume that $X(\vartheta) \sim P_\vartheta$ and $X_1 = X(\vartheta) - EX(\vartheta) + \mu \sim \widetilde{P}_\mu$ where ϑ is known and μ is unknown. In the product-shift-experiment $\left[R^n, \mathscr{B}^n, \{\widetilde{P}_\mu^{\times n}, \mu \in R\}\right]$, expectation estimation and testing may be done as in Sections 2 and 3, respectively, where $g_1 = B'''(\vartheta)/(B''(\vartheta))^{3/2}$ and g_2 allows a similar representation.

Another problem which can be dealt with is to test the hypothesis $H_0 : \vartheta \leq \vartheta_0$ versus the alternative $H_{1n} : \vartheta \geq \vartheta_{1n}$ if one assumes that the expectation function $\vartheta \to B'(\vartheta) = E_\vartheta X$ is strongly monotonous. For this case, we finally present just the following particular result which applies to both estimating and testing.

Proposition 1. *If conditions* (1) *and* (2) *are satisfied for* $\gamma \in \left(\frac{1}{6}, \frac{1}{4}\right]$ *then*

$$P_{\vartheta_0}^{\times n}\left(\sqrt{n}\frac{\overline{X}_n - B'(\vartheta_0)}{\sqrt{B''(\vartheta_0)}} > z_{1-\alpha(n)} + \frac{B'''(\vartheta_0)}{6\sqrt{n}(B''(\vartheta_0))^{3/2}} z_{1-\alpha(n)}^2\right) \sim \alpha(n), n \to \infty.$$

5 Sketch of proofs

Proof of Theorems 1 and 2. If condition (2) is satisfied then $x = z_{1-\alpha(n)} = o(n^{\gamma}), n \to \infty$ for $\gamma \in \left(\frac{1}{6}, \frac{3}{10}\right]$, and if (1) then, according to (Linnik 1961; Nagaev 1965), $P_{\mu}(T_n > x) \sim f_{n,s}^{(X)}(x), x \to \infty$ where $f_{n,s}^{(X)}(x) = \frac{1}{\sqrt{2\pi}x} \exp\left\{-\frac{x^2}{2} + \frac{x^3}{\sqrt{n}} \sum_{k=0}^{s-1} a_k \left(\frac{x}{\sqrt{n}}\right)^k\right\}$ and s is an integer satisfying $\frac{s}{2(s+2)} < \gamma \leq \frac{s+1}{2(s+3)}$, i.e. $s = 1$ if $\gamma \in \left(\frac{1}{6}, \frac{1}{4}\right]$ and $s = 2$ if $\gamma \in \left(\frac{1}{4}, \frac{3}{10}\right]$. Here, the constants $a_0 = \frac{g_1}{6}$, $a_1 = \frac{g_2 - 3g_1^2}{24}$ are due to the skewness g_1 and kurtosis g_2 of X. Note that $\frac{g_1 x^2}{6\sqrt{n}} = o(x)$ because $x = o(n^{1/2})$, thus $x + \frac{g_1 x^2}{6\sqrt{n}} = o(n^{\gamma})$, and $P_{\mu}\left(T_n > x + \frac{g_1 x^2}{6\sqrt{n}}\right) \sim f_{n,1}\left(x + \frac{g_1 x^2}{6\sqrt{n}}\right)$. Hence, $P_{\mu}\left(T_n > x + \frac{g_1 x^2}{6\sqrt{n}}\right) \sim 1 - \Phi(x)$. Similarly, $P_{\mu}(T_n > z_{1-\alpha(n)}(s)) \sim \alpha(n)$, $s = 1, 2$. Further, $P_{\mu_{1,n}(s)}(T_n \leq z_{1-\alpha(n)}(s))$

$$= P_{\mu_{1,n}(s)}\left(\frac{\sqrt{n}}{\sigma}(\bar{X}_n - \mu_{1,n}(s)\right) < z_{1-\alpha(n)}(s) - \frac{\sqrt{n}}{\sigma}(\mu_{1,n}(s) - \mu)) = P_0\left(\frac{\sqrt{n}}{\sigma}\bar{X}_n < z_{\beta(n)}(s)\right).$$

The latter equality holds because $\{P_{\mu}, \mu \in (-\infty, \infty)\}$ is assumed to be a shift family. It follows that $P_{\mu_{1,n}(s)}(T_n \leq z_{1-\alpha(n)}(s))$

$$= P_0\left(\frac{\sqrt{n}}{\sigma}(-\bar{X}_n) \geq z_{1-\beta(n)} + \frac{-g_1}{6\sqrt{n}} z_{1-\beta(n)}^2 + I_{\{2\}}(s)\frac{3g_2 - 4g_1^2}{72n} z_{1-\beta(n)}^3\right).$$

Note that $-g_1, g_2$ are skewness and kurtosis of $-X_1$. Thus,

$$P_{\mu_{1,n}(s)}\left(T_n \leq z_{1-\alpha(n)}(s)\right) \sim f_{n,s}^{(-X)}(z_{1-\beta(n)}(s)) \sim \beta(n), n \to \infty.$$

Because $P_{\mu}(ACI^u$ *does not cover* $\mu) = P_{\mu}(T_n > z_{1-\alpha(n)}(s))$ and $P_{\mu_{1,n}(s)}(ACI^u$ *covers* $\mu) = P_{\mu_{1,n}(s)}(T_n \leq z_{1-\alpha(n)}(s))$, the theorems are proved.

Proof of Remark 1. The first statement of the remark follows from

$$P_{\mu}\left(\mu > \bar{X}_n + \sigma z_{1-\alpha(n)}^{-}/\sqrt{n}\right) = P_{\mu}\left(\sqrt{n}(-\bar{X}_n + \mu)/\sigma > z_{1-\alpha(n)}^{-}\right)$$

and the second one from

$$P_{\mu_{1,n}^{-}(s)}\left(\mu < \bar{X}_n + \sigma z_{1-\alpha(n)}^{-}/\sqrt{n}\right) = P_0\left(\bar{X}_n > \mu - \mu_{1,n}^{-}(s) - \sigma z_{1-\alpha(n)}^{-}/\sqrt{n}\right)$$
$$= P_0\left(\sqrt{n}\bar{X}_n/\sigma > z_{1-\beta(n)}(s)\right).$$

Proof of Theorem 3. We start from the well known relations

$$\alpha = 1 - \Phi(z_{1-\alpha}) = \left(1 + O\left(\frac{1}{z_{1-\alpha}^2}\right)\right) \frac{1}{\sqrt{2\pi} z_{1-\alpha}} e^{-\frac{z_{1-\alpha}^2}{2}}, \ \alpha \to 0.$$

The solution to the approximative quantile equation $\alpha = \frac{1}{\sqrt{2\pi}x} e^{-\frac{x^2}{2}}$ will be denoted by $x = x_{1-\alpha}$. Let us put

$$x e^{\frac{x^2}{2}} = \frac{1}{\sqrt{2\pi}\alpha} =: y. \tag{3}$$

If $x \geq 1$ then it follows from (3) that $y \geq e^{\frac{x^2}{2}}$, hence $x^2 \leq \ln(y^2)$. It follows again from (3) that $y^2 \leq \ln(y^2) e^{x^2}$, thus $x^2 \geq \ln\left(\frac{y^2}{\ln y^2}\right)$. After one more such step,

$$\ln\left(\frac{y^2}{\ln y^2}\right) \leq x^2 \leq \ln\left[\frac{y^2}{\ln\left(\frac{y^2}{\ln y^2}\right)}\right].$$

The theorem now follows from

$$x^2 = \left\{\ln y^2 - \ln 2 - \ln\ln y\right\}\left\{1 + O\left(\frac{\ln\ln y}{(\ln y^2)^2}\right)\right\}, y \to \infty.$$

Let us remark that the inverse of the function $w \to we^w$ is called the Lambert W function. An asymptotic representation of the solution of (3) as $y \to \infty$ can therefore be derived from the more general representation (4.19) of W in (Corless et al. 1996) if one reads (3) as $we^w = y^2$. Our derivation of the particular result needed here, however, is much more elementary than the general one given in the paper just mentioned.

Proof of Theorem 4. Recognize that if $g_{n,s}(x) = o\left(\frac{1}{x}\right), x \to \infty$ then $f_{n,s}^{(+/-)(X)}(x + g_{n,s}(x)) \sim f_{n,s}^{(+/-)(X)}(x), x \to \infty$. Let us restrict to the case $s = 1$. According to (Linnik 1961),

$$P_{\mu_0}(T_n(1) > z_{1-\alpha(n)}) \sim P_{\mu_0}\left(\frac{3\sqrt{n}}{g_1} > T_n(1) > z_{1-\alpha(n)}\right).$$

The function $f_n^{(1)}(t) = t - \frac{g_1 t^2}{6\sqrt{n}}$ has a positive derivative, $f_n^{(1)'}(t) = 1 - \frac{g_1 t}{3\sqrt{n}} > 0$, if $g_1 t < 3\sqrt{n}$. Denoting there the inverse function of $f_n^{(1)}$ by $f_n^{(1)^{-1}}$, it follows $f_n^{(1)^{-1}}(x) = x + \frac{g_1 x^2}{6\sqrt{n}} + O\left(\frac{x^3}{n}\right)$ and $f_n^{(1)}\left(f_n^{(1)^{-1}}(x)\right) = x + o\left(\frac{1}{x}\right)$. Thus,

$$P_{\mu_o}(T_n(1) > z_{1-\alpha(n)}) \sim P_{\mu_o}(T_n > z_{1-\alpha(n)}(1)) \sim \alpha(n).$$

Moreover, $P_{\mu_{1n}(1;\alpha(n),\beta(n))}(T_n(1) \le z_{1-\alpha(n)}) = P_{\mu_{1n}(1;\alpha(n),\beta(n))}\left(T_n \le \left(f_n^{(1)}\right)^{-1}\left(z_{1-\alpha(n)}\right)\right) =$

$$P_{\mu_{1n}(1;\alpha(n),\beta(n))}\left(\sqrt{n}\frac{\bar{X}_n - \mu_{1n}(1)}{\sigma} \le z_{1-\alpha(n)}(1) + \frac{z_{1-\alpha(n)}^2 g_1}{6\sqrt{n}} + O\left(\frac{z_{1-\alpha(n)}^3}{n}\right) - \sqrt{n}\frac{\mu_{1n}(1) - \mu_0}{\sigma}\right) \sim$$

$$f_{n,1}^{(-X)}\left(-z_{\beta(n)}(1) + O\left(\frac{z_{1-\alpha(n)}^3}{n}\right)\right) \sim 1 - \Phi(z_{1-\beta(n)}) = \beta(n).$$

Proof of Proposition 1. Because

$$P_{\mu_0}(\textit{reject } H_0) = P_{\mu_0}(ACI^u(1) \textit{ does not cover } \mu_0)$$

it follows by Theorem 1 that

$$P_{\mu_0}(\textit{reject } H_0) = P_{\mu_0}\left(\bar{X}_n - \frac{\sigma}{\sqrt{n}} z_{1-\alpha}(1) > \mu_0\right) = P_{\mu_0}\left(\sqrt{n}\frac{\bar{X}_n - \mu_0}{\sigma} > z_{1-\alpha}(1)\right).$$

With $P_{\mu_0} = P_{\vartheta_0}^{\times n}, \mu_0 = B'(\vartheta_0), \sigma^2 = B''(\vartheta_0)$ and $B'''(\vartheta_0)/(B''(\vartheta_0))^{3/2} = g_1$, the proof of Proposition 1 is finished.

Acknowledgements

The author is grateful to the Reviewers for their valuable hints and declares no conflicts of interest.

References

Brown, LD: Fundamentals of statistical exponential families. IMS, Lecture Notes and Monograph Series. Hayward, CA (1986)

Corless, RM, Gonnet, GH, Hare, DEG, Jeffrey, DJ, Knuth, DE: On the Lambert *W*-Funktion. Adv. Comp. Math. **5**, 329–359 (1996)

Ibragimov, IA, Linnik, YW: Independent and stationary sequence. Walters, Nordhoff. Translation from Russian edition, 1965 (1971)

Linnik, YV: Limit theorems for sums of independent variables taking into account large deviations. I-III. Theor. Probab. Appl. **6** (1961). 131–148, 345–360; 7 (1962), 115–129
Nagaev, SV: Some limit theorems for large deviations. Theory Probab. Appl. **10**, 214–235 (1965)
Osipov, LV: Multidimensional limit theorems for large deviations. Theory Probab. Appl. **20**, 38–56 (1975)
Petrov, VV: Asymptotic behaviour of probabilities of large deviations. Theor. Probab. Appl. **13**, 408–420 (1968)
Richter, W-D: Skewness-kurtosis controlled higher order equivalent decisions. Open Stat. Probability J. **7**, 1–9 (2016). doi:10.2174/1876527001607010001

Permissions

All chapters in this book were first published in JSDA, by Springer; hereby published with permission under the Creative Commons Attribution License or equivalent. Every chapter published in this book has been scrutinized by our experts. Their significance has been extensively debated. The topics covered herein carry significant findings which will fuel the growth of the discipline. They may even be implemented as practical applications or may be referred to as a beginning point for another development.

The contributors of this book come from diverse backgrounds, making this book a truly international effort. This book will bring forth new frontiers with its revolutionizing research information and detailed analysis of the nascent developments around the world.

We would like to thank all the contributing authors for lending their expertise to make the book truly unique. They have played a crucial role in the development of this book. Without their invaluable contributions this book wouldn't have been possible. They have made vital efforts to compile up to date information on the varied aspects of this subject to make this book a valuable addition to the collection of many professionals and students.

This book was conceptualized with the vision of imparting up-to-date information and advanced data in this field. To ensure the same, a matchless editorial board was set up. Every individual on the board went through rigorous rounds of assessment to prove their worth. After which they invested a large part of their time researching and compiling the most relevant data for our readers.

The editorial board has been involved in producing this book since its inception. They have spent rigorous hours researching and exploring the diverse topics which have resulted in the successful publishing of this book. They have passed on their knowledge of decades through this book. To expedite this challenging task, the publisher supported the team at every step. A small team of assistant editors was also appointed to further simplify the editing procedure and attain best results for the readers.

Apart from the editorial board, the designing team has also invested a significant amount of their time in understanding the subject and creating the most relevant covers. They scrutinized every image to scout for the most suitable representation of the subject and create an appropriate cover for the book.

The publishing team has been an ardent support to the editorial, designing and production team. Their endless efforts to recruit the best for this project, has resulted in the accomplishment of this book. They are a veteran in the field of academics and their pool of knowledge is as vast as their experience in printing. Their expertise and guidance has proved useful at every step. Their uncompromising quality standards have made this book an exceptional effort. Their encouragement from time to time has been an inspiration for everyone.

The publisher and the editorial board hope that this book will prove to be a valuable piece of knowledge for researchers, students, practitioners and scholars across the globe.

List of Contributors

Rainer W. Alexandrowicz
Department of Psychology, Applied Psychology and Methods Research Unit, Universitaetsstr. 65, 9020 Klagenfurt, Austria

Clemens Draxler
The Health and Life Sciences University, EWZ 1, 6060 Hall/Tirol, Austria

Matthias Fischer
Department of Statistics & Econometrics, FAU Erlangen-Nürnberg, Erlangen, Germany

Kevin Jakob
Department of Statistics, University of Augsburg, Augsburg, Germany

Wolf-Dieter Richter
Department of Mathematics, University of Rostock, Rostock, Germany

Jayme Pinto and Nikolai Kolev
Department of Statistics, University of São Paulo, São Paulo, Brazil

Gauss M. Cordeiro and Maria do Carmo S. Lima
Departamento de Estatística, Universidade Federal de Pernambuco, 52171-900 Recife, PE, Brazil

Antonio E. Gomes and Cibele Q. da-Silva
Departamento de Estatística, Universidade de Brasília, 70910-900 Brasília, DF, Brazil

Cícero R. B. Dias, Gauss M. Cordeiro and Pedro Rafael Diniz Marinho
Department of Statistics, Federal University of Pernambuco, Recife, Brazil

Hemílio Fernandes Campos Coêlho
Department of Statistics, Federal Univeristy of Paraíba, João Pessoa, Brazil

Morad Alizadeh
Department of Statistics, Persian Gulf University, Bushehr, Iran

Alan D. Hutson
Roswell Park Cancer Institute, Department of Biostatistics and Bioinformatics, Elm and Carlton Streets, Buffalo, NY 14263, USA

Muhammad H Tahir
Department Statistics, Baghdad Campus, The Islamia University of Bahawalpur, 63100 Bahawalpur, Pakistan

Gauss M. Cordeiro
Departamento de Estatística, Universidade Federal de Pernambuco, Cidade Universitária, 50740-540 Recife, PE, Brazil

Giovana O. Silva
Departamento de Estatística, Universidade Federal da Bahia, Av. Adhemar de Barros, s/n°, 40170-110 Salvador, Brazil

Edwin M. M. Ortega
Departamento de Ciências Exatas, Universidade de São Paulo, Av. Pádua Dias 11, 13418-900 Piracicaba, SP, Brazil

Shahedul A. Khan and Saima K. Khosa
Department of Mathematics and Statistics, University of Saskatchewan, Wiggins Road, Saskatoon, SK S7N 5E6, Canada

Wolf-Dieter Richter
Institute of Mathematics, University of Rostock, Ulmenstraße 69, Haus 3, 18057 Rostock, Germany

Index

www.ingramcontent.com/pod-product-compliance
Lightning Source LLC
LaVergne TN
LVHW080722210926
848485LV00003B/505